Second Edition

AUTOMATED PROCESS CONTROL SYSTEMS: Concepts and Hardware

Ronald P. Hunter, P.E.

Federal Products Corporation
Providence, Rhode Island

PRENTICE-HALL, INC., *Englewood Cliffs, New Jersey 07632*

Library of Congress Cataloging-in-Publication Data

HUNTER, RONALD P., (date)
Automated process control systems.

Includes index.
1. Process control. 2. Automatic control. I. Title.
TS156.8.H86 1987   629.8   86-12281
ISBN 0-13-054479-5

Editorial/production supervision and
interior design: **Kathryn Pavelec**
Cover design: **Lundgren Graphics, Ltd.**
Manufacturing buyer: **Gordon Osbourne/Carol Bystrom**
Cover photograph courtesy of
Cincinnati Milacron, Industrial Robot Division

To my wife, Jean

Printed in the United States of America

10 9 8 7 6 5 4 3 2

ISBN 0-13-054479-5 025

PRENTICE-HALL INTERNATIONAL (UK) LIMITED, *London*
PRENTICE-HALL OF AUSTRALIA PTY. LIMITED, *Sydney*
PRENTICE-HALL CANADA INC., *Toronto*
PRENTICE-HALL HISPANOAMERICANA, S.A., *Mexico*
PRENTICE-HALL INDIA PRIVATE LIMITED, *New Delhi*
PRENTICE-HALL OF JAPAN, INC., *Tokyo*
PRENTICE-HALL OF SOUTHEAST ASIA PTE. LTD., *Singapore*
EDITORA PRENTICE-HALL DO BRASIL, LTDA., *Rio de Janeiro*
WHITEHALL BOOKS LIMITED, *Wellington, New Zealand*

CONTENTS

8   The Operational Amplifier: A System Black-Box Component   181

9   Calculus, Operational Amplifiers, and Analog Computers   210

10   The Operator Interface   233

Preface
to the Second Edition

This book is intended to serve as a textbook for electronics, electromechanical, biomedical, industrial instrumentation, or automatic process control technology students in a third- or fourth- (or both) semester survey course of automatic control systems. It also may well serve as a handbook for engineering and technical personnel who find themselves faced with the need for a basic familiarization with the concepts of automatic process control systems, but not the depth of knowledge that would be implicit in the taking of a series of specific formal courses.

This book is specifically designed to provide a short, easily-read, nonmathematical and self-contained introduction to and survey of automatic control systems. The entire text is written from a "systems" point of view.

We begin with a chapter on automatic control system theory and then delve into the hardware necessary to implement these systems. Next, sensors, measurement techniques (a topic that has been dramatically expanded in the second edition), and signal-conditioning equipment are covered. Operational amplifiers (and the conversion between analog and digital quantities) and system controllers are covered next. Simple on–off controllers, programmable controllers, single-loop analog controllers, and analog computers and digital computers as controllers are all presented, including hardware, software, and interfacing considerations, and the theory behind how each can be made to perform as a controller. The second edition includes an additional chapter devoted to the operator interface to process control systems.

The text then discusses various methods of causing the output from a controller to be converted into action, specifically the action whereby process conditions are altered in accordance with the controller's command. Finally, the techniques used to analyze and troubleshoot large automatic control systems are discussed.

Each category of equipment is discussed in relation to its importance and contribution to the overall system capability and performance. The material is presented in enough detail that typical principles of operation of these various classes of hardware are covered, along with comparative advantages and applications of each. Interfacing considerations are also dealt with in adequate detail for the reader to understand the more important criteria for their use in a system. The second edition has expanded coverage of automatic process control to cover the important field of robotics and its place in the process control industry.

The educational background assumed includes a familiarity with basic electrical concepts, some previous familiarity with semiconductors, and a basic knowledge of physics. Only very basic knowledge is assumed. Wherever additional knowledge is required, a special chapter or appendix (which may be either covered or skipped, as desired) is included. Examples are:

1. The appendix on integrated physics may precede study of the chapters on sensors and transducers. This will also help those who are weak in electrical fundamentals.

2. The appendix on the binary number system may precede study of the chapters on digital signal conversion and digital computers.

3. A chapter covering the principles of operation of digital computers precedes the chapter on digital computers used as process controllers. It specifically refers to the operation of typical minicomputers and microcomputers.

4. Where special chapters are not considered necessary, new technical terms are defined where they are introduced, and a glossary of terms is included at the end of each chapter.

Barrington, R.I.   RONALD P. HUNTER

INTRODUCTION TO AUTOMATIC CONTROL SYSTEMS

The purpose of this text is to fill in the knowledge and experience gap between components and specific functional devices, and useful, practical operating equipment as interconnected into *systems*. It is aimed at imparting a basic understanding of the identifiable *functional subsystems* normally encountered in process control systems, and of the principles that apply to the interaction of these functional subsystems as they operate together in a complete *automatic control* system.

This text is aimed primarily at a familiarization with systems concepts and automatic control principles, but it will also introduce (and survey) typical classes of hardware that are used to implement process control systems, including robots. An understanding of the basic principles of how to break a system (any system) down into functional subsystems is absolutely necessary in order to understand, troubleshoot, and repair even moderately complicated equipment. The same principles carry over to large and very complex systems.

However, before getting involved with the intricacies of process control systems, we should discuss the need for and applications of automatic process control.

PURPOSE OF AUTOMATIC CONTROL

Originally, automatic control was envisioned primarily as a means of reducing manufacturing costs by reducing payroll expenses and increasing production rates. In many systems significant reductions in the number of personnel on production lines have been realized, especially since robots were introduced to industry. However, in many other systems the savings from the reduced number of production-line workers has been offset by systems acquisition costs and the need for qualified personnel to maintain the more complex *automation* systems (other considerations have made the systems cost-effective). Therefore, savings in the area of payroll reductions were not always as great as were originally expected.

Increased production rates were realized, as expected; however, another reason for increased production was also realized, more unexpectedly, and that was a reduction in scrap. Automatic control systems, once operating and producing acceptable end product, continue to operate without the errors that often result from the Monday morning "blahs," inattention by production workers, and normal human errors; therefore, great savings have been realized. Many operating systems can justify the purchase and installation expenses simply on the basis of the economies realized by reduction of reject (out-of-tolerance) end product (scrap).

Automatic control systems can be adjusted to produce end product to closer tolerances than is consistently available using human operators; therefore, improved quality of end product is frequently achieved. Many systems must be automated because the required quality of the end product is simply not economically possible using human operators.

While speaking of processes that must be automated because the process itself has features that render it beyond the capability of human operators, we must include processes that operate in extremes of isolation (e.g., on the moon or on the bottom of the ocean); in extremes of environment (e.g., areas high in nuclear radiation); in total darkness, where the process reacts at extremes of speed (e.g., control of high-speed military aircraft); where the degree of complexity of control is excessive (e.g., transcontinental natural gas pipeline flow); and where certain control features are too critical to trust to human capabilities.

This list is not complete, however; left out is one very important reason for automation—convenience. Automatic speed control in an automobile is primarily for the convenience of the driver. Vending machines, home heating and air-conditioning systems, and many remotely controlled servomechanisms are also primarily for convenience.

Whatever the justification for automatic control systems, one extremely important consideration is, and always has been, the cost of automation. Many recent technological advances have resulted in the development of new hardware which is suitable for use in process control systems (e.g., microcomputers and robots), in significant reductions in the cost of computers and some other hardware items used for process control, and in increased capability for many process-control system hardware items.

Taking into account all these justifications and the availability of both the technology and the hardware, one of the fastest-growing fields in modern times has been industrial process control *automation*. The large majority of production systems designed and installed over the past few years have either been initially designed to be automated or at least designed such that automation can be easily added at a later date. Therefore, industrial automation will in one way or another affect the careers of most engineers and engineering technicians, so the basic concepts of automation are a necessary element in their mental "toolboxes."

HOW AN INDUSTRIAL CONTROL SYSTEM IS IMPLEMENTED

An industrial control system is implemented as this text is organized—in the order "from *sensor* to *servo*," including all the major identifiable functional subassemblies normally included in such systems.

Therefore, to preview what is to come, we shall first discuss sequentially the topics that must be considered in automating a system (in general terms), and then examine a common example of an automatic system. The text continues by leading into the specific chapters, which individually survey the types of hardware used to implement process control systems, and discuss the systems concepts, theories, and considerations involved in the actual implementation of systems.

The first piece of hardware in any process control system is the device that measures the *variables* (quantities) which must be considered in making a decision as to how to adjust the process to produce the desired end product. These measurements are made by hardware devices called *sensors*, which measure the desired variables and convert the information from one energy system to another if necessary (Fig. 1-1). For example, an electronic control system cannot deal directly with the speed of rotation of a mechanical shaft. Therefore, the sensor must measure the shaft rotational speed and convert (transduce) that information from a mechanical system measurement to an electrical one for use by the *controller*.

The text begins with a survey of the most common types of sensing (measuring) devices encountered in process control, and with systems considerations involved in selection of specific devices and use of these devices. The next topic that is considered relates to the handling (interfacing) of output signals from the sensors.

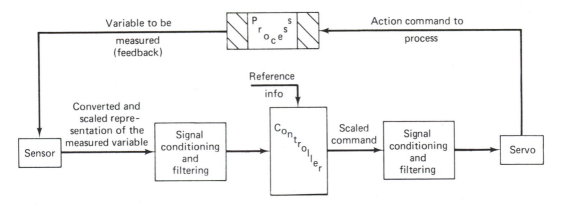

Figure 1-1 Feedback process control system.

Subsequent chapters discuss what the signal means (scaling), how much faith can be placed in the signal (accuracy, sensitivity), how to handle inaccuracies (noise) present in the signal, and how to keep from increasing inaccuracies or errors present in the signal. Then the final signal conditioning required to provide the measured information in acceptable form for various types of controllers is discussed.

The text next discusses the most common types of controllers used in process control systems. How they are used, how they handle information internally, and how they process input information in order to calculate the required command signals to *control* the process are also covered.

The signal conditioning necessary for processing of the command signals from the controller is then discussed along with the various types of hardware devices (*servos*) which ultimately convert the controller commands into action. Finally, techniques commonly used for understanding and troubleshooting of automatic systems are covered.

Process Control Example

Based upon the foregoing procedure, let us now investigate the process of automating an oil-fired circulating-hot-water home heating system. Figure 1-2 illustrates such a system with no automation at all. To automate this system, some variable measurements must first be made. The initial selection of variables is made based upon the final end product, which is room air temperature. Therefore, room temperature must be measured and information supplied to the system to enable the controller to make a decision to turn the system on or off.

Once the controller makes a decision to turn the system on, first the arc must be energized and then the fuel pump turned on, because if the pump is energized first there may be an explosion in the furnace when the arc is finally energized. Therefore, there is a requirement (a safety requirement) that the controller must

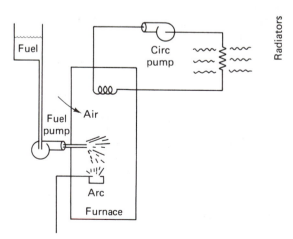

Figure 1-2 Basic heating system.

know that the arc is, in fact, established prior to turning on the fuel pump. This establishes the need for a sensor capable of sensing the presence or absence of the arc.

With the flame established in the furnace, the water in the system will be heated. Once it reaches an acceptable level, the circulator pump must be energized, both to circulate the hot water to the various radiators in the house and to cool the heat exchanger tubes in the furnace. Therefore, the temperature of the water in the furnace heat exchanger tubes must also be sensed.

Three sensors are therefore required: room temperature, arc presence, and water temperature in the furnace heat exchanger. The outputs from all these sensors must be processed until they are in a proper form (electrical in this case) for the controller. The controller must be designed to accept these signals and make the proper individual decisions as to when to command the arc, fuel pump, and circulator pump to start. Therefore, its output (or command) signals must be processed (conditioned) as necessary to accomplish the end control. Figure 1-3 shows the system as automated for home use.

If this system was to provide heating for an apartment building or a factory, several additional parameters would necessarily be measured: the room temperature in each space to be heated and, in addition, possibly the level of fuel in the tank, the level of water in the circulating system, and the pressures at the output of each pump. These measurements would also have to be properly conditioned for the controller to use them to accomplish whatever purpose it is wired for.

It should be obvious that even for a simple system, the control strategy and interconnections for automatic control can become numerous and complicated. When automation is designed for more complicated industrial process control systems, it is common for several hundred variables to be measured and their values

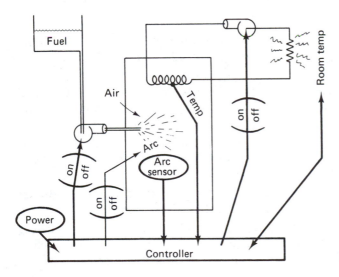

Figure 1-3 Automated heating system.

passed to the controller. The controller must then be capable of assimilating this quantity of information and processing it to correctly calculate the necessary output commands. The design of such systems is relegated to a specific engineering field called *automatic control theory.*

INTRODUCTION TO AUTOMATIC CONTROL THEORY

Most electronics and instrumentation technicians and engineers have been educated in the fields of circuit components and basic circuit theory and analysis, both in analog and digital circuits. This text is designed to build on this background and survey the systems applications of circuits and components studied previously. In fact, these circuits are *not* stand-alone devices; their primary use is in conjunction with other components, many of which are *not* electrical in nature. As they are connected together to realize a higher utility than they were individually designed for, they become portions of a larger, whole system. They become components or subsystems in this larger system.

The techniques for determining which subsystems or components are necessary, specific specifications for each of the individual subsystems, overall performance specifications for the system, interfacing requirements, interactions (both desirable and undesirable) among the subsystems, and adjusting the entire system to get the desired end result all fall within the responsibility of automatic control systems engineers. The body of knowledge available to accomplish these tasks constitutes automatic control system theory. A knowledge of the basics of this theory is therefore every bit as necessary as a knowledge of transistors, digital logic, computers, and circuit theory. More specifically, the field of automatic control theory includes *the application of scientific knowledge and engineering principles to the automatic feedback control of time-variant processes.* The key concepts in this definition include:

1. *Application of scientific knowledge:* that is, principles from the various pure sciences, such as physics, chemistry, biology, and mathematics, which have been discovered and published.
2. *Application of engineering principles:* this has the connotation of practical applications experience and *empirical* knowledge as compared to purely theoretical knowledge.
3. *Feedback:* this is a necessary part of this definition, since most of the problems that must be solved involve determination of proper feedback-loop characteristics and parameters.
4. *Control:* however it is defined, it is the key word in the definition—how easy to state but how difficult to achieve; it is the ultimate goal.
5. *Time-variant processes:* nearly all automatically-controlled processes are controlled based upon the time variation of their parameters. (This excludes "batch processing"; however, that will also be discussed later.)

Note that this definition has specifically excluded *open-loop* systems (which have no feedback of any form). There are many applications for open-loop control but they are of little interest to automatic control theory because there is no possibility of true *control*.

Goal of Automatic Control Theory

Engineers and scientists in this field have as their goal the empirical and mathematical determination of the time-based characteristics of various processes for which automatic control is deemed desirable. Furthermore, they strive to discover the basic laws of nature that govern biological and other feedback systems and to apply these basic principles to the control of processes. They implement various theories and techniques, observe their effects on process control systems, and attempt to refine the basic theories and techniques based upon the results.

Origin of Feedback Control

As people progress through the ages in their knowledge of the behavior and underlying causes of behavior of all naturally occurring phenomena, they invariably discover feedback. Feedback seems to be a necessity for all self-regulating systems; and all naturally occurring processes of life and evolution have unmistakably demonstrated the presence and the effects of feedback. The presence of feedback has always caused a self-regulation effect; and it is the altering of this effect that we are calling control.

Basic Characteristics of Feedback Control

First, we must define feedback. In *feedback control,* cause (input) and effect (output) are compared and the difference is used to alter the effect. It therefore requires historical knowledge of what has happened before it can be felt in a system. The effect that it has on the output can be basically divided into two opposed types of effects: positive and negative.

Positive feedback always adds to the difference between the original cause and effect. It therefore has an unstabilizing effect on the system, causing the difference between input and output to become greater until finally, in any physically realizable system (with finite output limits), the output will saturate (settle at its maximum possible values). It has definite uses in such devices as oscillators; however, it is not a useful concept in so far as automatic control systems are concerned, except as a condition to be avoided.

Negative feedback, on the other hand, decreases the difference between the desired effect and the actual effect, exactly the opposite of positive feedback. If the difference between the desired effect and actual effect is viewed as an undesirable condition, the term *error* seems to naturally apply. In schematic form, then, the negative *feedback control system* will look as shown in Fig. 1-4.

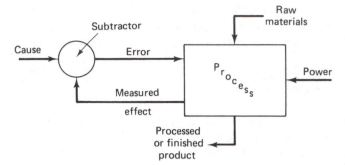

Figure 1-4 General feedback control system schematic.

The *cause* is the reason for running the process, the desired condition of the finished product. In a normal system there may be many desired conditions of the processed or finished product; each one must be specified and controlled. Each specified value or desired condition of the finished product is technically referred to as a *setpoint* of the process.

The subtractor, which determines the error by comparing the setpoint and feedback, is technically referred to as a *comparator*. Therefore, the system schematic in Fig. 1-4 is more correctly drawn as illustrated in Fig. 1-5. Also, note in Fig. 1-5 that the "raw materials," "power," and "finished product" arrows have been omitted. This is customary in practice, where they are understood to be necessary to the process and are therefore not shown.

Figure 1-6 shows two changes from Fig. 1-5. The least significant change is the symbol for the comparator; other symbols are also used inside the circle, but the circle shape is universal and so is its position in the schematic diagram. The second and most significant change is the addition of a controller. The successful operation of a system by simply using the unprocessed "error" signal as the input command to the process is almost impossible. The degree of control over the output of a process depends primarily upon how the controller "processes" the error signal before issuing a "command" to the process. Therefore, we will call the error signal an "actuating error," since it is the signal upon which the controller acts.

The physical construction and theory of operation of the most common types of controllers (including analog and digital computers) are covered in later chapters. Also, in the chapters on controllers, some of the more basic techniques for processing the error signal in order to develop the *optimum* command signal are discussed.

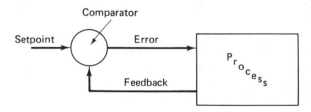

Figure 1-5 General feedback control system schematic.

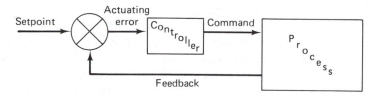

Figure 1-6 Standard general feedback control system schematic.

To a very large extent, the field of automatic control theory deals with the processing of the error signal and design of the controller. The balance of its area of concern is with the overall system dynamics, additional processing needed for the feedback signal, selection of points in the process where the feedback signal is to be measured, position where the command is to be applied, and interactions among various portions of the total system, which may be individually (and simultaneously) under the influence of many controllers.

The only techniques available for determining these complex interrelationships and for design of compensating and control circuits involve higher mathematics. Since most of these relationships and variables are time-dependent, the major mathematical tools are calculus and statistical theory.

To delve into the complex mathematical aspects of automatic control theory would be entirely out of place in this text. It is not even necessary to understand the operation of a system, only to design one. However, some of the concepts that must be considered, and can be discussed from a qualitative point of view, are pertinent here and will be discussed.

Basically, we have been talking about time-variant systems, systems whose parameters (variables) vary with time. This is not to exclude the fact that these same parameters may also vary with respect to other parameters besides time; this is one of the subtleties that adds to the problems of control systems engineers. There are other systems which have parameters that do not appreciably vary with time—in which, instead, the time processing of the systems' variables is of relatively little importance. These systems are commonly called *batch systems*.

Batch systems consist of mixing up batches of ingredients (based upon some recipe) to develop their individual end products. The more important parameters are proper proportions of ingredients and the addition of these ingredients under the exact conditions necessary for the end product to be acceptable. Batch processes are therefore frequently open-loop, having no need for feedback at all during the operation of the process, even though the recipe may be altered at the end of the batch based on observed results. Not infrequently these ingredients must be "cooked" and the recipe may require time-sequential addition of ingredients (they are still referred to as batch processes, however). When control is exercised over the flow rates, pressures, temperatures, and so on, these portions of the batch processes become closed-loop processes.

Another feature used to distinguish among types of processes is whether or not they are continuous. Examples of a continuous process are hot or cold rolling of sheet metals, paper production, film coating, and petrochemical production. All of these are continuous time-variant processes. On the other hand, examples of discontinuous processes include chemical mixing, mixing of pastry dough, and food processing; all of these are basically batch processes.

Continuous processes are of major interest in this text; therefore, we shall direct primary emphasis toward them. The intent of this text is to survey the maximum amount of equipment commonly used in systems, and all the equipment used in batch processing is also used in continuous (time-variant) processing. The most notable exception is the controller itself, and open-loop batch-processing-type controllers will be discussed in Chapter 11.

Effects of Negative Feedback

In general terms, negative feedback has the efect of reducing the sensitivity of overall system performance to variation in the parameters of any of the components or subsystems within the closed-loop portion of the system. Furthermore, it causes

Figure 1-7 Typical industrial process control system control room. Specifically, this photo is of Foxboro Electric Consotrol Instrumentation featuring backup control, as installed in the Humble Oil refinery, Benicia, Calif. (Courtesy of Foxboro Company.)

stabilization of the overall gain characteristics of the closed-loop portion of the system. Frequency-response characteristics are normally also improved by the application of feedback.

The application of negative feedback is, however, affected by control system dynamics in that the properties of the system itself cause a shift in the phase of the feedback signal as compared to the application of a corrective command signal. When the phase difference approaches 180°, the command is completely out of phase with the disturbance upon which the command is based; the end result is positive feedback even though the control system was theoretically designed with negative feedback. The prediction of and compensation for these phase differences is a major source of concern for process control systems engineers. Mathematical tools such as Bodie plots, root-locus diagrams, and Nyquist criteria are available to them for prediction and corrective design efforts.

Other more modern and some more exotic techniques, such as predictive control schemes, adaptive control schemes, distributed control techniques, hybrid control hardware designs, interactive control schemes, multi-computer backup control designs, and others are becoming very popular. However, the basic systems design considerations always must come first, and the same principles of basic automatic control system theory apply to them all. Figure 1-7 is a photograph of a typical industrial process control system embodying many of these concepts.

QUESTIONS

1. How does the emphasis of this text differ from the emphasis of texts on electrical components or electrical circuits?
2. Why is automatic control used?
3. Why is automatic control more important today than ever before?
4. Why will industrial automation eventually affect most engineers and engineering technicians?
5. How is an industrial control system implemented?
6. Describe an example of a feedback control system that you are personally familiar with. Be sure to identify the sensor, signal conditioning units (if any), controller, and servomechanism.
7. What is the goal of automatic control system theory?
8. Why is the field of automatic control system theory primarily dedicated to feedback control systems and not to open-loop systems?
9. In your own words, define automatic control theory.
10. What does the field of automatic control theory try to achieve?
11. What is the primary use of feedback?
12. In your own words, define feedback.
13. What is the basic difference between positive and negative feedback?
14. What is the error signal in an automatic control system?

15. What is the setpoint?

16. What is the function of the comparator in an automatic control system?

17. What is the function of the controller in an automatic control system?

18. What is the function(s) of the signal conditioning units in a feedback control system?

19. List several significant differences between open- and closed-loop control systems.

20. When we speak of a controller for a process, to what portion (hardware portion) of the system are we referring?

21. Make a list of the system effects of open-loop control versus the system effects of closed-loop control.

22. Sketch the standard feedback control system block diagram, labeling each data path and functional block.

23. Why is the field of automatic control system theory (like all branches of engineering) so involved with higher mathematics?

24. What are batch systems?

25. How do batch systems (normally) basically differ from time-variant systems?

26. What is one of the most serious problems introduced into automatic control systems by the use of feedback?

27. In a single word, what is the most basic difference between open-loop and closed-loop systems?

28. Continuous processes
    A. Consist of mixing up fixed quantities of ingredients to develop their end product.
    B. Are processes where control of the time variation of their variables is of prime consideration.
    C. Are never used with negative feedback.
    D. Do not fit the definition of process control systems.

29. The presence of negative feedback in any system
    A. Reduces overall sensitivity to variations in that part of the system enclosed by the feedback path.
    B. Decreases the overall gain of the system.
    C. Stabilizes the overall system's performance characteristics.
    D. Normally increases the frequency response of the system.
    E. All of the above are correct responses to this question.

30. The primary difference between open- and closed-loop control systems is
    A. The use of a controller.
    B. The former is analog in nature and the latter is digital.
    C. The former is always more expensive than the latter.
    D. The presence (or absence) of a feedback loop.
    E. All of the answers above are correct.

31. Which of the following are valid justifications for putting a process control system under automatic control?
    A. Where the degree of complexity of control is beyond human capabilities.
    B. Where the process must operate in extremes of environment.
    C. Where the process reacts at extremely fast rates.
    D. Where certain control features are too critical to trust to human capabilities.
    E. Any or all of the above are valid justifications.

32. A feedback control system is one in which what currently exists is continually compared with desired conditions and the difference between the two is used to alter the process so as to reduce this difference. True or False?

33. Which of the following statements describe properties inherent in an open-loop control system?
    A. Output has no effect on input.
    B. Inherently stable.
    C. "Controller" has no way of knowing if its command was executed.
    D. "Controller" does not care whether its command was executed.
    E. All of the statements above describe an open-loop control system.

34. The function of a signal conditioner is
    A. To sense the signal.
    B. To prepare the signal for succeeding electrical units.
    C. To amplify the signal.
    D. To modify the signal.

GLOSSARY

Actuators   Hardware devices which differ from servomechanisms primarily in that they have no feedback; otherwise, they are the same.

Algorithm   As used in this text, it refers to the prescribed set of rules or equations by which the solution to the automatic control problem will be calculated. The automatic control system controller implements these rules or equations in the process of calculating signals to be sent to process actuators.

Automatic Control   Control of a manufacturing process or communication system by self-operating controlling equipment according to the principles of automatic control system theory.

Automation   *See* Automatic Control.

Control   Exercising automatic feedback control.

Controller   The hardware device which accepts both the information from the system sensors and desired condition commands at its input, calculates the correct response to be issued to the system servos in order to exercise automatic control, and issues these commands in proper format. The controller contains the control strategy algorithm.

Empirical   Factual information obtained solely as the result of experimentation or observation, without reliance on scientific theory.

Feedback   The use of a measured process variable to provide current information of the state of that variable to a controller in order for it to exercise automatic control.

Feedback Control System   A system comprising one or more feedback loops which compare the controlled signal(s) (as measured by sensors) with the setpoint and uses the difference between them (the error signal) to compute the correction necessary to drive that error to zero.

Functional Subassembly   *See* Subsystem.

Open-Loop   A system having no feedback upon which to base control decisions.

Optimum   The best choice; the most favorable selection or condition; the best obtainable set of circumstances.

Sensor   The hardware piece of control system equipment which measures system variables. Normally, this piece of equipment also transduces the measured quantity into another energy system. Sensors provide the feedback signals in process control systems by detecting the current condition, state, or value of a process variable and providing the system with a signal which reflects that condition, state, or value.

Servomechanism (Servo)   Hardware devices which accept control system commands at their inputs and (normally) convert (transduce) these commands into mechanical motion. They are characterized as being actuators connected with feedback for their own control. In process control systems, servomechanisms are the devices which actually cause the system variables to change value in accordance with control system controller computations.

Subsystem   Each of the functional hardware pieces of equipment which are assembled and interconnected with other functional elements (or subsystems) into a system.

System   An integrated assembly of functional hardware devices so selected and interconnected as to perform a higher level function (which is the system design objective) than any of the individual devices could perform alone.

Time-Variant System   As used in this text, it refers to a process control system whose variables change value continuously (primarily) as a function of the passage of time.

Variables   Process control system quantities which must be measured or controlled, and whose values must be considered in making process control decisions. These quantities may vary in value as a function of time, or as a function of other system variables, or both.

2

MECHANICAL PARAMETER SENSORS

INTRODUCTORY CONCEPTS

To understand what a sensor is, we must first define a transducer. A *transducer* is a physical piece of hardware that has been designed to *proportionally* transform (or convert) variable information present in one energy system to any other energy system (normally, a more convenient energy system).

For example, the normal automobile gas-tank-level mechanism converts liquid-level information into an electrical signal that is proportional to the level of gas in your tank (a continuously variable parameter); therefore, the mechanism (located in the gas tank itself) is a liquid-level-to-electrical signal transducer. Another example is an ordinary liquid-in-glass thermometer, which converts temperature information to height of a level of liquid in the thermometer tube; therefore, it is a temperature-to-liquid level transducer. There are numerous other examples, many of which will be discussed later in the following chapters.

If that is a transducer, what is a sensor? A sensor is, first and foremost, a transducer. The one fact that differentiates a sensor from a transducer is its application or use. A sensor is a transducer that is used to *sense* (or measure) the condition of a process control system variable and provide an *input* to the rest of the process control system. It is the first piece of hardware in a process control system loop. It measures directly the condition (or value) of a process variable and converts this information into an energy system that is more convenient to deal with automatically.

In our gas-tank-level problem, the device described is not only a transducer (in that it directly converts gas-tank liquid level into an electrical signal), but it is also a sensor which is an "input" to be used by the driver to make decisions. We use such a device because an electrical signal is more convenient or easier to use in the rest of the system than a mechanical one. You cannot argue with the fact that it would be possible to get out and dip a ruler into the gas tank periodically to check the level of the gas inside, but that is not very convenient. By using the sensor described we can run a pair of small wires up to a gage on the dashboard calibrated to read, not in electrical units, but in the electrical equivalent of the level of gas in the gas tank.

By the way, isn't this gas tank *gage* also a transducer? It converts the gas level (which is a variable electrical signal at this point) into the mechanical rotation of a pointer; therefore, it is an electrical-to-mechanical transducer (or, more simply, an *electromechanical* transducer). However, is it a sensor? No! It does not directly measure or sense the gas tank level itself but only indicates the value of the output from the sensor. Therefore, it is called an indicator.

Some of the reasons for using sensors must have become apparent by now. They are used for convenience, for providing an input to indicators, recorders, or alarms, and for providing an input to automatic process control systems. As an example, a digital process control computer cannot directly understand the temperature of a batch reactor (a big cauldron full of some chemical mixture which must be cooked to get a particular end product). We *must* use a temperature-to-electrical sensor and then (possibly) an analog-to-digital converter to be able to feed this temperature information directly to the computer so that only then can the computer decide whether or not it (the temperature) is at the proper value and initiate the proper correction sequence of operations if it is not within acceptable tolerances.

Sensors are generally categorized by type of output (electrical, etc.), by whether the output is *analog* (continuous) or *digital* (discrete steps or values), *and* by the general type of measurement made. The following text will discuss sensors under general categories of the type of measurement made. Only a few representative types of sensors for each category will be discussed, so this is neither an individually comprehensive nor an all-inclusive listing. To accomplish such a listing would result in a book measured by the pound, not by the number of pages. There are many good references which cover more sensors in more detail, and no attempt is made here to compete with them. That depth of understanding is inappropriate in a systems survey text such as this one. Furthermore, we shall deal primarily with sensors that provide an electrical signal at their output, since in this text we shall be dealing largely with electrical controllers.

Throughout the rest of this chapter it will be necessary (and desirable) to introduce commonly used instrumentation terms and definitions. Some of these will be defined in the text itself. The remainder are simply used and it is left up to the reader to consult the glossary of technical terms included at the end of the chapter.

These terms are common in electronics and instrumentation and *must* be completely understood by all students. It is very embarrassing to be caught using a technical term incorrectly. It is disastrous to purchase a piece of equipment without understanding the terms used in the *specifications,* which would alert you to the fact that the instrument will *not* work in your application.

MOTION AND POSITION SENSORS

Linear Motion Sensors

Probably among the simplest of sensors are the *position* or *motion sensors* (where motion is simply considered to be change of position). These sensors are also called *displacement transducers,* where displacement is defined similarly to motion. In our context, they both refer to a limited change of position.

The linear *potentiometer* is probably the simplest linear motion-to-electrical transducer. It consists of a slider or wiper (Fig. 2-1A), which moves across a *resistance* element. The resistance element would be fixed to one surface and the arm fixed to another surface. The electrical output (variable resistance) would then be proportional to the relative positions of these two surfaces.

Direction of relative motion can be decided by increase or decrease in resistance values. They are made with maximum possible relative motions in excess of 6 inches. The resistance element itself can be made to vary *linearly* or *nonlinearly,* as specified. The *resolution* is determined by the spacing of the turns of wire (Fig. 2-1B) on the wirewound type. For film-type resistance elements, the resolution can be essentially infinite. Hard stops must be built in to prevent the device from being

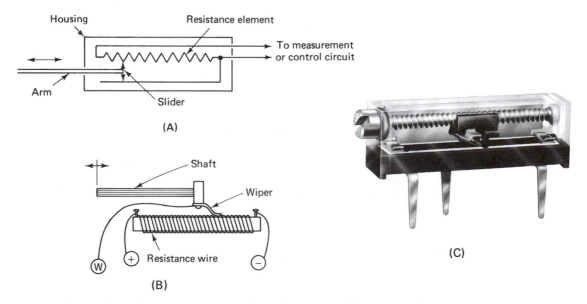

Figure 2-1 Linear motion potentiometer. [(C) Courtesy of Dale Electronics, Inc.]

overranged and thereby physically destroyed. The potentiometers, which must be used where environmental conditions are favorable, require that significant forces overcome friction and have limited mechanical life.

A simple variable linear motion inductor would serve a similar purpose. It consists essentially of a coil and a core (Fig. 2-2A), each fixed to a different surface. Relative motion between the two surfaces will result in a change in the value of coil *inductance,* which is proportional to the amount of relative motion. This is the inverse of the common audio speaker (which is provided with an electrical signal and converts that signal to mechanical motion). Ordinarily, the coil would be part of an *oscillator* circuit (Fig. 2-2B) whereby changes in core position cause changes in inductance which in turn causes a change in the *frequency* of the oscillating circuit. This device is more rugged than the linear motion potentiometer and would have less friction and a greater mechanical life expectancy. It also has a tremendous advantage in that the input and output are *electrically isolated* from each other. This is a very serious system consideration, as will be discussed later. Furthermore, compared to the potentiometer, this device would not be as sensitive to mechanical vibrations, wear, shock, and adverse environmental conditions.

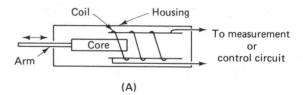

(A)

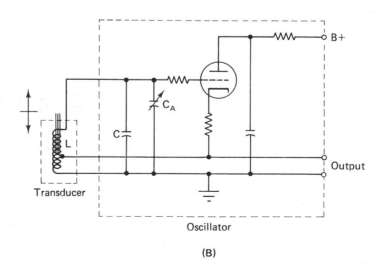

(B)

Figure 2-2 (A) Linear motion variable inductor; (B) inductive displacement transducer in oscillator tank circuit.

The simple linear inductor does not find as much practical use as does its cousin the *linear variable differential transformer* (LVDT). This device (Fig. 2-3) has all the advantages of the simple linear variable inductor (except cost—the LVDT is more expensive) and has several additional systems advantages. First, it is much more sensitive to relative motion: LVDTs can be purchased that can sense fractions of millionths of an inch of motion. Other models can be purchased which can measure in excess of 6 inches total range of motion. Furthermore, the electrical output can be in the form of polarity sensitive *dc voltages,* which are normally considerably easier to deal with in a system.

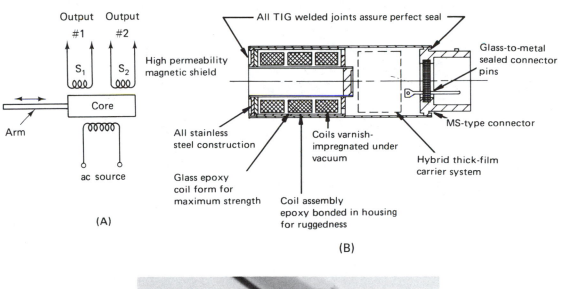

(A)

(B)

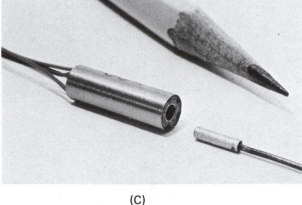

(C)

Figure 2-3 (A) Linear variable differential transformer; (B) hermetically sealed dc-operated LVDT; (C) subminiature LVDT. [(B) and (C) Courtesy of Schaevitz Engineering.]

To explain briefly the electrical operation of an LVDT, consider the primary coil to be connected to a sinusoidal *ac* input source. The two secondaries are mechanically and electrically as identical as economically possible (one of the reasons for the expense of this device) and are symmetrically located with reference to the primary. First, we shall analyze the circuit operation when the polarity of the ac source is as indicated in Fig. 2-4A.

On the positive half-cycle of the primary voltage, secondary coil *currents* flow as indicated through R_1 and R_2. Note the polarities of the voltage drops across R_1 and R_2 ($R_1 = R_2$ in value).

If the core is exactly (electrically) centered between S_1 and S_2, then I_1 will equal I_2 and the voltage drops across R_1 and R_2 will be equal but opposite in polarity; therefore, e_{out} will be zero.

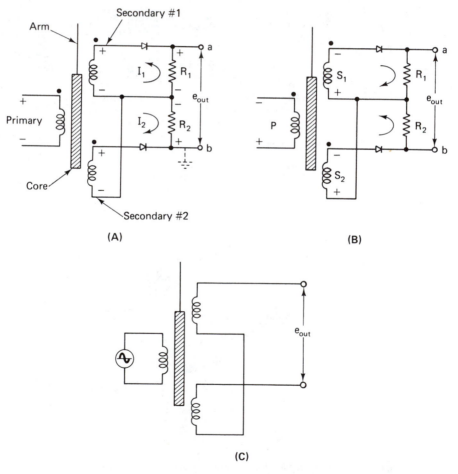

Figure 2-4 LVDT signal-conditioning circuits.

If the core is positioned more toward S_1, then I_1 will be greater than I_2; therefore, the voltage drop across R_1 will be greater than the drop across R_2 and point a will be relatively more positive than point b; therefore, e_{out} will be positive with respect to common.

If the core is positioned more toward S_2, then I_2 will be greater than I_1; therefore, a larger voltage drop (voltage difference) will appear across R_2 than across R_1 and point b will be relatively more positive than point a; therefore, e_{out} will be negative with respect to common.

The magnitude of the output voltage will be proportional to the position of the core with respect to the center position; the farther from center (either way), the greater the output voltage. The polarity of the output voltage indicates the direction from center.

When the primary excitation is the opposite polarity (as in Fig. 2-4B), I_1 and I_2 are blocked by the diodes and there is no output current or voltage. Remember: these are ac signals; therefore, the descriptions above are "snapshots" of the currents, polarities, and voltages at specific instants in time.

The electrical output from Fig. 2-4A will be a half-wave-rectified ac; however, by including a capacitor across points a and b, bipolar dc output voltages will be obtained.

The LVDT output is dependent upon the characteristics and mechanical placement of the two secondary coils relative to each other. They must be (electrically) as near to identical as possible and must be placed symmetrical to the primary in order to get a linear output. The output characteristics are also very dependent upon the *excitation* frequency and the values of resistance in the secondary circuit.

It may be desirable to simply use an ac output rather than to rectify it. In this case the windings are connected in series opposition (Fig. 2-4C), and the output will be a phase-sensitive variable-peak-voltage ac signal at the same frequency as the excitation (input) frequency. Figure 2-5 illustrates one common industrial application of precision LVDT's.

Another group of linear position sensors operate on the principle that the value of *capacitance* of an electrical capacitor is dependent on (among other things) the distance between its plates. Figure 2-6 shows a simple variable capacitor position sensor. In practice, the values of capacitance from such a transducer are very small, and the changes in capacitance for changes in plate separation are only a small percentage of the total value of capacitance; therefore, the circuitry associated with use of this type of sensor is fairly complex. Normally, like the linear variable inductor, it would be included as a component in an oscillating circuit, the frequency of oscillation being proportional to the separation of the plates. All these sensors are carefully designed such as to minimize the effects of unwanted changes in output due to temperature, humidity, aging, and so on.

All the sensors previously discussed are analog or continuous in nature; that is, they go *smoothly* from one value to the next. There are several more recent developments in mechanical linear motion sensors which are digital in nature, or have only a very finite number of discrete output values regardless of the changes

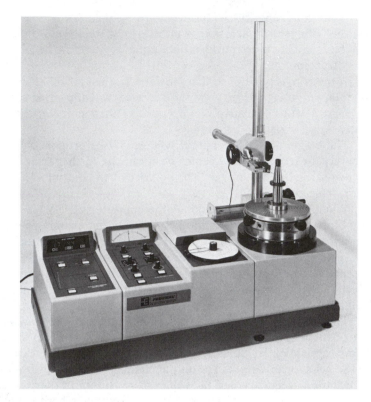

Figure 2-5 Circular geometry measuring machine. The circular table at the right rotates the workpiece. A stationary head, consisting basically of a very precise LVDT with a contact probe, is brought into contact with the workpiece. Any surface irregularities of form (from a perfectly circular form) will be sensed and recorded on the polar chart recorder. (Courtesy of Federal Products Corporation.)

in input (position changes) and the output switches or *jumps* from one value to the next. The difference between analog and digital, and the concepts surrounding digital representations of analog quantities, will be discussed in greater detail later in this text. Of primary importance now is the fact that the "direct digital sensors" are more compatible with digital computers, digital readouts, and digital displays, all of which are finding increased use with process control systems, and therefore their use may be more desirable in many systems. This will be covered more completely later in the book, but we must start somewhere.

The direct digital position (motion) sensors are more commonly (technically anyhow) called *encoders*. Therefore, technical literature for a direct digital position or motion sensor may read "linear position encoder," but we now know that it is a sensor. Encoders can generally be divided into two categories, the *incremental* type and the *absolute* type. First we shall discuss the differences between the two types. All the sensors discussed up to this point are *absolute,* in that the output can

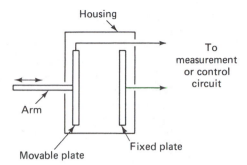

Housing

To
measurement
or control
circuit

Arm

Movable plate

Fixed plate

Figure 2-6 Linear motion variable capacitor.

be correlated directly to the input at any position. Therefore, in order to make a position measurement, all that is necessary is to directly measure the output from this type of sensor.

The *incremental type* operates quite differently; the output consists only of a series of pulses, each one representing some *increment of* position *change*. Therefore, to make a measurement it is necessary first to "zero" the sensor and then count (and remember) the number of pulses occurring since the time the zero position was left. An example may help the reader understand this type of sensor.

Refer to Fig. 2-7A. The *range* (and *span,* since it starts at zero) for this *incremental direct digital linear position encoder* is 0 to 1 inch (in.), and it has a resolution of 0.05 in. The inch markings on the scale are for the purposes of this example only and would not be present in a commercial transducer. The scale, arm, and counter (one piece) must initially be zeroed. Then as the input (movable) arm moves, for each mark passed on the scale there will be a digital pulse sent to the counter. For the position indicated in Fig. 2-7A the counter has received and registered a (net) total of 11 pulses. Since each pulse represents 0.05 in. of movement, the "absolute" position would be the resolution (0.05 in.) times the number of pulses (11); or 0.55 in. from the zero point.

Note that the counter must also have the capability to distinguish whether each pulse it receives from the encoder is to be added to the net previous total (movement to the right or increasing distance) or subtracted from the net previous total (movement to the left or decreasing distance). Therefore, there must be a "direction" output from this sensor also, and the counter must be able to count "bidirectionally" (up and down).

There are several quite different techniques employed in the physical construction of the scales and in the (normally photosemiconductor) marking detectors, which generate the pulses from this sensor. These manufacturing techniques are beyond the scope of this text, but they make this type of sensor quite expensive to manufacture.

The *absolute direct digital linear position encoder* is designed quite differently. First, the scale is marked completely differently; actually there are quite a number of commercially available marking schemes used, but we shall use a straight *binary-count* marking scheme for our example (Fig. 2-8).

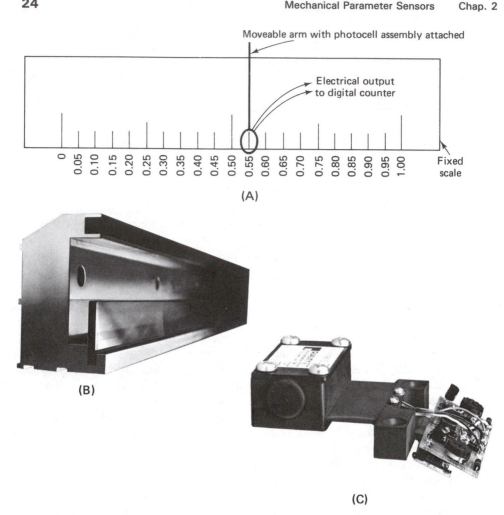

Figure 2-7   (A) Linear digital incremental position encoder scale schematic; (B) linear digital position encoder scale; (C) sensing head. [(B) and (C) Courtesy of Quality Measurement Systems, Inc.]

In Fig. 2-8 consider the dark areas (only) on the scale to be electrically conductive areas, three tracks wide. Imagine three fixed brass brushes, one over each track, and that the scale can be moved relative to the brush holder. At any position of the scale (relative to the brushes) there will either be electrical continuity between each of the brushes and the conductive strip(s) on its track, or there will be no electrical continuity. For example, when the brushes are in position 1 there will be electrical continuity between the brushes and conductive strips for tracks 1 and 3; track 2 will register no electrical continuity. In position 2 there will be continuity for the brush on track 2 and no continuity for the brushes on tracks 1 and 3.

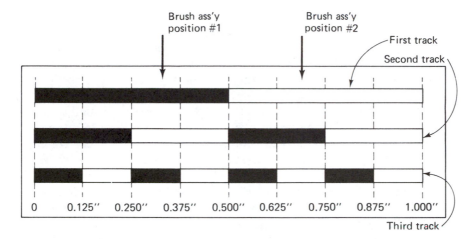

Figure 2-8 Linear absolute position encoder scale schematic.

Therefore, by "looking" (electrically) at the brushes and comparing them you can determine the actual absolute position of the brushes on the scale. Note that the resolution of this scale is 0.125 in. and it is necessary to deal with three separate brushes. In order to obtain the same resolution as for the incremental encoder in the previous example, a total of five tracks and brushes must be dealt with in the associated electronic circuitry. Therefore, the circuitry is more complex than for the incremental type and the scale is much more complex (and expensive) to make.

In practice, there would normally be photocell pickups for each of the tracks (instead of brushes) as for the incremental type. Each type has its advantages and its disadvantages as compared to each other and as compared to analog sensors. The decision as to which one to use in a particular situation must be made based on the peculiarities of each application.

We have covered many of the principles of operation used in linear position (motion) sensors. Although the list is in no way all-inclusive, it is sufficient for the purposes of this text. Next, we shall treat other basic measurement sensors in the same manner, to give the reader a survey of the different types of sensors available. We shall refer back to these devices when we start to introduce analog and digital automatic system controllers.

Rotary Motion Sensors

Rotary position, angular motion, or angular displacement sensors are the next category. In general terms, there is a rotary (angular) position sensor which operates on a principle very similar to each of the linear position sensors. The following text will frequently reference the linear displacement analog to the rotary displacement sensor being discussed. There are also a couple of angular motion sensors which have no linear displacement counterparts.

The rotary (angular) position variable resistance (potentiometer) sensor is probably the most common transducer in commercial use today. How frequently during the course of a normal day do you twist a radio volume control or some other knob which is attached to a rotary position-to-electrical transducer? The electrical signal appearing at its terminals is proportional to the position marks on the scale of the knob. In industry there are innumerable applications for such a sensor.

The rotary potentiometers (Fig. 2-9) may be wire-wound or of the film-composition type. They normally are restricted to less than 360 degrees of rotation, although multiple (2, 5, and 10)-turn potentiometers are also commercially available. Some special potentiometers are made to have nonlinear resistance elements; logarithmic and exponential tapered potentiometers are common examples. They can be purchased in sealed cases but still suffer the same physical environmental and applications limitations as were noted for linear resistance position transducers.

Angular displacement sensors operating on the variable inductance and variable capacitance principles are also available. They operate on identical principles as do their linear motion counterparts; they are simply constructed differently.

The *rotary variable differential transformer* (RVDT) is commercially available and operates on almost exactly the same principle as the LVDT. The primary difference is that the core rotates (see Fig. 2-10A) rather than moving in a linear fashion. The same type of electronic circuitry would be used as for the LVDT, depending upon the electrical output desired. The range of motion for this device is normally limited to approximately ±45° from the center position. There is also a two-coil (rather than the three for the RVDT) version called a *variable reluctance angular position transducer* (Fig. 2-10B).

An ordinary AM-radio tuning capacitor will suffice as an example of a variable capacitance angular position sensor (Fig. 2-11). All other systems considerations (environment, electronics, etc.) would be identical to the linear capacitance displacement sensor.

There is an entire group of sensors which have seen widespread use in the military for years but have only recently been used much in industry. They are normally considered in pairs, which results in an electromechanical servomechanism (see Chapter 16). However, they are also very capable (although relatively expensive)

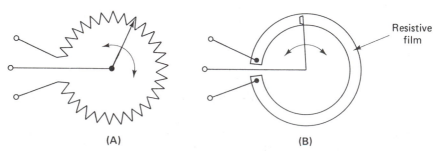

Figure 2-9 Two rotary potentiometers: (A) wire-wound; (B) film composition.

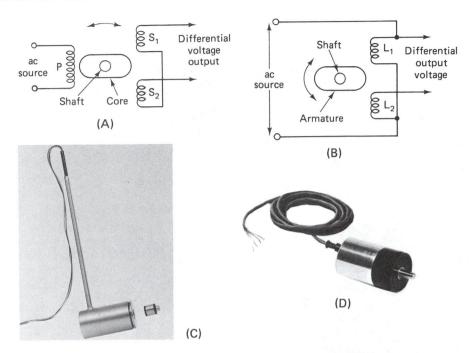

Figure 2-10 (A) Rotary variable differential transformer; (B) variable reluctance angular position transducer; (C) and (D) commercial RVDTs. The particular models shown in parts (C) and (D) each have the oscillator and demodulator built into the RVDT housing; therefore, they will be powered from a dc source and the electrical output will be 0 to 10V dc (linearly corresponding to shaft position). [(C) and (D) Courtesy of Pickering and Co., Inc.]

rotational position sensors. They are called *synchros* and they resemble ac motors in physical construction (Fig. 2-12A and B).

They have a rotating armature (rotor) which is connected to an ac excitation source (Fig. 2-12C). The stationary field (stator) consists of three separate windings, each one physically located 120° from the others. Note that the three stator coils are connected together on one end and the other ends of each are available. There will be two brushes on the rotor to supply excitation to the rotor coil.

The currents induced in each of the stator coils will depend upon the position of the rotor coil relative to each of the three (stator) windings. It can be shown that by comparing the currents in each of the stator windings, the position of the rotor can be precisely determined. Also the direction of rotation can be determined by comparing the *changes* in currents in each of the stator windings.

These are excellent devices, though somewhat more complicated and expensive; however, technical literature in the last year or so has shown a marked commercial revival of interest in these devices.

There is a purely mechanical type of rotary *motion* (motion, not position) sensor, the *flyball governor* (or centrifugal governor) device. As rotary motion sen-

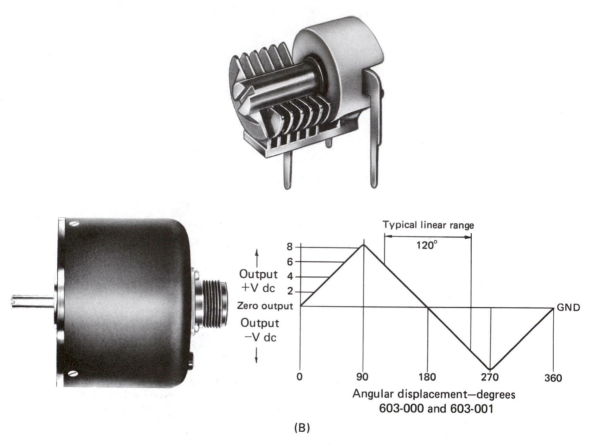

Output +V dc

8
6
4
2

Zero output

Output −V dc

Typical linear range
120°

GND

0   90   180   270   360

Angular displacement—degrees
603-000 and 603-001

(B)

Figure 2-11 (A) Angular motion variable capacitor; (B) precision differential capacitor angular displacement transducer having oscillator and demodulation circuitry built into the transducer housing. [(A) Courtesy of E.F. Johnson Co.; (B) courtesy of Trans-Tek, Inc.]

sors go, this is undoubtedly one of the oldest known sensors. Referring to Fig. 2-13A notice the weights, there may be from one to four of them. As the shaft spins, the weights are thrown outward by centripetal forces. Each weight is fastened to a mechanical arm which is pivoted at point *a*, and therefore they are free to move up and out as the speed of the rotating shaft increases.

There is a sliding sleeve around the shaft which, even though it rotates at the same speed as the shaft, is free to slide (linearly) up and down the rotating shaft itself. The amount of up or down motion is controlled directly by the weights, as the sliding sleeve is directly connected to the same arm the weights are fastened to, through pivots *b* and *c*. Therefore, as the shaft rotates, the weights are thrown outward, and as they move outward, the sliding sleeve is forced up the shaft toward the pivots at *a*.

(A)

(B)

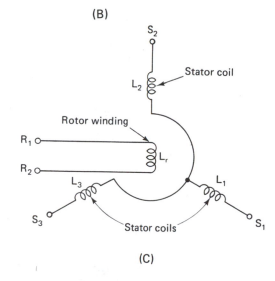

(C)

(D)

Figure 2-12 (A) Typical synchro motor (generator); (B) cutaway view illustrating rotor and stator; rotor brushes are barely visible in the base section; (C) schematic diagram of the synchro; (D) (synchro) resolver with electronics to convert shaft angle to digital format. [(A) and (B) Courtesy of Kearfott Division, Singer Co.; (D) courtesy of Computer Conversions Corp.]

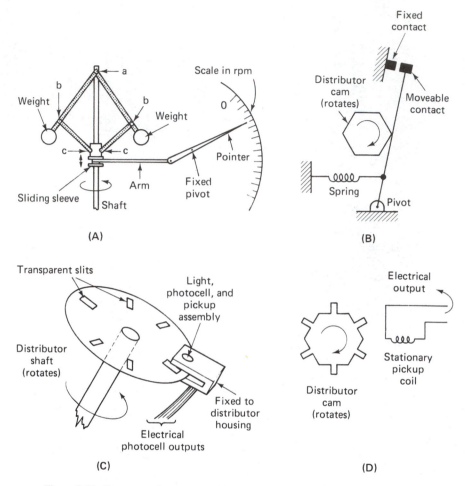

Figure 2-13 Rotary motion sensors: (A) centrifugal governor; (B) mechanical automotive-type points; (C) optical (pointless) auto ignition system; (D) magnetic (reluctance) auto ignition system.

Therefore, thus far we have a rotary motion (rotational speed)-to-linear mechanical position transducer. The rest of the mechanism shown in Fig. 2-13A is simply an indicator which moves across a scale calibrated directly in rotational shaft speed (i.e., r/min). Any of the position transducers discussed thus far could be used in place of the pointer to give an electrical output and thus make this sensor compatible with electronic controllers.

This mechanism is much in use today, but probably more as a stand-alone device than providing an input to the rest of the system. Examples of its use include the regulators in clock mechanisms, mechanical speed controllers, and emergency trip-out devices on high-speed turbines.

Other rotary motion sensors include small ac and dc motors used as generators. The voltage out of either, or the frequency of the ac waveform generated, will all be proportional to the speed (r/min) of the rotating shaft. When used in this type of application these generators are called *tachometers*. They are in very common use and are excellent angular-rate-of-motion sensors. Two commercial examples are illustrated in Fig. 2-14.

Furthermore, the frequency of the waveform generated by interrupting a circuit with mechanical points or by interrupting an electro-optical circuit as in Fig. 2-13B and C is proportional to the speed of the rotating shaft. The photocell circuit illustrated is a schematic of a pointless automobile ignition system, where the pulses generated are proportional to the engine r/min, but also each pulse is generated at a specific angle of crankshaft rotation (this sets the timing of the automobile engine).

A different development along the lines of pointless ignition systems in automobiles made use of a principle used for years in industry. A gear (possibly with magnetized teeth) is rotated past a stationary pickup coil (Fig. 2-13D). As each tooth on the gear passes near the pickup coil, it causes a current to be induced in the coil;

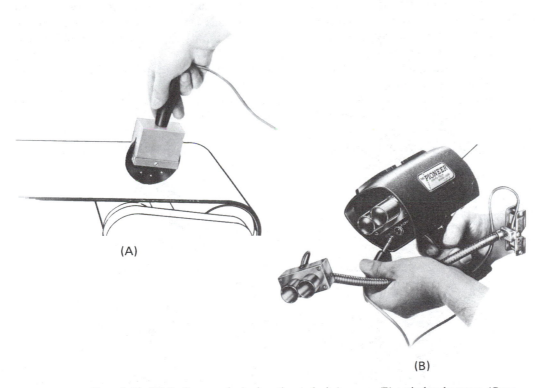

(A)

(B)

Figure 2-14 (A) Surface speed wheel motion (velocity) sensor; (B) optical tachometer. (Courtesy of Pioneer Electric and Research Corp.)

and the frequency of these induced currents is proportional to the shaft speed. This is not altogether different from the principle used in magnetic cartridges for record players; where the vibrations of the stylus cause the reluctance of a coil to vary, thereby causing a voltage to be generated at its output.

So much for analog rotary motion sensors. As with linear motion digital sensors, there are both incremental and absolute rotary digital shaft angle encoders. Basically, the only difference between the linear and rotary versions is that in the rotary versions, the scale is printed on a rotating disc rather than on a linear scale.

Figure 2-15 compares the rotary and linear digital incremental position encoder scales. All comments and considerations noted for the linear incremental encoder also apply to the rotary models. Both the rotary and the linear scales are in very widespread commercial use today. They are available in a multitude of resolutions, and many different techniques are used for generating the incremental pulses. Electronics similar to those used for the linear incremental encoders would also be required with these devices.

Absolute digital shaft angle encoders are also in widespread commercial use. Again, all comments made about the theory of operation of the absolute linear position encoder apply to the absolute shaft angle encoder also. Normally, in both linear and shaft angle absolute encoders, a minimum of four tracks is necessary and 10 tracks are available. The number of tracks sets the resolution if you will remember. A 10 track absolute encoder has a resolution of approximately 1/1000 of the span of the encoder.

Figure 2-16A is a cutaway illustration of a 4-bit (four-track), optical-coupled, absolute shaft angle encoder. Note that a separate photoelectric unit is required for each track and that instead of alternating conductive and nonconductive strips, there are transparent and opaque areas patterned on the rotating disc.

Some encoders use more than one disc, geared together to get fine and coarse readings. This is more practical for higher-resolution units than putting all the tracks

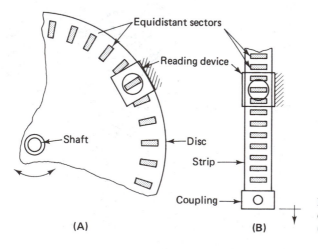

Figure 2-15 Basic incremental digital displacement transducers: (A) angular; (B) linear.

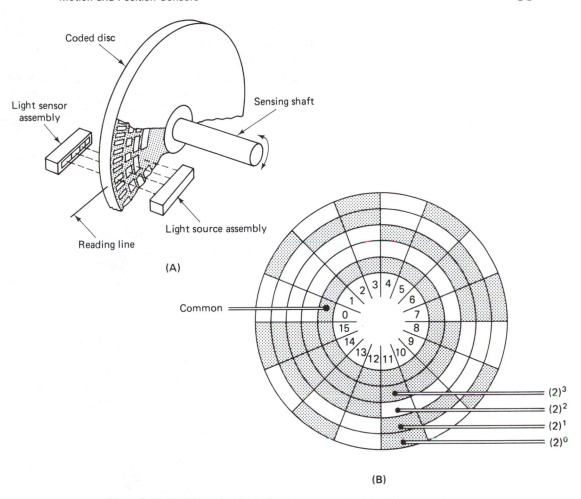

Figure 2-16 (A) Photoelectric shaft-angle encoder (basic); (B) binary code disc.

on one disc. The higher-resolution tracks would have marks so close to each other that it would be very difficult to make the optical sensing heads work correctly.

Proximity and Limit Sensors

The final category of position-sensing devices to be discussed cannot be categorized as either linear or rotary sensors—they are simply known as *proximity sensors*. Proximity sensors operate on a variety of principles, each having advantages in particular applications. However, they all are designed to sense the physical closeness, presence, or proximity of objects.

All proximity sensors are digital, since their outputs represent the presence or absence of an object in the proximity of the sensing head. The sensing heads can be designed to either physically contact the objects or to be noncontact sensors.

Contact-type proximity sensors normally have a lever of some sort sticking out of the device. The sensor is designed such that when the lever is moved by contact with an object, an electrical set of contacts are activated (Fig. 2-17A). Non-contact proximity sensors frequently employ *ultrasonics,* magnetic fields (Fig. 2-17B), light beams (Fig. 2-17C), and even airstreams to detect the presence (or absence) of objects within the measuring range of the sensing heads.

The proximity sensor is also widely used as a limit detector, to detect the extremes of motion of physical devices. In this application, it is normally called a *limit*

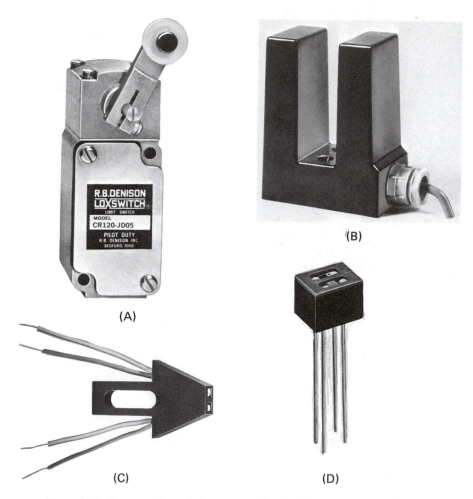

(B)

(A)

(C)

(D)

Figure 2-17 Commercial proximity sensors: (A) physical contact type (solid-state output); (B) magnetic type; (C) and (D) optical types. [(A) and (B) Courtesy of R. B. Dension, Inc.; (C) and (D) courtesy of Optron, Inc.]

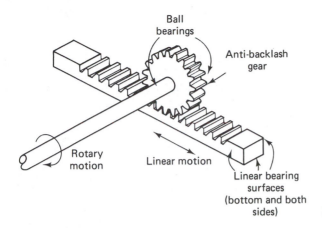

Ball bearings

Anti-backlash gear

Rotary motion

Linear motion

Linear bearing surfaces (bottom and both sides)

Figure 2-18 Rack and pinion.

detector, or simply a *limit switch* (since most proximity sensors used in this type of application are of the contact type). An example of an application for these devices is the mechanisms that stop the movement of automatic overhead garage doors when they are fully up or down.

Any of the linear displacement transducers can be used for sensing rotary motion, and vice versa, by using a linear-to-rotational motion transducer. Figure 2-18 is an example of a very common linear/rotational motion conversion device, the *rack and pinion.*

In any given application there may be systems constraints that require the use of such a device; this is not uncommon. Of course, every time a new element is added in a system, the overall conversion accuracy suffers; rack and pinions are no exception. Slop (*backlash*) between the gear teeth has a disastrous effect on system performance, therefore, special gears are used, called *antibacklash gears,* which pretty much eliminate the slop between gears but restrict the power available through the gear system and increase system friction. As mentioned before, this is only done when system considerations require it.

Also, there is frequently the need for amplification or attenuation of motion in order to apply a given sensor to a particular application. Levers and gearing arrangements are mechanical amplifiers and are frequently used. However, each time one is used, there is added friction, frequent lubrication (maintenance) requirements, and slop as the mechanical parts wear. Unfortunately, it is the inclusion of such mechanical parts in a system (because they are necessary, not by choice) that normally sets the maintenance requirements and failure rate of a given system.

FORCE SENSORS

This may seem to be an odd place to discuss force sensors, but most force sensors use linear position sensing devices. First, the force is converted to a relatively small motion by bending, stretching, compressing, or twisting a piece of metal; then a

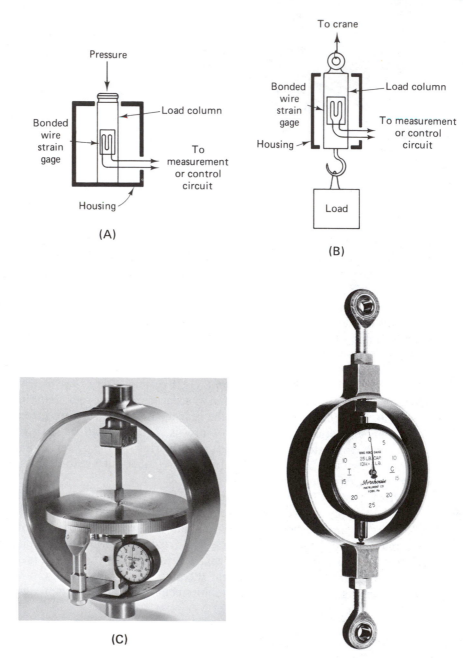

Figure 2-19 (A) Load cell; (B) crane scale; (C) commercial proving ring (calibrates compression load sensors); (D) commercial ring force gage (calibrates tension load sensors). [(C) and (D) Courtesy of Morehouse Instrument Co.]

linear motion transducer senses the amount of distortion of that piece of metal, which is, of course, proportional to the amount of force (stress and strain) the metal is subjected to.

Figure 2-19A and B illustrates the technique used for measuring forces of compression and tension. The metal load columns are precisely machined such that their deformation under load is predictable. The only problem left is to find a linear displacement sensor which is sensitive enough to measure the deformations—of the order of magnitude of tens-of-thousandths to millionths of an inch of deformation (frequently under very adverse vibration, shock, and environmental conditions).

The *bonded wire strain gage* is one of the devices capable of meeting these requirements. The device operates on the principle that the resistance of a length of wire (conductor) depends upon (among other parameters) its cross-sectional area and its length. A given piece of wire will change its resistance as it is stretched, (made longer and thinner) or compressed (made shorter and thicker); and this is exactly what is done.

A piece of wire conductor is bonded to a piece of plastic (Fig. 2-20) such that there is no relative motion between the two. The piece of plastic is then bonded (by very specially formulated bonding agents) to the load column in question (Fig. 2-19) such that there is no relative motion between it and the load column.

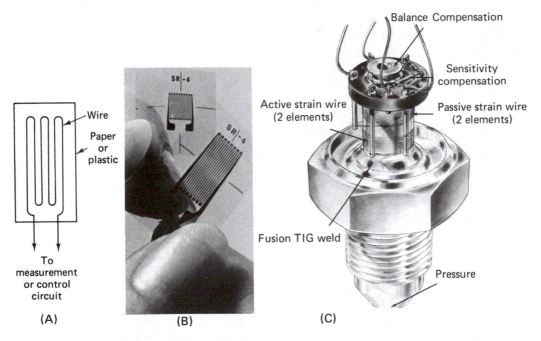

Figure 2-20 (A) Bonded wire strain gage; (B) commercial bonded wire strain gages; (C) commercial bonded strain gage pressure transducer. [(B) Courtesy of BLH Electronics; (C) courtesy of Standard Controls, Inc.]

Therefore, as the load column is subjected to forces (stresses), it deforms (strains), and this distortion (tension or compression) is transmitted directly to the piece of conductor, which also distorts (becomes longer or shorter). By measuring the amount of resistance of the piece of wire, the amount of distortion and thereby the amount of distorting force can be directly determined.

Bonded wire strain gages have excellent resistance to vibration and shock and can be covered with protective coatings such that they can operate exposed to the elements or even submerged in water. They have excellent life expectancies as long as they are not distorted (strained) beyond their elastic limit, which can be easily ensured by proper design.

The electrical output (small variable resistance) from these devices becomes a problem, however. The total resistance of the element is normally around 100 ohms (Ω), and the changes in resistance due to deformation amount to only a few per cent of the total. Therefore, special measurement techniques have been developed for use with bonded wire strain gages. These techniques will be discussed in Chapter 7.

Figure 2-21B illustrates the application of a bonded wire strain gage to mea-

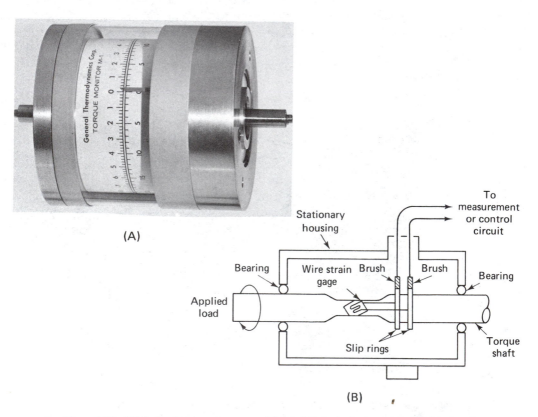

Figure 2-21 (A) Indicating torque sensor; (B) bonded strain torque sensor and its brush assembly. [(A) Courtesy of General Thermodynamics Corp.]

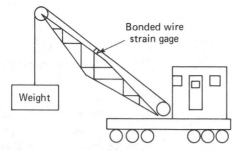

Figure 2-22 Bonded wire strain gage application.

sure the twisting of a shaft, which will be proportional to the torque transmitted by the shaft. Note the additional complexity introduced by the slip rings and brushes. The (electrical) resistance of the brushes is also an appreciable percentage of the total resistance of the bonded wire strain gage.

Figure 2-22 illustrates the application of a bonded wire strain gage to measure the amount of bending in a derrick's boom, another common application.

There is also a device called an *unbonded wire strain gage,* which as you might suspect from its name, is simply a piece of wire stretched between two movable pieces. As the pieces move relative to each other (Fig. 2-23), the resistance of the wire changes proportionally. This device is not in as common usage as its bonded counterpart because of the additional design problems involved.

Greater *sensitivity* can be achieved in all these applications by substituting a specially designed piece of semiconductor material for the bonded wire strain gage. This piece of semiconductor material has been specially designed such that its end-to-end resistance varies over orders of magnitude as the piece is subjected to strain. This particular concept has been used in many commercial products but has not replaced the bonded wire strain gage in most applications. The semiconductor piece

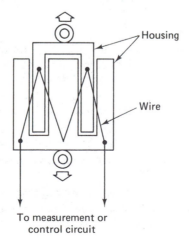

Figure 2-23 Unbonded strain gage.

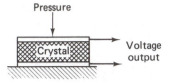

Figure 2-24 Piezoelectric crystal pressure transducer.

is more limited in its application as a result of its costs and susceptibility to vibration shock and temperature; otherwise, it is an excellent device.

One final force measurement device must be discussed, the *piezoelectric crystal*. It is a crystal device (made of Rochelle salt or quartz, for example) which exhibits the properties of a force-sensitive voltage source. As the crystal is distorted (Fig. 2-24), a voltage is internally generated and appears across two of its surfaces. This voltage appears, but the crystal has such a low power-generating capability that the voltage will not hold for any appreciable time; therefore, it is more correctly termed a charge rather than a voltage.

The application for such a sensor is to rapidly changing forces, where the charge does not have much time to leak off. Vibration measurement is a primary use. Also it is much used as the sensor in ceramic or crystal-type phonograph cartridges, where the needle causes distortion of the crystal and the voltages generated are directly amplified by the phonograph amplifiers.

In industrial applications special *charge amplifiers* are used to reduce the loading effects on such crystals and thereby increase their outputs.

QUESTIONS

1. In your own words, define a transducer.
2. In your own words, define a sensor.
3. In your own words, explain the relationship between the terms *transducer* and *sensor*.
4. List several applications for sensors.
5. How are sensors categorized?
6. Make a table, listing in one vertical column each of the sensors covered in this chapter. Then make four adjacent vertical columns, labeling them: Variable measured, Principle of operation, Advantage, and Disadvantage. Attempt to fill every blank space in the table.
7. Briefly, and in your own words, differentiate between analog devices and digital devices (or signals).
8. In your own words, differentiate between absolute and incremental digital signals.
9. What types of electrical outputs would you expect from incremental and from absolute digital position encoders?
10. Differentiate between position and motion as used in this chapter.
11. List the systems advantages and disadvantages of analog sensors and digital sensors.
12. Differentiate among accuracy, resolution, and sensitivity.

13. Differentiate between range and span.

14. List several of the advantages and disadvantages of electrical and mechanical transducers.

15. What does the term *proximity* mean (as used in this chapter to describe proximity sensors)?

16. Differentiate among the applications of position sensors, force sensors, and torque sensors.

17. Most force sensors rely on the deformation of a carefully machined "load column" which is then measured by a linear motion to electrical transducer. True or false?

18. What type of electrical output should be expected from an incremental digital shaft angle encoder?
    A. A series of digital pulses (a pulse train) over a single pair of output wires.
    B. Several parallel wires, each one with a digital voltage level, which must be interpreted together to indicate shaft angle.
    C. A variable resistance analog signal.
    D. A bipolar dc voltage.

19. Absolute digital position (or angle) encoders have an electrical output which can be used *directly* to indicate the exact (within the measurement accuracy of the sensor) position of the sense head relative to the digital scale. True or false?

20. Which of the following parameters can a bonded wire strain gage best be used to measure?
    A. Speed of rotation of a shaft.
    B. Closeness of an object.
    C. Deformation of a metal piece.
    D. Relative position of two linear motion surfaces.
    E. Instantaneous position of a rotating shaft.

21. Which of the following parameters can a proximity sensor best be used to measure?
    A. Speed of rotation of a shaft.
    B. Closeness of an object.
    C. Deformation of a metal piece.
    D. Relative position of two linear motion surfaces.
    E. Instantaneous position of a rotating shaft.

22. When a dc generator is used as a sensor to determine the speed of rotation of a shaft, it is correctly termed
    A. An angular shaft angle encoder.
    B. A flyball governor.
    C. A bourdon tube.
    D. A tachometer.

23. Which of the following phenomena is commonly used in industry to sense minute (very small) changes in the physical dimensions of a load (force) column?
    A. The proportionality between liquid level and pressure.
    B. The attenuation of nuclear radiation by solid materials.
    C. The variation of resistance of wire as it is deformed.
    D. The sensitivity of hair to moisture.
    E. The fact that where hydraulic flow velocity is high, the corresponding pressure will be low, and vice versa.

24. Incremental digital position (or angle) encoders have an electrical output which can be used *directly* to indicate the exact (within the measurement accuracy of the sensor) position of the sense head relative to the digital scale. True or false?

25. Which of the following types of process variables can be measured using a magnetic pickup type of proximity sensor?
    A. Speed of rotation of a shaft.
    B. Quantity of metal objects moving along a production line.
    C. Fluid flow through a turbine flow meter.
    D. Breaks in a continuous (ferrous) metal wire or tube.
    E. All of the answers above are correct.

26. How could the resolution of a linear motion wire-wound potentiometer be increased?
    A. Increase the number of turns of wire over the same linear distance.
    B. Decrease the diameter of the wire used.
    C. Decrease the spacing between the adjacent turns of wire.
    D. Redesign the shape of the wiper.
    E. Combination of all of the above will result in increased resolution.

27. Which of the following parameters can a centrifugal (flyball) governor best be used to measure?
    A. Speed of rotation of a shaft.
    B. Closeness of an object.
    C. Deformation of a metal piece.
    D. Relative position of two linear motion surfaces.
    E. Instantaneous position of a rotating shaft.

28. The common LVDT is
    A. A differential transformer.
    B. A mechanical position-to-electrical transducer.
    C. A sensor.
    D. An inductive electromechanical transducer.
    E. All of the answers above are correct.

29. Which type of linear digital position encoder can economically be made to be the more sensitive?
    A. An absolute linear digital position encoder.
    B. An incremental linear digital position encoder.
    C. They are equally as precise.
    D. Neither can be made to be as sensitive as the other.

30. A piezoelectric device exhibits what electrical characteristics?

31. Which would be the best application for a piezoelectric transducer?
    A. Very low slowly changing pressures.
    B. Very high slowly changing pressures.
    C. Rapidly changing pressure measurement.
    D. High-temperature pressure measurement.

32. The bonded wire strain gage exhibits temperature-sensitive characteristics which must be controlled and has relatively low resistance so that lead length must be compensated for. True or false?

33. Which parameter do bonded wire strain gages actually *directly* measure?
    A. Stress.
    B. Deformation.
    C. Pressure.
    D. Torque.
    E. Force.

34. Torque is a mechanical system parameter which is sensed by measuring
    A. Force only.
    B. Distances only.
    C. Deformation of a shaft.
    D. Speed of rotation only.
    E. All of the above are used to sense torque.

35. How can the sensitivity of the bonded strain gage be increased *without* sacrificing any other parameters?
    A. Use one in conjunction with a piezoelectric cell.
    B. Double or triple the applied voltage.
    C. Double or triple the applied frequency.
    D. Use two, in a differential configuration (one in tension, one in compression).
    E. All of the techniques above are acceptable in every instance.

36. "In order to make a measurement it is first necessary to 'zero' the sensor and then count (and remember) the number of pulses occurring since the time the zero position was left." This quote from the text describes _____.

GLOSSARY

Absolute   (as compared to incremental, when referring to digital quantities)   Absolute systems provide information at their output which represents the absolute position of the sensing head relative to the zero on the measurement scale as well as each incremental change in the input (within the resolution of the system).

Alternating Current (ac)   current which periodically reverses its direction of flow. Normally it alternates between two maximum values, one negative and the other positive. The type of current provided by alternators and oscillators.

Analog   (continuous as compared to digital)   Representation of naturally occurring system variables by an unlimited set of values having negligible separation between adjacent values.

Backlash   The "slop" or "play" between mechanical parts which are machined for a close fit.

Binary   A characteristic of a system or device involving choice in which there are only two possible distinguishable (or acceptable) states or solutions. Based on the binary number system.

Capacitance   That property of an energy system which permits temporary static storage of energy and can provide that energy back to the system upon demand.

Current   The medium of energy transfer in an electrical circuit. Modern theory accepts as fact that extremely tiny (almost massless) negatively charged electrons which move from areas of higher concentration (more negative) to areas of lesser concentration (less negative or more positive) constitute the flow substance called current. A voltage difference between two points in a circuit is necessary for current to flow.

Digital   (discrete as compared to continuous or analog)   Representation of naturally occurring system variables by a limited number of discrete values; having a discrete separation between adjacent values with no interim values possible.

Direct Current (dc)   current which can flow in one direction only. The type of current provided by a battery.

Electrical Isolation   The process of ensuring that there are no possible paths for electrical current to flow between two (isolated) circuits. Normally provided by transformer action, optical coupling, or other magnetic coupling between the two circuits.

Electromechanical   A particular class of transducers which accept electrical signal inputs and convert this electrical signal to mechanical motion.

Excitation (Electrical) Stimulus   An external electrical force which adds energy to a system and which, when applied, will cause a response.

Frequency   The number of times a given event occurs as compared to some standard unit of time.

Incremental   (as compared to absolute when referring to digital quantities)   Incremental systems provide information at their output which represents (only) the occurrence of each single incremental change in their input. No information is contained in the output which relates each incremental change to the initial, or zero, position.

Inductance   That property of an energy system which tends to maintain the flow of energy even after the source of energy has been removed.

Linear   Straight-line relationship, related by a constant ratio.

Nonlinear   Relationship between two (or more) quantities which cannot be described by a linear approximation.

Oscillator   An electrical circuit which provides a time-variant output which continuously varies between two extreme values. The frequency of oscillation and characteristics of the electrical output are determined by circuit design.

Phenomenon   An observable effect which can be described scientifically.

Piezoelectric   A phenomenon exhibited by certain crystalline substances wherein they generate an electrical charge when mechanically deformed and, conversely, will mechanically deform when subjected to a properly applied voltage.

Potentiometer (Pot)   An electromechanical variable resistance device.

Proportional   Having a defined, repeatable, and predictable relationship.

Proximity   Closeness, nearness to.

Range   A statement of the full limits over which an instrument is effective. A statement of the largest and smallest (inclusive) values over which an instrument can be used within its stated accuracy specifications.

Resistance   That property of electrical components which offers linear opposition to the flow of electrical current through the component.

Resolution (of an instrument)   A statement of the largest incremental change in the input which will produce no detectable change in the output. The degree to which a system can distinguish between adjacent values.

Sensitivity   The ratio of the percent variation in one quantity as compared to the percent variation in a second quantity which causes the first to change. The ratio of the magnitude of the output as compared to the magnitude of the measured quantity.

Span (of an instrument)   The difference between the largest and smallest values which can be tolerated to result in useful information within stated accuracy specifications.

Specifications   Detailed descriptions defining the performance capabilities and requirements of equipment.

Transducer   A hardware piece of equipment which converts variable information from one energy system to another (normally a more convenient) energy system. Sensors and actuators are transducers.

Ultrasonic   An adjective describing a piece of equipment which operates using ultrasonic frequencies. Ultrasonic frequencies are those which lie above the audio frequency range.

Voltage   A specifically defined measurement of the intensity of the driving force (in an electrical circuit) which causes electrical current to flow against the effects of resistance (capacitance and inductance) in a circuit.

3

HYDRAULIC
AND PNEUMATIC
SENSORS
(FLUID SENSORS)

Before commencing presentation of a new type of sensor, we must start out by defining exactly what we mean by hydraulics and pneumatics. *Hydraulics,* as used here, is intended to include all liquids but specifically to exclude flowable slurries such as cement. Any liquid that will readily seek its own level in a container will be included. *Pneumatics,* as used here, includes common, safe (nonexplosive, noncorrosive), commercially used gases: normally, air.

From the basic knowledge of physics that you already have (or received from Apendix A, you realize that many of the physical phenomena occurring for liquids also occur for gases (pneumatics); therefore, it should not be difficult to accept the fact that many of the same sensors will work in both energy systems (and this is the case). Therefore, we shall discuss them together, pointing out differences between sensors used in hydraulic or pneumatic systems where they exist.

In discussing these sensors there are basically only three general categories of measurements to be considered. They are *pressure* (in both systems), *level* (hydraulic system only), and *flow* (both systems). There are many other very important parameters to be measured, but we shall not deal with them specifically here; examples include humidity, density, pH, and composition. Some of these will be dealt with independently at the end of Chapter 5. Others are more limited in their application and not in common enough usage to be included in a survey; therefore, they are only mentioned.

46

PRESSURE SENSORS

We shall begin our discussion by considering pressure sensors. *Pressure* is defined as a force spread over an area. Therefore, its dimensions are those of force over those of area: pounds per square in. (psi), grams per square centimeter (g/cm^2), and so on.

Another very common way to measure pressure, primarily very low pressure (less than a few psi), is by noting the ability of the pressure to push up (force up) a column of liquid against the effects of gravity. The common manometer (Fig. 3-1) is the device used to indicate this type of pressure measurement. Pressure is piped to one of the connections for the manometer and the other end is open to *ambient* (atmospheric or room) *pressure*. The liquid will move in the manometer in response to the *difference* in pressures in the two tubes. In other words, this is a *differential pressure* measuring device, where one pressure is measured compared to another pressure. When an unknown pressure is measured as compared to ambient pressure, we call the pressure measurement *gage pressure*.

In the measurements of Fig. 3-1 the gage pressure measurement for Fig. 3-1A is in terms of the height of a column of mercury (Hg) and h_m is 2.04 in. of mercury. The gage pressure measurement for Fig. 3-1B is in terms of the height of a column of fresh water (H_2O) and h_w is 27.67 in. of water. Figure 3-1C is a common pressure gage (we shall discuss the principle of operation of these gages later) and it reads 1.0 psi. In fact, all three pressure sensors indicate exactly the same pressure reading. Some conversion constants for converting the various methods of pressure indication are given in Fig. 3-2. There are others, but they are not used as commonly.

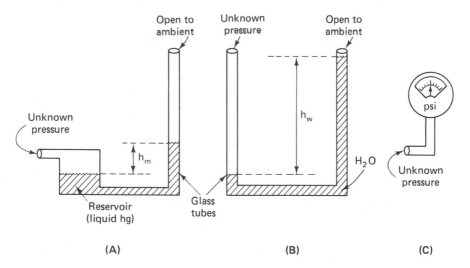

Figure 3-1 Pressure measurement gages: (A) single-leg manometer; (B) U-tube manometer; (C) common pressure gage.

Atmospheric pressure = 14.7 psi or 29.92 in. Hg or 34 ft. H_2O

or $1.033 \frac{kg}{cm^2}$ or 406.8 in. H_2O or 1 atmosphere

From these, various conversion constants can be derived.

i.e.: $\frac{29.92 \text{ in Hg}}{14.7 \text{ psi}} = 2.04 \frac{\text{in Hg}}{\text{psi}}$ OR $\frac{406.8 \text{ in } H_2O}{14.7 \text{ psi}} = 27.67 \frac{\text{in } H_2O}{\text{psi}}$

Figure 3-2 Pressure unit conversion.

Without really introducing it, we have just covered the simplest pressure measurement devices for both hydraulics and pneumatics. There is a system disadvantage to the use of manometers in that they do not have an electrical output, which is normally required for systems. Several techniques are available for converting the manometer levels to electrical parameters using linear position transducers in addition to the manometers.

Figure 3-3A shows a simple inductive coil wrapped around the glass tube of a mercury manometer. Since mercury is conductive, the self-inductance of the coil will change as the level of mercury changes *within the range of the coil.* Figure 3-3B shows a manometer (using any liquid) that has a float and magnetic core piece in the measurement leg. A specially designed set of LVDT windings surrounds the glass tube of the manometer, and therefore we get an LVDT type of output for our system.

Both techniques are used but the range of pressure measurements is quite small and these devices may be fragile. A very common use for a device such as illustrated in Fig. 3-3B is for measuring the difference between two pressures, where ambient

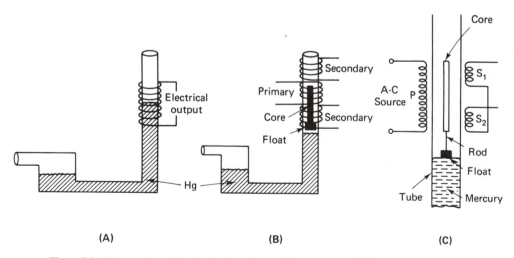

Figure 3-3 Manometer to electrical conversion techniques: (A) variable-reluctance sensor; (B) LVDT sensor; (C) cutaway view of LVDT sensor.

pressure is not one of the pressures. In this case, very small differences between very large pressures can be accurately sensed. In this application the sensor would be technically referred to as a *differential pressure sensor* (D/P sensor). Remember that the term "D/P sensor" is not applied to measurement of one pressure with respect to ambient (atmospheric) pressure. Although this is actually a differential pressure measurement, too, that measurement is referred to as a *gage pressure measurement*. From physics, can you remember what a pressure measurement as compared to true zero pressure is called? It is an *absolute pressure measurement*.

Nearly all other pressure measurement sensors rely on the measured pressure acting over a surface area and the resulting force causing some type of mechanical deflection. For electrical output a secondary mechanical motion (linear or rotational)-to-electrical transducer is used.

Let us now discuss some of the more popular pressure-to-mechanical motion sensors, realizing that most of the position sensors discussed previously can be applied as secondary transducers to obtain a desired form of electrical output.

Probably the most common mechanical pressure sensor is the *bourdon tube* (Fig. 3-4). The bourdon tube itself is made of a thin, springy metal, formed in the shape of a somewhat flattened tube, with one end sealed and the other end opened. The tube thus made is then mechanically formed into any of several different shapes, each shape having advantages in particular situations. The most common shape is the C shape shown in Fig. 3-4.

As illustrated, pressure is applied to the open end of the bourdon tube. As

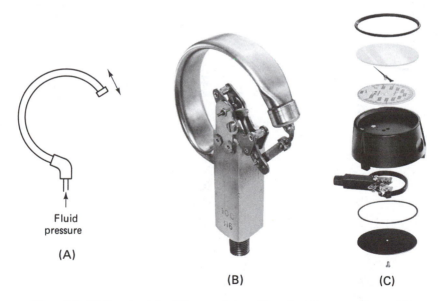

Fluid
pressure

(A)

(B)

(C)

Figure 3-4 (A) Bourdon tube; (B) uncased view of bourdon tube pressure gage; (C) exploded view of complete bourdon tube pressure gage. [(B) and (C) Courtesy of Marsh Instrument Co.]

the pressure acts against the top and bottom surfaces (of the inside of the tube), it acts against unequal areas (the top or outside area being larger than the bottom or inside area). A net force is generated which has the effect of straightening out the tube, much like a New Year's Eve blow toy. The output is therefore a quasi-linear motion, and it is shown to be connected to a gear system and a pointer in Fig. 3-4B. It returns to its original shape upon removal of the pressure, owing to the springiness of the metal. This mechanism (Fig. 3-4B) is by far the most common of all pressure-measuring indicators in use over the past century. It is inexpensive, can be made to be quite accurate, wears very little, and is extremely rugged and reliable.

In order to increase its sensitivity or to fit into different-shaped containers, the bourdon tube can be formed into other shapes: for example, the flat spiral shape (Fig. 3-5A), the helical-spiral shape (Fig. 3-5B), or even a twisted shape (Fig. 3-5C). Each of these has its own systems advantages, although they are more expensive.

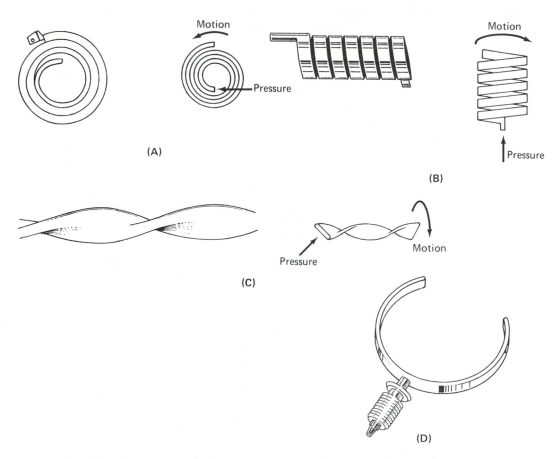

Figure 3-5 (A) Flat-spiral bourdon tube; (B) helical-spiral bourdon tube; (C) twisted bourdon tube; (D) U-shaped bourdon tube.

Bourdon tubes come in various sizes to measure pressures from a few psi up to many thousands of psi. Also, they can be purchased in pairs, with the free ends connected together. This configuration is designed such that the free ends fight, or oppose, each other's motion with increasing pressure in the tubes; therefore, the resulting motion of the free ends (connected together) is proportional to the difference between the pressures applied to each of the tubes. In this configuration we have *bourdon tube differential pressure sensors.*

For measurement of very high pressures it is not necessary to form the tube in any special shape at all. Therefore, some high-pressure sensors are simple closed-off ends of tubing made of appropriate materials, and the distortion of the tubing resulting from high internal pressures is measured. Figure 3-6A shows a strain tube, and a specially designed strain gage is attached to measure the distortion of the tube produced by internal pressures. Figure 3-6B shows a modification of this where the tube is simply flattened on one end. Again, a strain gage would be used to measure the distortion. Some examples of bourdon tubes connected with secondary electromechanical transducers are illustrated in Fig. 3-7.

The *bellows* is another common pressure-to-mechanical motion transducer. It is made of a springy material which has been (first) formed in the shape of a thin-

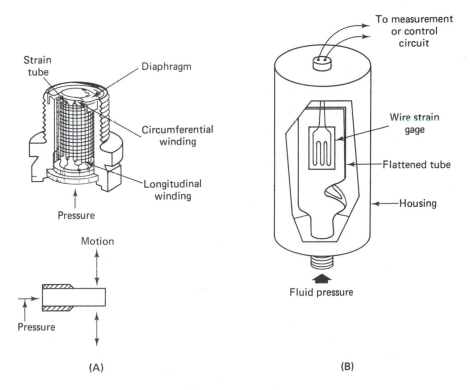

(A)   (B)

Figure 3-6 Examples of strain tubes.

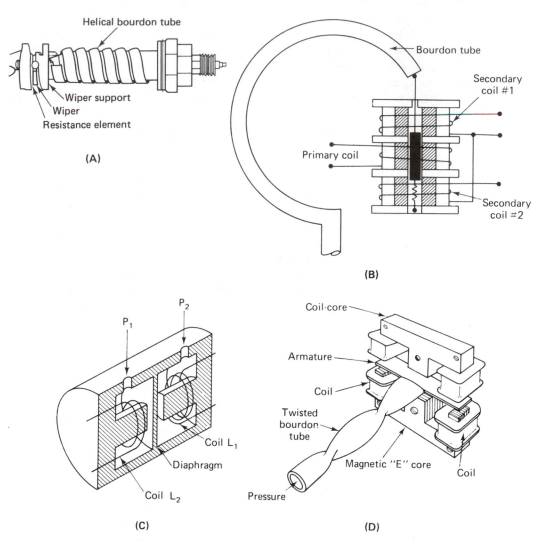

Figure 3-7 (A) Resistance change measured by a helical Bourdon tube. Reluctance change measured by: (B) core motion in differential transformer; (C) diaphragm deflection; (D) angular deflection of bourdon tube.

walled tube. This tube is then worked and deep convolutions are formed in it (Fig. 3-8). This mechanical deforming of the metal in order to form a bellows has a tendency to make the material brittle and failure-prone. Therefore, this device is not as reliable as the bourdon tube, and is considerably more expensive, both primarily because of the extra working necessary to form it. Once the bellows has been successfully made and tested, it will probably have a very long service life; the high failure rate is only upon initial manufacture.

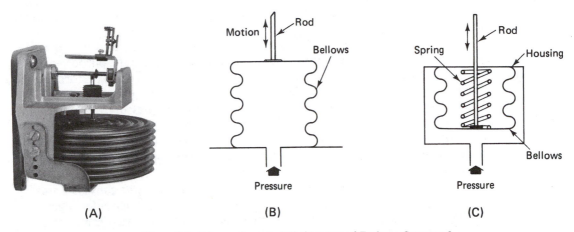

Figure 3-8 Bellows element. [(A) Courtesy of Foxboro Company.]

After being formed, the bellows has one end sealed and the other end equipped with a pipe fitting for connection to the pressure to be measured. Referring to Fig. 3-8B, as the pressure increases inside the bellows, the bellows is forced to lengthen. By sensing the amount of elongation using a linear position transducer, this motion can be converted to an electrical signal for systems use.

The bellows is an excellent pressure-sensing device. It is generally more sensitive than a bourdon tube and can sense lower pressures. It can be used to sense somewhat higher pressures if an auxiliary spring is included to oppose excessive elongation, as illustrated in Fig. 3-8C; for smaller pressures the springiness of the metal itself is sufficient.

The bellows can be connected into various configurations in order to sense differential pressures. Permitting a second pressure to be admitted to the inside of the bellows as in Fig. 3-8C is one configuration and Fig. 3-9 illustrates another commonly used configuration.

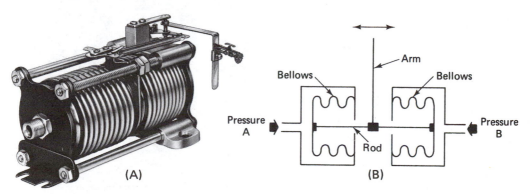

Figure 3-9 Differential-pressure sensor employing bellows elements. [(A) Courtesy of Foxboro Company.]

Two additional devices are somewhat similar to the bellows, the diaphragm and the capsule.

The *diaphragm* (Fig. 3-10) is a thin circular disc made of a springy material which is stamped or pressed into the shape illustrated. It serves the purpose of sealing one chamber from another in a transducer and at the same time performing the function of a pressure-to-mechanical motion transducer. As pressure is applied to one side, the center portion is pushed up away from the pressure (remember, the edges are sealed to the case and cannot move). The amount of deformation at the center of the diaphragm is proportional to the pressure applied. By sealing the diaphragm between two cases and admitting pressures to both sides, it can be made into a very effective differential pressure sensor.

An adaptation of the diaphragm is the capsule. The *capsule* consists primarily of two diaphragms (of opposite curvature) welded together at their outer edges (Fig. 3-11A). Several capsules can be connected together for increased sensitivity, and they can also be designed into differential pressure sensors.

Diaphragms and capsules are easier to manufacture than bellows, and are use-

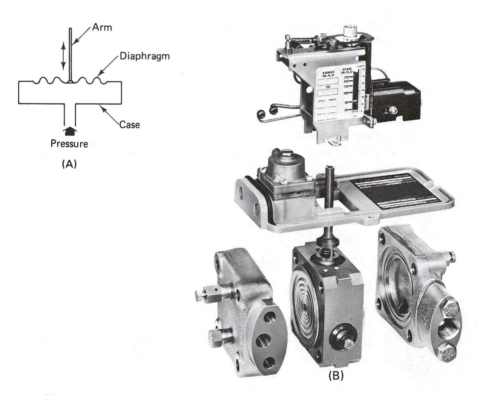

Figure 3-10 (A) Diaphragm pressure gage schematic; (B) commercial diaphragm differential pressure instrument. [(B) Courtesy of Taylor Instrument Process Control Division/SYBRON Corp.]

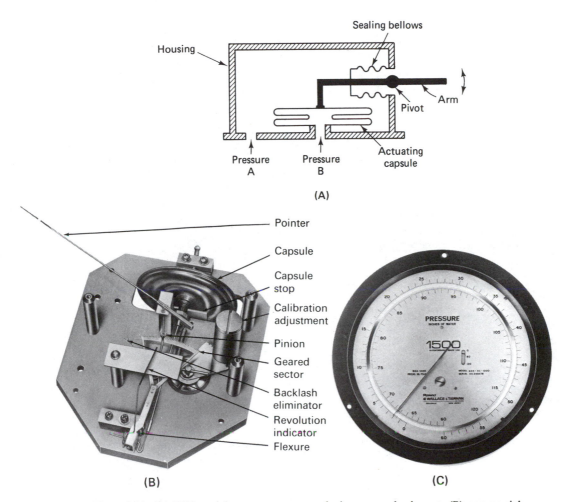

Figure 3-11 (A) Differential-pressure sensor employing a capsule element; (B) commercial differential pressure sensor using the capsule element; (C) commercial precision pressure gage using the capsule element shown in part (B). [(B) and (C) Courtesy of Wallace & Tiernan Division, Pennwalt Corp.]

ful primarily in the measurement of low pressures. They are all in very common commercial use.

Figure 3-12 illustrates the application of a variable capacitance device to the measurement of pressure. Pressure sensors using this principle do not require a secondary mechanical position-to-electrical transducer and thereby do have one advantage over the others. In the sensor illustrated, a diaphragm is connected as one plate of a capacitor.

Referring back to the section on force sensors when we talked about a piezoelectric force transducer, you will remember that it responded to changing pres-

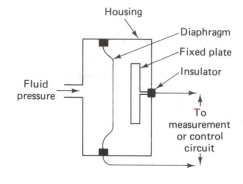

Figure 3-12 Capacitor pressure transducer.

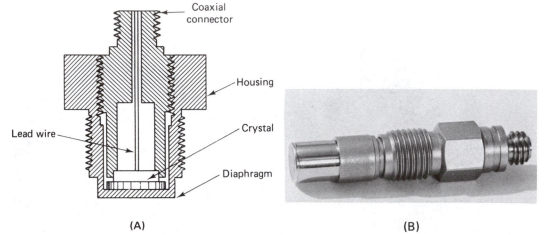

(A)   **(B)**

Figure 3-13 (A) Piezoelectric pressure transducer; (B) typical commercial quartz pressure transducer. [(B) Courtesy of Sundstrand Data Control, Inc.]

sures. Therefore, in an application where pressure *change* measurement is required, this may be an excellent choice of sensor. Figure 3-13 illustrates a commercially available transducer using a piezoelectric crystal for a primary transducer.

This has been a representative sampling of commercially used pressure sensors. There are many others but they are not used as commonly as those presented.

LEVEL SENSORS

Let us now turn our attention to some representative liquid-level sensors. There are two basic types of applications for liquid-level (of course, gases have no "levels") sensors; the first is the need to know the analog value or continuous (precise) level at all times, and the second is simply to know when the actual liquid level is at a predetermined value (digitally).

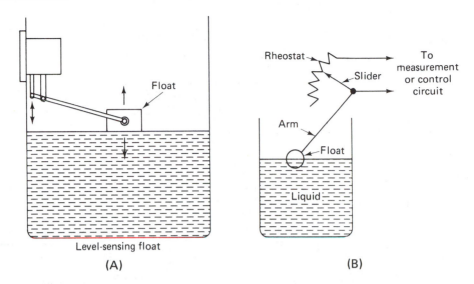

Level-sensing float

(A)

(B)

Figure 3-14 (A) Level-sensing float; (B) liquid-level sensor employing a float-operated rheostat.

First, we shall discuss the type that continuously monitors the level of a liquid in a tank. The simplest type of liquid-level sensor is a common float (Fig. 3-14). The float is actually a liquid level-to-mechanical motion transducer, and therefore a secondary transducer (or transducers) must be used to get an electrical output that will be proportional to the level.

A second method of sensing level is to sense the pressure at the bottom of the tank. The relationship between depth of a liquid and the pressure created by the weight of the liquid above that depth is: depth is equal to pressure divided by the density of the liquid. Therefore, by measuring the pressure at the bottom of the tank, and knowing the density of the liquid in the tank, the depth can be determined easily. Any of the pressure sensors previously mentioned could conceivably be used. Figure 3-15 illustrates the use of this type of level sensor.

By immersing two electrodes in the tank (Fig. 3-16), the liquid level can be sensed by either measuring the *conductivity* of the liquid between the two electrodes

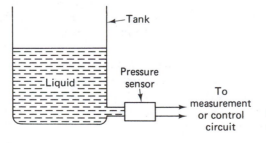

Figure 3-15 Using a fluid-pressure transducer to sense liquid level.

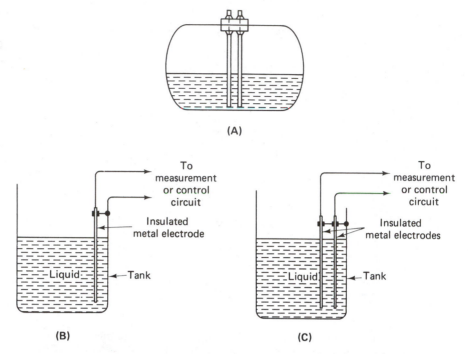

Figure 3-16 (A) Conductivity liquid-level sensor (non-metal tank); (B) capacitive liquid-level sensor for a metallic tank; (C) capacitive liquid-level sensor for a non-metallic tank.

(the higher the liquid level, the greater the total conductivity of the measuring system), or by using the two electrodes as plates of a capacitor and monitoring the change in capacitance as the dielectric (the liquid in the tank) changes value. Both methods are used; however, they are only usable in circumstances where the properties of the liquid being sensed will not foul the electrodes or produce a hazardous situation.

Other sensors are designed to sense liquid level based on the fact that at the surface of the liquid in a tank there exists a sharp change in density (a change from liquid to gas). Ultrasonic waves are reflected by changes in density, so an ultrasonic transceiver can be located either in the bottom of the tank pointed upward, or in the top of the tank pointed down, and the relative time delay of the reflected signal will be proportional to the level of liquid in the tank. This principle (Fig. 3-17) is also widely used in industry.

Figure 3-18 illustrates the use of a nuclear radiation sensor to detect the level of liquid in a tank. The nuclear radiation emitted by the source is assumed to be constant and the liquid attenuates the quantity of radiation reaching the *radiation* sensor (the Geiger-Müller tube). Therefore, the intensity of radiation sensed by the G-M tube will be proportional to the liquid level.

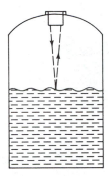

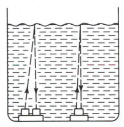

Figure 3-17 Sonic sensing of liquid level.

Figure 3-19 illustrates several commercial-level sensors operating on the principles previously discussed.

Of course, we can always weigh a tank with liquid in it. The weight will be directly proportional to the level of liquid in the tank. This is an extremely accurate method and is commonly used in calibration tests of hydraulic flow devices (to be discussed later).

There is another very widely used analog liquid-level measurement technique, the *purge bubbler* system (Fig. 3-20). In this technique a relatively constant flow of air is permitted to flow down the air tube and bubble out of the bottom. The pressure must be adequate to "blow," or force, the liquid out of the tube. The higher the level of liquid in the tank, the more pressure will be required to keep the liquid from entering the bottom of the air tube.

Therefore, the pressure of the air in the tube is proportional to the level of liquid in the tank. If you look at this system correctly, it actually is a modified manometer measuring system, set up to measure the pressure at the bottom of the tank, which is proportional to the level of the liquid in the tank.

Now let us turn our attention to applications where there is no necessity to know the exact level at all times, the only requirement being to determine if the

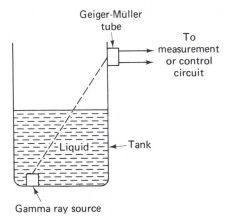

Figure 3-18 Gamma-ray liquid-level sensor.

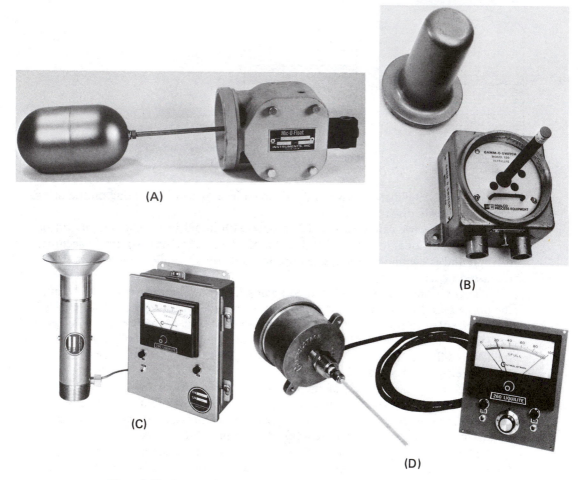

Figure 3-19 Commercial level sensors: (A) float type; (B) nuclear type; (C) ultrasonic type; (D) capacitance type. [(A) and (B) Courtesy of C-E INVALCO, Combustion Engineering, Inc., Tulsa, Okla.; (C) and (D) courtesy of Controlotron Corp.]

liquid is (at least) up to a predetermined level. In all these *level-limit sensors,* the sensing device is physically located at the level in question. There are innumerable methods to accomplish this sensing, a few of which are shown without further explanation in Fig. 3-21.

FLOW SENSORS

A wide variety of physical principles are used in the design of the flow sensors currently available. Some are for liquids, some for a liquid with chunks in it, and

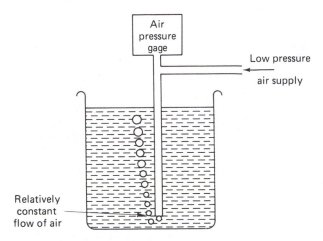

Relatively
constant
flow of air

Figure 3-20 Purge bubbler level measuring system.

others for gas flow. Even with this variety of sensors, however, flow measurement is one of the most acute problem areas in modern systems design. True, many physical devices are available, but relatively few have found widespread acceptance either because of limitations of the sensor itself or of cost. We shall discuss the systems limitations of each type as it is presented.

Let us begin by discussing one of the more common, simpler and older flow sensors. The *nutating disc meter* (common household water meter) is one of the very few flow sensors which actually directly measures the *quantity of liquid flow* through a pipe. Nearly all the others measure not the quantity of flow but the *velocity* of flow. Of course, there is a direct relationship between the two ($Q = AV$), but there are problems in obtaining accurate velocity measurements, as will be discussed later; but now back to the nutating disc sensor.

Refer to Fig. 3-22. The chamber is an enclosed cylindrical-shaped chamber except for openings at the input and output where the pipes are connected. Inside this chamber is a flat circular disc which seals against the walls of the chamber. It is mounted on an off-center pivot and is forced to remain always in a tilted position such that one edge touches the bottom of the chamber and the opposite edge of the disc touches the top of the chamber. Therefore, the disc effectively divides the chamber into two equal parts, each sealed from the other. Furthermore, the inlet and outlet pipe openings in the chamber are physically located such that no matter what the position of the disc, it *always* blocks flow straight through from inlet to outlet. Also, the disc itself is not permitted to rotate.

In the position illustrated in Fig. 3-22, as liquid flows into the chamber (from the right) the disc forces it to fill (only) the top portion of the chamber. As this chamber fills, the full pressure of the incoming liquid is felt all over the top of the disc. If the disc were mounted on the center, it would block flow and that would be the end of the flow; however, it is mounted off-center. Therefore, there is a greater area on one side of the pivot point than the other, and the disc begins a peculiar wobbly, oscillating, or tilting motion (technically referred to as a *nutating*

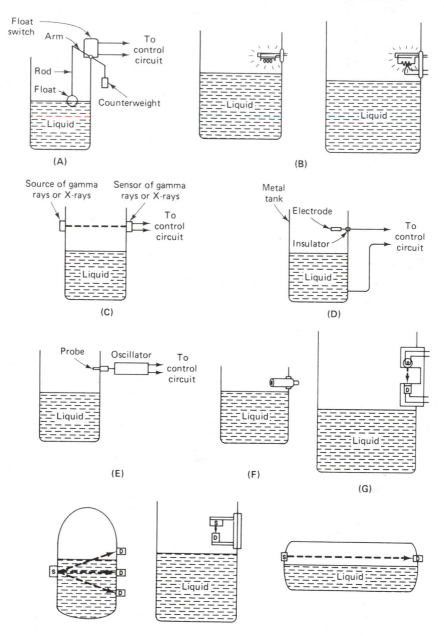

Figure 3-21 (A) Float switch employed as a level-limit sensor; (B) level sensing by heat transfer rate; (C) level-limit sensor employing gamma rays or X-rays; (D) conductivity level-limit sensor; (E) ultrasonic level-limit sensor; (F) level sensing by dielectric variation; (G) level sensing by optical means; (H) radiation sensing of liquid level.

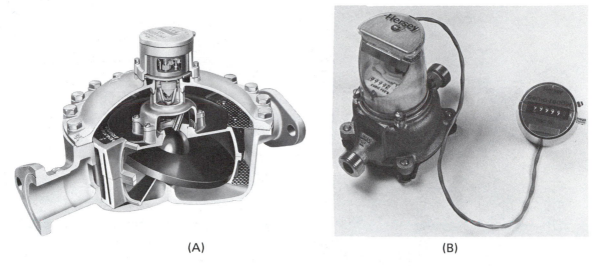

(A)   (B)

Figure 3-22 Nutating disc flow meter. [(A) Courtesy of Neptune Measurement Co.—a subsidiary of Neptune International Corp.; (B) courtesy of Hersey Products, Inc.]

motion), which effectively causes the quantity of liquid just admitted and trapped above the disc to be transferred to the outlet portion of the chamber. At this point the disc will be tilted with the left edge against the bottom of the chamber and the right edge against the top, thereby exposing the underside of the disc to liquid entering and simultaneously permitting the liquid trapped on top of the disc to exit the chamber at the outlet.

This process continues, each oscillation of the disc transferring a fixed volume of water from the inlet to the outlet. By counting the number of oscillations, the total quantity of liquid having passed through the meter can be calculated. To facilitate this counting function there is a pin projecting from the top of the nutating disc assembly. The pin either physically touches, or is magnetically coupled to, a counting mechanism. In the common household water meter the counter is geared such that the counter actually registers gallons of flow directly.

The nutating disc is an excellent and reliable device, capable of accurately measuring flow from approximately zero flow through its design maximum. However, as the quantities of flow become greater, larger devices are required, and these become cumbersome and expensive. Furthermore, they are mechanical devices subject to friction and wear and must be used only with "clean" (no lumps) liquids having some minimum lubricating qualities. Also, nutating disc meters are capable of handling only limited velocities of flow. Therefore, this type of flow sensor has rather limited application in industry.

We focus our attention now on devices that can only measure quantity of flow indirectly, some of the oldest and simplest of which are illustrated in Fig. 3-23. All these devices place calibrated restrictions in the line of flow and measure the pres-

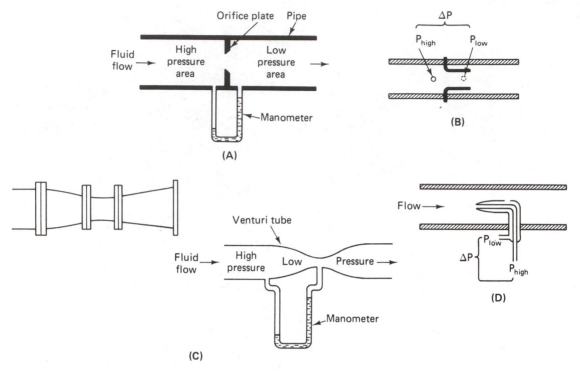

Figure 3-23   Rate-of-flow sensors employing: (A) orifice plate; (B) nozzle; (C) venturi tube; (D) Pitot tube.

sure drop across the obstruction. This pressure drop occurs because of a law of physics which states generally that where velocity is relatively low, pressure is higher; and conversely, that where velocity is relatively higher, pressure is lower.

The velocity of flow is less (therefore, the pressure is higher) on the upstream side of the obstructions, and the velocity of flow is considerably higher through the reduced cross-sectional area of the obstruction itself. Therefore, there is a pressure drop, and the magnitude of this pressure drop is proportional to the velocity of flow through the obstruction. However, the relationship between pressure drop and velocity of flow is very nonlinear. Furthermore, the obstruction must be designed for a specific range of flows (velocities), and the lower velocities will not register a pressure drop at all. In practice, all these devices are usable over relatively small spans of flow. Furthermore, they cannot be used with liquids that may have "chunks" in it or the openings may become plugged up, and the pressure drop is affected by other phenomena, such as temperature and viscosity of the liquid, which means that they must be selected for each specific application.

Each of the types of sensors illustrated has an advantage over others, but only a few will be mentioned here. The *orifice plate meter* is the least expensive, but the least accurate and has the most limited span of measurements. The *Pitot tube* is more susceptible to getting plugged up and is therefore used more with pneumatic systems than with hydraulic systems. It is commonly used as an airspeed sensor for aircraft and small boats, since it really measures velocity of flow. The *venturi* and *nozzle tubes* are more sophisticated (and more expensive) versions of the orifice plate. They are more accurate and are usable over wider spans of flow than the orifice plate meter, and are less susceptible to some of the perturbations that affect all these sensors.

In practice, all are used extensively in industry. They all suffer from maintenance problems, though, in that they must be disconnected from the system, and the system drained, before repairs are possible. In most cases bypass piping and extra valves are used so as to minimize the replacement problems.

There are several other types of flow-measurement sensors which rely on measurement of the pressure drop across an obstruction in the flow path. The *float* or *variable area meter* (Fig. 3-24A) places a calibrated weight (dimensions are also carefully controlled) in a transparent flow tube (normally glass or clear plastic) which changes internal diameter from smallest at the bottom to largest at the top. This meter must be mounted in a vertical position, and as flow enters from the bottom it pushes the float up the tube until exactly enough area is exposed (between the edges of the float and the sides of the tube) for the fluid to squeeze by. Therefore, the height of the float itself is an indication of the velocity (and therefore the quantity) of flow. These devices can measure quite accurately, but normally are used for visual indication only, not in automated systems. They are widely used as visual flow indicators to give a quick check as to whether the automated flow device is working properly.

The *target meter* or *cantilever beam meter* places a flat plate across the flow stream (Fig. 3-24B). The plate is fastened to the end of a springy bar, and the amount of distortion of this bar is an indication of the velocity of flow. They are not in widespread commercial use.

The *turbine flow meter* was designed specifically for aerospace and petrochemical flow measurements. In these applications the fluids to be measured had good lubricating qualities and were quite free of chunks. The requirement was for flow measurement over wider spans and with increased measurement accuracy as compared to the orifice and similar meters.

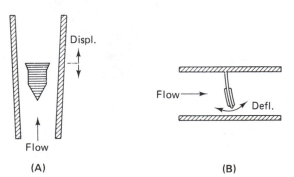

Figure 3-24 (A) Float (variable area); (B) cantilever vane (target meter).

It operates on the same principle as a child's pinwheel. As a liquid (or gas) passes a propeller-type blade, it causes that blade to rotate. The higher the velocity of flow, the faster the propeller blade spins. At very low flow rates, the blade will not spin; therefore, this sensor also has a lower flow cutoff rating.

In practice, the turbine (propeller with many blades) is suspended on very special bearings (extremely low friction) in the center of the pipe, the blades rotating close enough to the inside of the pipe to minimize leakage past the edges of the blades. The blades themselves are either magnetic or of a magnetic material such that as they pass a coil (which is sealed from the flowing liquid by a nonmagnetic seal, such as stainless steel) they induce a pulse in the coil (Fig. 3-25). Therefore, the electrical output from this type of device is a series of *pulses,* each pulse representing the passage of one of the turbine blades. The frequency of the pulses is therefore proportional to the velocity of the turbine itself, which is in turn proportional to the velocity of flow past it.

This type of sensor is also a maintenance headache, and is quite expensive; however, it is one of the best devices available for many applications (especially petrochemical applications) and is therefore widely used. One of the largest sources of error in the measurement of flow by this device is *turbulence* in the flow stream before (and immediately after) encountering the turbine. Therefore, flow-straightening vanes are necessary to assure a minimum of turbulence at the turbine itself.

One of the newer and more successful innovations in hydraulic flow sensors

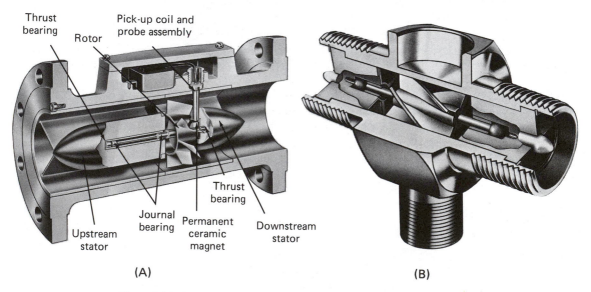

(A)   (B)

Figure 3-25 Cutaway photos showing the internal construction of commercial turbine flow meters. [(A) Courtesy of Neptune Measurement Co.—a subsidiary of Neptune International Corp.; (B) courtesy of Special Products Division of Halliburton Services, a Halliburton Company.]

is the *magnetic flow meter.* This device does not place any obstruction in the fluid flow path and is therefore useful for measuring flow that has chunks in it (i.e., sewage flow into a treatment plant). Furthermore, it gives a direct electrical output, which is a great systems advantage.

The typical magnetic flow meter basically consists of a straight section of non-magnetic pipe with an electrically insulating liner on the inside of that section of the pipe. It makes use of the fact that a moving *conductor* placed in a magnetic field will experience a voltage appearing across its opposite ends as it moves through the field. In the magnetic flow meter the liquid itself is used as the moving conductor. This places the restriction of at least a minimum of conductivity on the liquid itself, the only restriction on the liquid for this sensor.

Referring to Fig. 3-26A, try to imagine isolated cylindrical sections of the liquid as though they were detached from the main flow stream, moving at the velocity of the flow stream and oriented in a direction such that they cut the lines of flux created by an external (very powerful) electromagnet. Since these "conductors" are inside the insulated section of the pipe, it is necessary to insert two probes through the wall of the pipe (and insulated from it) and through the insulating lining on the inside of the pipe. These probes are essentially even with the inside surface of the pipe and therefore do not obstruct the flow stream at all. They are there to simply make electrical contact with the two ends of the liquid "conductors" as they pass through the intense magnetic field and provide two terminals external to the pipe to measure this electrical potential.

In reality, there are not isolated sections of liquid as conductors; rather, the fluid acts as a continuous conductor, thereby making continuous contact with the two probes and providing a continuous electrical signal at the output. As long as the fluid has at least a minimum level of conductivity, there are essentially no other restrictions on the flowing liquid. Unlike all the flow sensors discussed up to this point, variation in consistency, content, temperature, viscosity, or even turbulence does not affect the electrical output. There is one restriction on the physical installation of a magnetic flow meter in that it must be mounted in a section of pipe where it remains full of liquid at all times; otherwise, there would be periods of time when there was no contact between the probes and the liquid, and the electrical output would be discontinuous.

This sensor has essentially no limit to the range of velocities that can be measured except as placed by the electrical measuring equipment (the lower the velocity, the lower the output voltage) and by the physical size of the pipe itself. The electrical output (with ac coil excitation) will be an ac voltage with magnitudes in the microvolt area. Therefore, fairly exotic electronics must be used to detect and amplify these low voltages to usable levels. The electrical signal is linearly proportional to the rate (and therefore the quantity) of flow. The primary disadvantage of using a magnetic flow sensor is its cost; they are quite expensive as compared to other flow-measurement devices. Also, the requirement for a minimum fluid conductivity excludes its application in the petrochemical industry. Figure 3-27 illustrates a typical commercial magnetic flow meter.

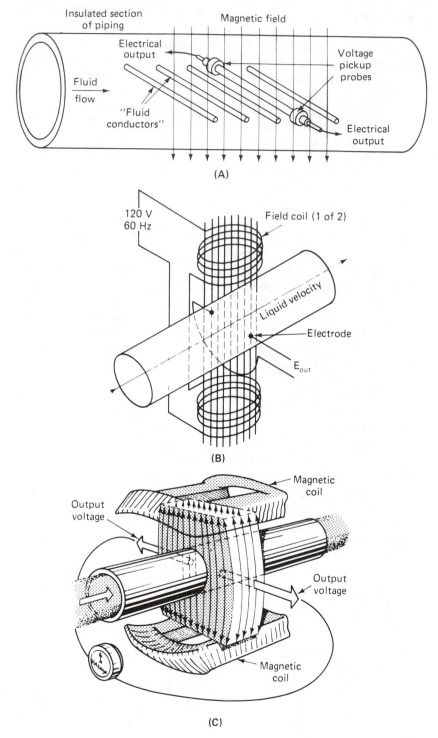

Figure 3-26 Magnetic flow meter schematics.

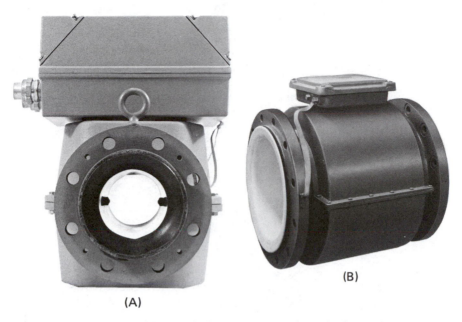

(A)

(B)

Figure 3-27 Commercial magnetic flow meters—note the probes in part (A). [(A) Courtesy of Taylor Instrument Process Control Division/SYBRON Corp.; (B) courtesy of Brooks Instrument Division, Emerson Electric Co.]

The *fluidic oscillator* is another relatively new and promising hydraulic flow sensor. It also places no obstruction in the line of flow, thereby causing a very small pressure loss, and has no moving parts to wear out. Referring to Fig. 3-28, the top sketch illustrates a flow stream coming out of a blunt opening. Note that on one side of the stream (bottom), there is a wall, whereas there is no wall illustrated on the other side (top) of the stream. As the liquid comes squirting out, it disperses, and much turbulence is created on the side of the stream nearer to the wall. Turbulent flow is high-velocity flow, so, as a result of the high-velocity turbulent flow between the main stream and the wall, an area of lower pressure is created. The area of relatively higher pressure on the other side (top side) of the stream forces the main stream to deflect slightly toward the wall. This causes increased turbulence and still lower pressure on the wall side of the stream, which maintains it in this deflected position. In the absence of any further disturbance, the flow stream will remain "attached" to this wall.

Now shift your attention to Fig. 3-28B. The housing is designed differently from Fig. 3-28A, but exactly the same phenomenon would occur as soon as the flow started. It will be attracted to the bottom wall and would remain there, the little cavities at *a* ensuring this. However, as the stream is deflected toward the bottom wall, it meets a very small obstruction at *b* which diverts a small portion of the main stream into the feedback passage. As the feedback passage fills with liquid, it creates

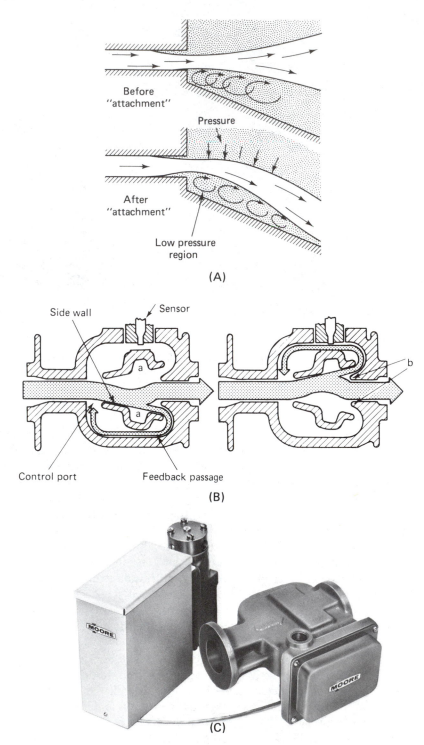

Before
"attachment"

Pressure

After
"attachment"

Low pressure
region

(A)

Side wall   Sensor

a

a

b

Control port   Feedback passage

(B)

MOORE

MOORE

(C)

Figure 3-28 Fluidic flow meter: (A) wall attachment principle; (B) principle of operation; (C) commercial fluidic flow meter. (Courtesy of Moore Products Co.)

an area of high pressure at the lower control port. This high pressure then deflects the stream toward the opposite wall (the top wall). As the main stream approaches the top wall, turbulence is created and it attaches itself to this wall exactly as it had done to the bottom wall. There is an obstruction on the top wall also, which diverts a small portion of the main stream into a feedback passage which eventually causes a high pressure at the top control port, and the stream is deflected toward the lower wall, where the process repeats itself.

The stream thus oscillates back and forth between the two walls. The frequency of oscillation is linearly related to the velocity (and quantity) of flow through the sensor. The manufacturer claims excellent accuracy for this device (available in many sizes, depending on expected flow rates) as long as a minimum flow rate is maintained. It must be used with relatively clean liquids so that the feedback and control passages do not get clogged up. Shock, vibration, changing temperatures, and changing viscosities do not affect this device. A pressure sensor is illustrated as detecting the pressure in the upper feedback passage, each high pressure pulse indicating a complete cycle.

The field of *ultrasonics* has also made very successful excursions into the area of hydraulic flow measurement in several versions. One of the more promising techniques is illustrated in Fig. 3-29. In this design a small obstruction is inserted through the pipe into the flow path. As the liquid flows past this obstruction it causes not just turbulence, but a very special kind of turbulence. This kind consists of swirling whirlpools, or *vortexes,* first off one side of the obstruction and then off the other side. These whirlpools are areas of high velocity and therefore low pressure, and they are swept along the pipe by the main flow.

A short distance down the pipe an ultrasonic transmitter is located on one side of the pipe and its receiver is located directly opposite it. There is a continuous transmission of an ultrasonic wave (through the fluid) across the pipe, but the vortexes interfere with the received quantity of the signal, causing level shifts in the magnitude of the received signal. The frequency of these level shifts is directly proportional to the frequency of vortex generation, which in turn is directly proportional to the velocity (and therefore quantity) of liquid flow.

There are definite restrictions as to type of liquid and rate of liquid flow for these devices. However, the calibration is dependent only on the physical characteristics of the obstruction placed in the flow path, so it maintains its calibration well.

A second ultrasonic (liquid) flow-metering sensor is the ultrasonic wave velocity flow meter. Some commercially available designs of this sensor can be clamped around any section of installed piping, or it can be built into special sections of pipe. One design consists of two heads which are located on opposite sides of the pipe. One head is an ultrasonic transmitter and it transmits an ultrasonic signal through the wall of the pipe into the fluid. The other head is an ultrasonic receiver and it is clamped on the opposite side of the pipe a fixed distance (along the pipe) away from the transmitter. This is illustrated in Fig. 3-30.

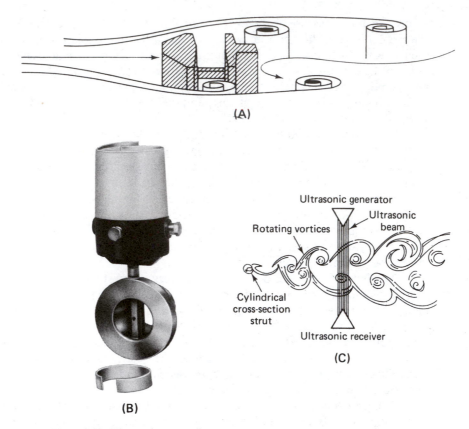

(A)

(B)

(C)

Figure 3-29 Ultrasonic vortex flow meter. (Courtesy of Fischer & Porter Co.)

The ultrasonic signal is transmitted into the fluid, and it takes the signal a finite amount of time to travel from the transmitter to the receiver. The amount of time depends upon the pipe and liquid characteristics (which are fixed quantities, not variables, and can therefore be calibrated out) and the velocity (and direction) of fluid flow through the pipe.

Therefore, by measuring the time differences between transmission and reception, the device can be calibrated in terms of velocity of fluid flow.

Another design has a schematic similar to Fig. 3-30A except that both heads are transponders (receivers and transmitters). Ultrasound bursts are sent from one to the other alternately. The alternate bursts yield a frequency difference (due to the Doppler effect) which is proportional to the velocity of flow through the pipe. Figure 3-30C illustrates a sensor operating on this principle.

There are several additional flow sensors which are used for measuring *pneumatic flow*. Pneumatic flow is always calculated from the cross-sectional area of

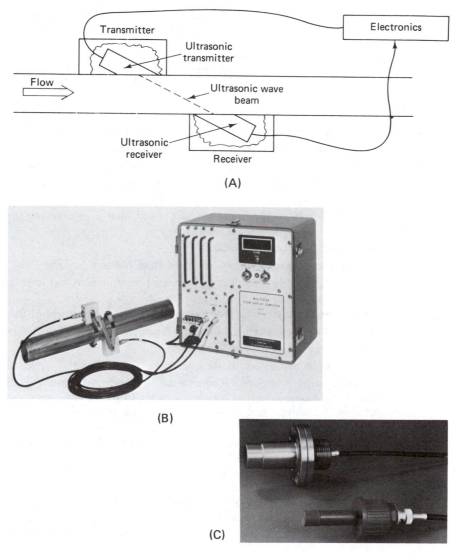

Figure 3-30 Ultrasonic wave velocity liquid flow meters. [(B) Courtesy of Controlotron Corp.; (C) courtesy of DuPont Company.]

the duct and a measurement of air velocity flow; there is no common direct volumetric pneumatic flow rate sensor. Pitot tube velocity-measuring devices are commonly used for this purpose. Another device consists of a resistance heating element with a temperature sensor attached. This device is lowered into a duct and the rate of cooling of the heating element is proportional to the velocity (and temperature) of the air flowing past it. There are many other gas (pneumatic) flow devices, but they are not commonly enough used to warrant space here.

QUESTIONS

1. Differentiate between hydraulics and pneumatics.
2. What general categories of hydraulic (and pneumatic) sensors are there?
3. Name at least two different sets of pressure measurement units and explain each.
4. Differentiate among absolute, gage, and atmospheric pressures.
5. How does a differential pressure measurement differ from a normal pressure measurement?
6. What is a bourdon tube, how is it constructed, and what is its basic principle of operation?
7. Repeat Question 6 explaining a bellows.
8. Generally, how are electrical outputs obtained from pressure sensors?
9. What is the difference between analog fluid-level measurement and level-limit sensors and their applications?
10. What is the most basic practical problem related to the measurement of fluid flow?
11. Which of the flow sensors actually (directly) measures quantity of fluid flow?
12. Describe the operation of turbine flow meters.
13. Describe the operation of magnetic flow meters.
14. Make a table, listing in one vertical column each of the sensors covered in this chapter. Then make four adjacent vertical columns, labeling them: Variable measured, Principle of operation, Advantage, and Disadvantage. Attempt to fill every blank space in the table.
15. Which of the following phenomena is commonly used in industry to sense the difference between two *pneumatic* pressures?
    A. The proportionality between liquid level and pressure.
    B. The attenuation of nuclear radiation by solid materials.
    C. The variation of resistance of wire as it is deformed.
    D. The sensitivity of hair to moisture.
    E. The fact that where hydraulic flow velocity is high, the corresponding pressure will be low, and vice versa.
16. *Gage pressure* refers to the pressure difference between an unknown and _____ ?
17. *Absolute pressure* refers to the pressure difference between an unknown pressure and _____ ?

Match the following numbered descriptions with the letter of the most appropriate device (a–e) named below. Each answer can be used only once.
For questions 18–22, in which device (a–e) is pressure indicated by _____ ?

18. Liquid level?
19. Movement of a curved tube of resilient metal, sealed on one end?
20. Elongation of thin-walled, pleated cylinder, sealed at one end?
21. Bulging pleated disc of thin metal?

22. Motion of two pleated discs of opposite curvature?
   a. Diaphragm
   b. Bellows
   c. Bourdon
   d. Capsule
   e. Manometer

23. Liquid level (in a tank) can be sensed using which of the following phenomena?
   A. The proportionality between liquid level and absorption of various types of radiation by the liquid.
   B. The difference in incoming and outgoing flows.
   C. Reflection of ultrasonic radiation from areas of sharp change in density.
   D. The proportionality between liquid level and pressure.
   E. All of the answers above are correct.

24. Which of the following hydraulic flow sensors actually measures volume of flow (not velocity)?
   A. Magnetic flow meter.
   B. Turbine flow meter.
   C. Ultrasonic flow meter.
   D. Nutating disc (water meter) flow meter.
   E. All of the flow sensors above actually directly measure volume of flow.

25. Variations in fluid consistency, content, temperature, viscosity, and turbulence do *not* affect the operation of which of the following measuring devices?
   A. Pitot.
   B. Turbine.
   C. Magnetic flow meter.
   D. Fluidic oscillator.
   E. Float (variable area) flow meter.

Match the following numbered descriptions with the letter (a–e) of the most appropriate device named below.

26. Transmitted sound intensity varies proportional to flow.

27. Transmission time of sound is proportional to rate of flow.

28. L-shaped tube points into fluid stream.

29. Pipe constricts to cause flow-dependent pressure difference.

30. Plate with hole causes pressure difference proportionate to flow.
   a. Pitot
   b. Orifice
   c. Venturi
   d. Ultrasonic vortex flow meter
   e. Ultrasonic wave velocity meter

31. What is the difference between a single-input pressure transducer and a differential pressure (D/P) transducer?
   A. A D/P transducer always measures higher pressures.
   B. A single-input pressure transducer can measure only absolute pressure.
   C. A single-input pressure transducer can measure two pressures relative to another pressure.
   D. A differential pressure transducer measures the difference between two pressures.
   E. All of the answers above are correct.

32. Which of the following is the principle on which the ultrasonic vortex flow meter works?
    A. Vortexes generate ultrasonic sounds with frequency proportional to flow.
    B. Vortexes generate ultrasonic sounds with intensity proportional to flow.
    C. Ultrasonic waves generate vortexes proportional to flow.
    D. Vortexes generated due to flow modulate transmitted sound.
    E. The frequency of the ultrasonic wave is changed proportional to flow.

Match the following numbered descriptions with the letter (a–e) of the most appropriate device named below.

33. Magnetically generated electric pulse train proportional to flow.

34. Voltage proportional to flow.

35. Height of indication proportional to flow.

36. Frequency of pressure pulses proportional to flow.

37. Measures flow by passing measured volumes of fluid.
    a. Fluidic oscillator
    b. Nutating disc flow meter
    c. Variable area float
    d. Magnetic flow meter
    e. Turbine

38. Which of the following hydraulic flow sensors offers the least resistance to flow (is the highest-impedance measurement system)?
    A. Vortex shedding.
    B. Orifice plate.
    C. Nutating disc.
    D. Magnetic flow meter.
    E. The statement of the problem is contradictory to itself.

39. A high-impedance hydraulic flow measurement system is one in which the actual sensor itself has a low impedance. True or false?

40. A high-impedance pressure measurement system is one in which the actual sensor itself has a low impedance. True or false?

41. Turbine flow meters are primarily used to measure flow of _____ fluids.
    A. Corrosive.
    B. Chunky.
    C. Viscous.
    D. Petrochemical.
    E. Turbine flow meters are normally used for all the liquids mentioned.

42. The bellows transduces
    A. Mechanical motion to pressure.
    B. Hydraulic flow to differential pressure.
    C. Temperature to pressure.
    D. Light intensity to voltage.
    E. Pressure to mechanical motion.

43. In applications using a magnetic flow meter, the fluid
    A. Must have magnetic properties.
    B. Deflects a magnetized "hinge."
    C. Must be a conductor.
    D. Must be an insulator.
    E. Flows through a pipe made from a magnetic conductor.

44. The venturi tube transduces
   A. Mechanical motion to pressure.
   B. Hydraulic flow to differential pressure.
   C. Temperature to pressure.
   D. Light intensity to voltage.
   E. Pressure to mechanical motion.

GLOSSARY

Absolute Pressure   Measurement of one pressure as compared to the pressure of outer space (which is accepted as being absolutely zero pressure).

Bourdon Tube   A thin, springy, cylindrically shaped tube, (usually) bent into a curved shape with one end sealed and the other end open. Used to measure pressures.

Conductivity   A measure of the ability of a material to conduct electron flow; the reciprocal of restivity.

Conductor   A material having atomic properties which permit the flow of electron current while offering relatively little opposition to that flow.

Diaphragm   A flexible membrane which has the effect of sealing one chamber from another, while responding to pressure differences between the two chambers by mechanically deforming.

Differential Pressure   Measurement of one pressure with respect to another pressure. The difference between the two pressures.

Gage Pressure   Measurement of one pressure with respect to the pressure that would be measured at sea level. Normally measured with respect to ambient pressure, rather than sea level pressure.

Hydraulics   That branch of physics which deals with the behavior of liquids.

Pneumatics   That branch of physics which deals with the behavior of gases.

Pressure   Weight or force distributed over an area. The driving force or cause of hydraulic or pneumatic flow in a system.

Pulses (Electrical)   An abrupt change in voltage from one level to another, remaining at that new level for a brief period of time, and then returning to the original level.

Radiation   The propagation of any of several forms of energy through space, or through a material. Electromagnetic fields, light, sound, and heat are examples.

Turbulence (Hydraulic)   A term which describes the flow of a liquid which is in agitated motion, as compared to smooth (laminar) flow. Normally this is an undesirable condition of flow.

4

TEMPERATURE SENSORS

HEAT AND TEMPERATURE

Temperature measurements are frequently made by default, when the real measurement desired is a *heat* measurement. This raises two questions: first, what is the difference between *temperature* and heat; and second, why don't we use heat sensors to measure heat directly? The answer to the second question is that no practical true heat sensor has ever been developed, so that is why they are not used.

The answer to the first question is somewhat more complicated and should have been covered in a previous physics course (or the reader is referred to Appendix A). But, as a quick review, let us think of a hydraulic pipe with liquid flowing through it into a tank. The rate of flow through the pipe (and into the tank) is dependent (among other things) on the pressure applied to the liquid to cause the flow—higher pressure results in faster flow, and vice versa. The pressure applied to cause the liquid to flow is analogous to temperature which causes heat to flow (temperature being analogous to pressure and heat flow being analogous to fluid flow). The total quantity of water in the tank (gallons collected) is analogous to the total heat which flowed (units are British thermal units, Btu). Therefore, temperature causes heat to flow, or a flow of heat through an object will cause a temperature difference between the points where the heat enters a system and the point where it leaves. Therefore, the primary difference is one of cause and effect; temperature (differences) cause heat to flow. Let us now discuss some commercially popular temperature sensors.

There are very few measurements made in industry for which such an excellent and relatively inexpensive selection of sensors exist as for temperature measurement.

Many different physical phenomenon can be taken advantage of as direct indications of the temperature of a substance, and many different commercially designed products are available using each of these phenomena. Let us start off with a couple of the older ones and then work into the newer developments.

FILLED-SYSTEM THERMOMETERS

The oldest temperature sensors relied on the volumetric expansion of liquids with increasing temperatures—the ordinary *liquid-in-glass thermometers.* In these thermometers a glass tube has a thin hole bored its entire length (Fig. 4-1), with a small reservoir at one end. The tube is filled and the open end is sealed under conditions such that there is no air in the tube at all, only the liquid.

There is therefore a fixed volume of liquid in a sealed (fixed, but larger volume) container. As the temperature of the liquid changes, so does its volume, and the only place the liquid can expand (by design) is up the bored hole. Here we have a temperature-to-volumetric change transducer *and* a volumetric change-to-level transducer simultaneously, because the volume changes are indicated in terms of level changes.

Note that the accuracy of this device depends only upon the uniformity of diameter of the bored hole and the scale markings. Over given temperature ranges, liquids can be used that experience linear-volume-change-for-temperature-change characteristics, and since there is no air in the tube to be compressed as the liquid level rises, this remains a linear relationship in practice. These devices can be made to be extremely accurate and are constantly used to calibrate the more sophisticated devices to be discussed later.

The liquid-in-glass thermometer does not easily lend itself to inclusion in a system; however, with some modifications we can come up with a more compatible sensor while using essentially the same principle. The liquid-filled (or gas-filled) bulb is such a device. This device (Fig. 4-2) consists of a bulbous reservoir on one end (the end exposed to the temperature to be sensed) which is connected to a pressure sensor on the other end by a special type of tubing. The entire system is filled either

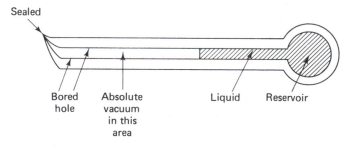

Figure 4-1 Liquid-in-glass thermometer.

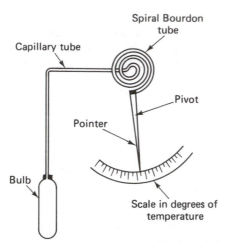

Figure 4-2 Filled bulb temperature sensor.

with a specific liquid or a specific gas and is sealed. The liquid (or gas) is therefore confined in a fixed volume, the bulb, tubing, and pressure sensor being of such a construction that they will not change internal dimensions significantly with pressure or temperature changes.

As the temperature of the liquid (gas) in the bulb changes, its volume also tries to change. Since it is confined in a fixed volume, the net effect is a change of internal pressure, hence the need for a pressure sensor. Therefore, this sensor is a temperature-to-pressure transducer. It has been widely used in industry for many years. Some readers may remember the automobile temperature gages used before the advent of "idiot lights." These temperature gages were of the filled-bulb type and used a bourdon tube pressure sensor under the dashboard to give a visual indication of engine temperature.

BIMETALLIC TEMPERATURE SENSORS

There is a second type of temperature sensor which has been around nearly as long as the liquid-in-glass thermometer, the *bimetallic-type temperature sensor*. These sensors make use of the physical phenomenon that all solid materials change their dimensions when their temperature is changed.

A bimetallic strip (Fig. 4-3A) is constructed by bonding two thin strips of two different metals together such that they cannot move relative to each other. Since all materials change their physical dimensions at different rates when undergoing the same temperature change, these two metal strips also try to change lengths with temperature change *but* at different rates. The net result is that the bimetallic strip *bends* with temperature change.

If one end of the bimetallic strip is fixed, the position of the other end (the free end) is a direct indication of the temperature of the strip. By using any of the

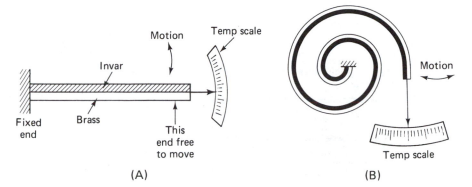

Figure 4-3 Bimetallic temperature sensors.

displacement sensors discussed previously as a secondary transducer with the bi-metallic strip, we can get an electrical signal out that is proportional to temperature. This device is simple, rugged, holds its calibration very well, and is quite inexpensive. Therefore, it is in extremely widespread use.

There are several modifications of the basic bimetallic strip which extend its usefulness. The first modification is to increase its sensitivity. For the bimetallic strip, this means a greater movement of the free end for the same temperature change. This is very simply accomplished by making the strip longer; but as it gets longer, it becomes unwieldy and awkward to handle. Therefore, this longer bime-tallic strip is formed in the shape of a spiral (Fig. 4-3B), and a very long length is compressed into a fairly small space. This spiral shape is also in extremely wide-spread use.

A further modification of the bimetallic strip consists of adding electrical contacts to the free end (Fig. 4-4) and including it as a part of an electrical circuit. As illustrated in Fig. 4-4, as the strips bend, they either make electrical contact and complete the circuit or they open the contacts, thereby breaking the electrical circuit. This device is in such common usage that it is given a special name, *thermostat*.

Thermostats similar to Fig. 4-4A are used in electric toasters, coffee pots, and irons. Thermostats similar to Fig. 4-4B are to be found in the majority of home heating systems. Thermostats of the design in Fig. 4-4A and B do have one problem, however. As the points open and close they are moving quite slowly, and it is common for them to arc and burn. There are several methods of modifying the thermostats so that the contacts open with a snap action. Figure 4-4C illustrates one of these, a bimetallic disc. Devices similar in construction to the bimetallic disc are commonly used to interrupt inductive circuits, such as thermal safety (overload) switches in electric motors.

When bimetallic devices are included as components in electric circuits, they can be designed to respond either to ambient temperature, in which case they are called thermostats; or they can be designed to respond more to electrical current

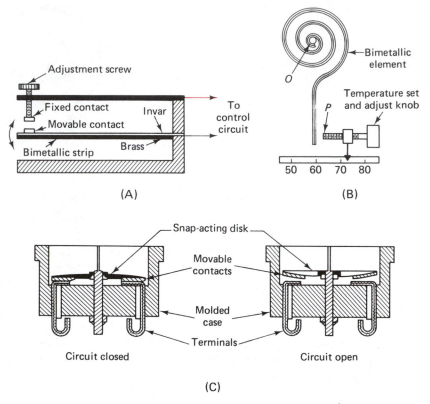

Figure 4-4 Bimetallic temperature sensors used in thermostats.

flowing through the circuit, in which application they would be called *circuit breakers*.

      Figure 4-5A illustrates common circuit-breaker construction, where heating elements are used to increase the sensitivity of the device. Figure 4-5B is even simpler in design, where the current flowing through the bimetallic strip causes self-heating

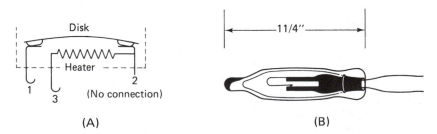

Figure 4-5 Bimetallic temperature sensors used in circuit breakers.

($I^2 R$-type heating) of the element. This latter sensor is commonly used in flashing Christmas tree lights.

RESISTIVE TEMPERATURE SENSORS

Another separate category of temperature sensors relies on the temperature-electrical resistance characteristics of metals (conductors) and *semiconductors.* First, we shall discuss sensors that rely on the temperature sensitivity of the electrical resistance of common metals.

The simplest of the *resistive temperature sensors* consist of a long piece of fine wire (copper, nickel, or platinum, for example) formed either into the shape of a coil or woven in a shape like a piece of very coarse (fabric) material. This wire sensor is then enclosed in a protective sheath, which in turn is in contact with the substance whose temperature is to be measured (sensed). Figure 4-6 shows a couple of commercial designs using this principle. The choice of wire element, sheath materials, and dimensions is made for each specific application; therefore, a great variety of these devices are available as stock items.

By changing the shape of the container (Fig. 4-7) the same principle can be used to measure the temperature of a surface. Here the wire-sensing element is effectively spread over a large area and can be fixed by an adhesive to the surface whose temperature is to be sensed. Resistance temperature sensors, especially platinum-wire elements, can be made to be so accurate and repeatable that replacement of defective sensors requires no recalibration of the system. Furthermore, platinum-wire temperature sensors are sufficiently accurate to be used as temperature measurement standards for calibration of other temperature sensors.

Use of these devices in a system presents significant problems to the electronics that are to interface with these resistive temperature sensors. This subject will be covered in detail in Chapter 6.

Now let us turn our attention to the semiconductor resistive temperature sensors. These devices are excellent temperature sensors and are called *thermistors,* to differentiate them from the other type of resistive temperature sensors. You may remember from past semiconductor courses that basic semiconductor materials are extremely sensitive to temperature (among other phenomena).

They, contrary to metallic conductors, exhibit an increase in resistance with decreasing temperature. Commonly available thermistors will exhibit changes of resistance from less than 100 Ω at high temperatures to hundreds of megohms of resistance at low temperatures. This makes the thermistor orders of magnitude more sensitive than the metallic resistive temperature sensors over the same temperature span.

The major problem hindering widespread commercial acceptance of thermistors has been quality control over the exact temperature/resistance characteristics of the commercially available devices. In the event of a sensor failure, a metallic resistive sensor can be commercially replaced with one with almost exactly the same

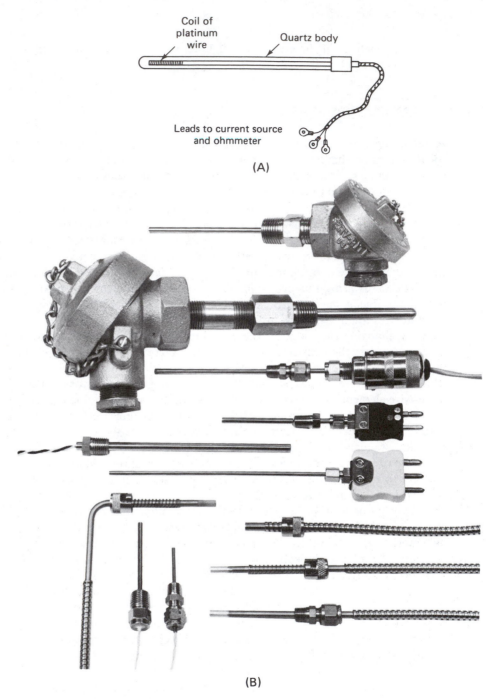

Coil of
platinum
wire

Quartz body

Leads to current source
and ohmmeter

(A)

(B)

Figure 4-6 (A) Schematic diagram of one form of platinum resistance thermometer; (B) group of industrial resistance temperature sensors. [(B) Courtesy of HYCAL Corp.]

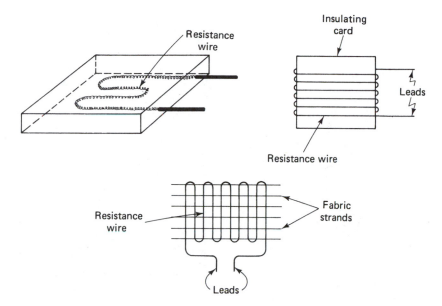

Figure 4-7 Resistance-type surface temperature sensors.

characteristics; therefore, recalibration of the temperature measurement system will not be required when replacing this type of sensor. Manufacturing tolerances on thermistors have not been good enough to make them competitive with metallic sensors until recently. Therefore, they were used more where sensitivity and/or size were more important than replaceability without system recalibration.

Modern thermistors have largely overcome this fault and they are in widespread commercial use. They are available in a multitude of sizes, shapes, and temperature ranges, some of which are illustrated in Fig. 4-8. They are smaller, more sensitive, and have generally faster response times than the metallic type; they may also be less expensive from a systems point of view because, owing to the larger changes in resistance for given temperature spans, less sophisticated measurement equipment may be acceptable.

THERMOCOUPLES

Another (older) type of temperature sensor is the *thermocouple*. Back in the early 1800s a man by the name of Seebeck discovered that if two wires, of different metals, are joined together to form a closed loop, and the two junctions are at different temperatures, an electric current will flow. The magnitude of the dc current will be directly proportional to the temperature difference between the two junctions, and the polarity is dependent upon the materials the wires are made of. This is an example of direct conversion of heat energy into electrical energy.

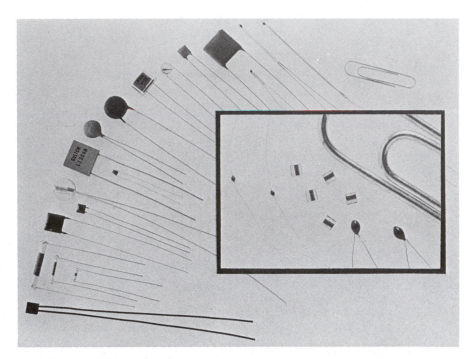

Figure 4-8 Group of industrial thermistors showing various sizes and shapes commonly available. (Courtesy of Gultan Industries, Inc., Piezo Products Division, Metuchen, N.J.)

If one junction is always maintained at a constant reference temperature (Fig. 4-9A), the magnitude of the current flowing will be directly proportional to the temperature at the other junction. Tables can be made up relating this current directly to temperature, the only variables in the tables being the temperature of the reference junction and the specific materials the two wires are made of. Because of their predictability, these devices, like resistance wire sensors, can be replaced in a system without the need for recalibration.

In practice, the devices are used with a voltmeter in place of the ammeter as shown in Fig. 4-9A. Therefore no current flows and the thermocouple acts as a voltage source, the voltage (which is down in the millivolt region) being proportional to the temperature sensed. This offers more predictable results, since the thermocouple junctions are capable of very little power, and measuring system lead length would otherwise affect the measured values.

Thermocouples are extremely simple and rugged and are in widespread use. There is a good selection of combinations of metals from which thermocouples are made, each covering a different range of temperatures to be measured. The measurement electronics required to measure the low power voltages generated by thermocouples are special and add significantly to system cost, however.

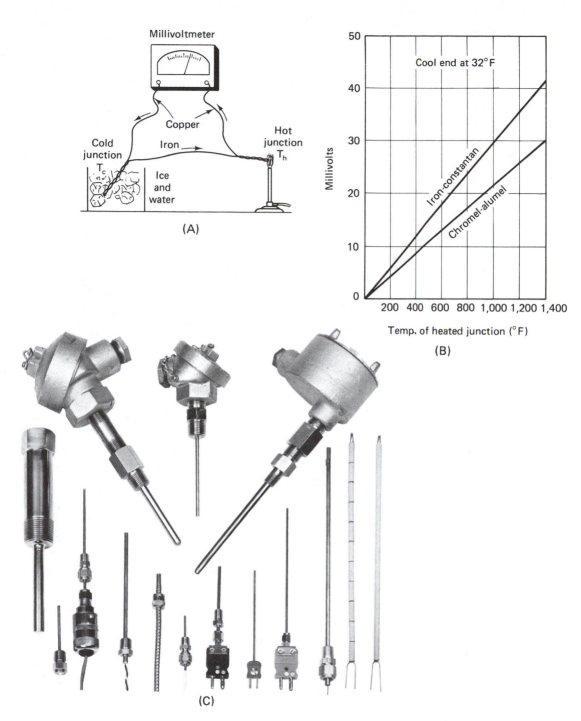

Figure 4-9 (A) Principle of the thermocouple; (B) graph showing temperature-voltage curves for iron-constantan and chromel-alumel thermocouples; (C) group of industrial thermocouples. [(C) Courtesy of HY-CAL Corp.]

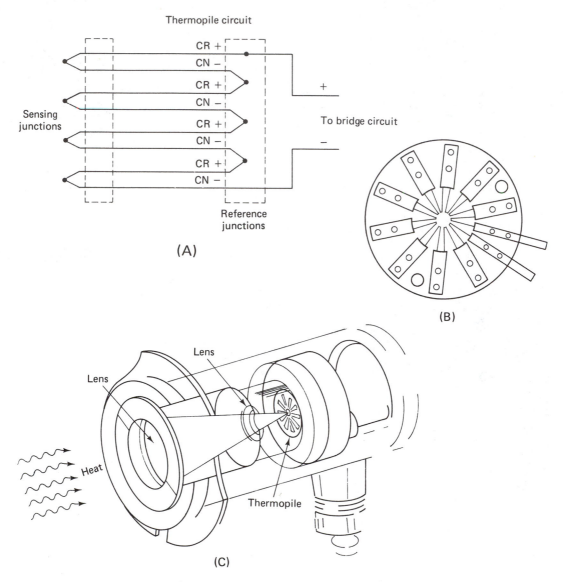

Figure 4-10 (A) Schematic diagram of thermopile; (B) phantom view of a thermopile; (C) phantom view of a pyrometer head.

The sensitivity of thermocouple temperature-measuring systems can be significantly increased by connecting several measuring junctions in series (Fig. 4-10). This configuration is technically referred to as a *thermopile*. It is commonly used in a device called a *radiation pyrometer* for measuring very high temperatures. Figure 4-11 is an abbreviated table of useful thermocouple temperature ranges.

Material	ISA Type	Range (°C)
Iron/constantan	J	0 to 1000
Chromel/alumel	K	0 to 1200
Copper/constantan	T	−190 to 380
Chromel/constantan	E	−190 to 1000
Platinum 13% rhodium	R	0 to 1790
Platinum 10% rhodium	S	0 to 1790

Figure 4-11 Abbreviated table of useful thermocouple temperature ranges.

ELECTRONIC ICE-POINT SIMULATION

In discussing and using thermocouples it should have become obvious that the actual use of ice baths for the reference junction is messy and impractical. Therefore, electronically simulated ice-point thermocouple reference junctions have been developed.

The electronic ice-point simulator (Fig. 4-12) is a battery-powered electronic circuit which generates the exact electromotive force (emf) that would be generated by a reference junction if it were submerged in ice at exactly 0°C. The simulator includes temperature-sensitive elements (internally) which compensate for the actual ambient temperature of the electronic circuit itself. Convenient terminals are normally provided which are designed such that secondary errors due to the connections at those terminals, and those errors due to lead wires from the sensor to the rest of the measuring system, are either compensated for or eliminated. Simulators are typically accurate to within 1°C or less, depending on the temperature range.

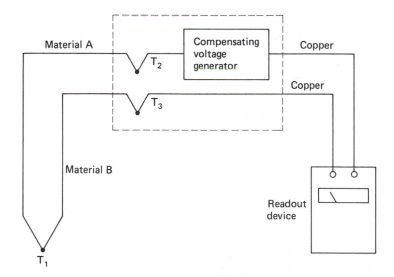

Figure 4-12 Schematic illustrating the proper connection of an electronic ice-point simulator into a thermocouple measurement system.

PYROMETERS

The radiation pyrometer operates on the principle that heat is a radiant form of energy and can be focused (at some distance from the source of heat) by a lens system. The heat is focused in the radiation pyrometer onto a thermopile, and the voltage generated by this thermopile is therefore proportional to the temperature being measured. These devices are used only for measuring such high temperatures as those of molten metals, filaments, and fires. The electrical output from the thermopile is compatible with automated systems. Figure 4-13 illustrates a hand-held version of a radiation pyrometer.

While on the topic of measuring extremely high temperatures, there is an optical device that is also used, the *optical pyrometer*. The optical pyrometer consists of a lens system and a very special light bulb. The optical system is oriented such that an observer looking through it (like a telescope) looks past the filament and at the object whose temperature is to be sensed (Fig. 4-14).

The heart of the system is the filament in that light bulb. There is a knob on the side of the pyrometer which is calibrated directly in terms of the temperature of that filament. Actually, it is a potentiometer which controls the current through the filament and therefore its temperature.

When an observer looks through the telescope, his eye perceives the filament as a dark line across the field of view. As he turns the knob, the filament heats up and turns color. The observer increases the temperature of the filament until it glows at exactly the same color as the substance in the background, which is the substance whose temperature is to be determined. At this point the filament can no longer be distinctly perceived by the eye and is at the same temperature (which can be read

Figure 4-13 Hand-held infrared radiation pyrometer. This type of pyrometer responds to the infrared radiation from a heat source and automatically indicates its temperature. The sensor in this device is a lead-sulfide detector; not a thermopile. (Courtesy of Pyrometer Instruments Co., Inc., Northvale, N.J.)

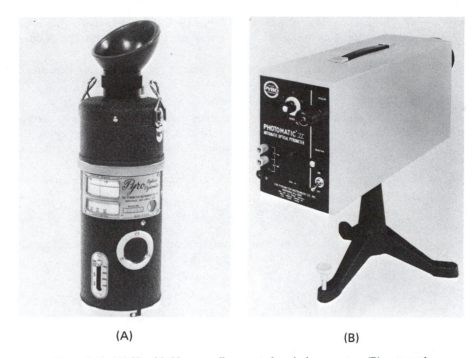

(A)   (B)

Figure 4-14 (A) Hand-held, manually operated optical pyrometer; (B) automatic optical pryrometer. This device has circuitry which electronically compares the color of the object whose temperature is to be measured with the reference filament, the temperature of which is known. It also has outputs capable of being used in a real-time process control system. [(A) and (B) Courtesy of Pyrometer Instrument Co., Inc., Northvale, N.J.]

from the dial) as the substance. This particular system is not easily adapted for automated systems but is in common industrial use.

This is a brief survey of those temperature sensors commonly in commercial use today. As mentioned previously, there is an exceptional selection of temperature sensors available, most of which are easily adaptable for automated systems use.

QUESTIONS

1. Differentiate between the terms *temperature* and *heat*.
2. Why are there no heat sensors discussed in this chapter?
3. How is heat measured when it is needed in a process control system?
4. List several different naturally occurring phenomena which can be taken advantage of for the design of temperature sensors.
5. Differentiate between *thermistors* and *thermocouples*.

6. Thermostats are in such common use that you should have a working knowledge of their principles of operation. In your own words explain the principle of operation of thermostats.

7. When you require a temperature measurement from a surface that cannot be touched, what phenomenon can be used to make that measurement?

8. Differentiate between optical pyrometers and radiation pyrometers.

9. Make a table, listing each of the sensors covered in this chapter in one vertical column. Then make four adjacent columns, labeling them: Variable measured, Principle of operation, Advantage, and Disadvantage. Attempt to fill in every blank space in the table with a (very) brief answer.

10. A thermocouple temperature sensor relies on what basic phenomenon for its operation?

11. "Heat" is sensed
    A. Directly through the use of heat sensors.
    B. Indirectly.

12. A high-impedance temperature measurement system is one in which the actual sensor itself has a low impedance. True or false?

13. The filled bulb temperature sensor actually directly transduces
    A. Pressure to flow.
    B. Temperature to flow.
    C. Temperature to mechanical position of a pointer.
    D. Temperature to volumetric expansion of the fluid.
    E. Temperature to current.

14. Which of the following lists includes types of sensors that exhibit a change in electrical resistance as a function of temperature?
    A. Coil of wire and thermistor.
    B. Instrinsic semiconductor material and coil of wire.
    C. Extrinsic semiconductor material and thermistor.
    D. Intrinsic and extrinsic semiconductor materials.
    E. All of the answers above are correct.

15. A thermistor is _____.

16. How is a thermocouple constructed?

17. The single most commonly made variable measurement in process control is _____ measurement.
    A. Pressure.
    B. Time.
    C. Temperature.
    D. Force.
    E. All of the measurements above are equally common.

18. The difference between temperature and heat is analogous to the difference between
    A. Pneumatic pressure and flow.
    B. Force and mechanical energy.
    C. Voltage and current.
    D. Hydraulic pressure and flow.
    E. All of the analogies above are correct answers to the question.

19. When you require information concerning the temperature of an object and cannot "touch" it, what phenomenon can you employ to obtain this information?

 A. Change in atmospheric pressure in the vicinity of the object.

 B. Measurement of air temperature and use of an equation.

 C. The fact that heat is a radiant form of energy and can be measured remotely.

 D. The fact that the surface emits *X-rays* and a radiation pyrometer will measure these rays remotely.

 E. None of the methods above will work; you're just out of luck.

GLOSSARY

Bimetallic   Consisting of two different metals which have been mechanically brought together for some useful purpose (i.e., sensing temperature).

Heat   The flow substance in a thermal system. A form of pure energy.

Semiconductor   A term commonly applied to any of several substances which have resistivities in the range between conductors and insulators. Modern science has discovered many of the unique electrical effects of these substances and designed them into many electronic devices which are collectively termed *solid-state* electronic devices.

Temperature   A measure of the degree of hotness or coldness of a substance. Also a measure of the driving force which has the ability to cause heat to flow between two bodies of different temperatures.

Thermistor   A semiconductor device whose electrical resistivity changes as a predictable function of its temperature.

Thermocouple   A device consisting of two wires, made of different metals, which are electrically joined at two points. When these two junctions are at different temperatures, an electromotive force (emf) will be thermoelectrically generated which is proportional to that temperature difference.

Thermopile   A grouping of thermocouples connected together in series (series aiding), commonly used to measure radiant energy.

Thermostat   A device designed to actuate a set of electrical contacts based upon a temperature measurement.

5

MISCELLANEOUS SENSORS

LIGHT SENSORS

Visible light is only a small portion of the spectrum of electromagnetic waves called *light*. Light includes infrared, visible light, and ultraviolet rays, which differ from each other in wavelengths. Since different materials will respond differently to the various wavelengths, it is possible to select materials that respond mainly to only narrow bands of wavelengths, even single colors in the visible spectrum. This must be kept in mind when specifying any of the light sensors described in this text. Many of the sensors commercially available can be purchased in versions that are selective as to sensitivity to the different wavelengths.

Phototubes are still in common commercial use and come in two general types. One type has an output which is an analog (proportional) output, and the second is more of a switching device. The proportional type is a vacuum tube similar in construction to a vacuum-tube diode in that it has two elements, the *cathode* and the *anode* (Fig. 5-1). The cathode is frequently a semicylindrical surface having its inner surface coated with a *photosensitive* material. The particular photosensitive material used on the cathode is selected because it exhibits the property of *photoemission;* that is, when light strikes the atoms of this photoemissive material, electrons are released from the material.

The anode is a thin rod physically located along the inner axis of the cylindrical cathode. If an electrical potential is applied across the cathode (negative) and the anode (positive), the electrons released by light incident on the photoemissive material (on the cathode) would be attracted to the anode. The amount of current

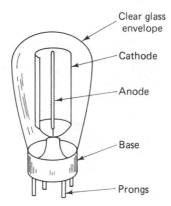

Clear glass
envelope

Cathode

Anode

Base

Prongs

Figure 5-1 Phototube.

flowing will be proportional to the quantity of electrons emitted from the cathode, which is, in turn, proportional to the intensity of light illuminating the cathode.

The physical construction of the cathode can be varied to suit different applications, and the particular photoemissive material used on the cathode can be selected to make the tube more sensitive to particular wavelengths of light.

The second type of phototube is somewhat similar in physical construction to the device just described. The primary difference is that instead of being a vacuum-type tube, this tube has the space around the anode and cathode filled with a gas. Thus, when the cathode emits electrons, as they accelerate toward the anode they collide with electrons associated with the gas atoms, knocking them out of their orbits, and these electrons are now free to add to the current flowing. Therefore, there are more electrons available to participate in the current flow to the anode.

This effect is quite rapid and dramatic and essentially causes the gas-filled phototube to "fire" or switch when illuminated by a minimum threshold of light. Therefore, it is much more sensitive than the vacuum-type phototube and would be used in situations where interruptions of light beams would trigger some device into action. The output might conceivably be connected to a counter, counting objects as they travel down a production line; or it might be across a doorway and sound an alarm as the beam is broken. When used in this type of application they are referred to as optical proximity sensors.

A second commercially available type of light-sensing device is the *photovoltaic cell*. The photovoltaic cell is constructed of materials that effectively create a situation where light energy is directly converted into electrical energy. They are commonly called *solar cells* and act as a source of voltage (emf), the dc voltage/current characteristics of which are dependent upon the intensity of illumination. These devices (Fig. 5-2) are frequently used in conjunction with ammeters whose dials are calibrated directly in terms of light intensity. These devices are then called *light meters* or (photographic) *exposure meters*. They are analog devices and have a very long service life. Some versions are used to power communication satellites while in orbit.

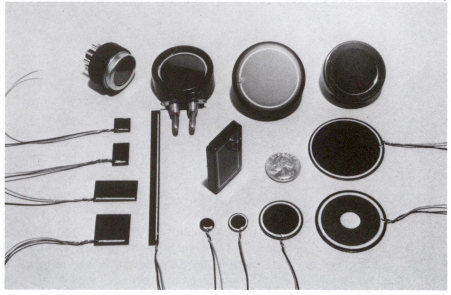

Figure 5-2 Industrial selenium photovoltaic cells. (Courtesy of VATEC, Inc.)

Photoconductive semiconductor devices comprise another very popular class of light sensors. Semiconductor materials are directly affected by radiation of all types (heat, light, and nuclear), and advantage can be taken of their light sensitivity to produce light-intensity sensors. Light radiation incident on semiconductor material has the effect of causing its resistance to decrease proportionately to increasing intensity.

Photoconductuve sensors (Fig. 5-3) can be made which respond to low levels of light intensity by using a lens built into the sensor housing to focus the light on

Figure 5-3 Cadmium-sulfide and cadmium-selenide photoconductive cells. (Courtesy of VATEC, Inc.

Figure 5-4 Inside view of a solid-state optical proximity sensor. The lens transmits light from the incandescent lamp and receives light from a retroreflective target, focusing the returning light on an internal phototransistor. The phototransistor operates a relay based on the presence or absence of reflected light. (Courtesy of MICRO SWITCH, a Division of Honeywell.)

the semiconductor material. These devices can be made to be very sensitive, small, rugged, and inexpensive, and are therefore in very common usage.

The sensitivity of photoconductive devices can be greatly increased by using specially designed *transistors* (Fig. 5-4) which have been enclosed in cases having lenses built into them to focus the light directly on either the base or the emitter. These devices, besides being more sensitive than normal photoconductive cells, can be made smaller and more rugged, since neither the relatively larger exposed surface area of semiconductor material nor its large transparent covering are required. These devices are also in very common commercial use. The electrical circuit symbols for these light sensors are illustrated in Fig. 5-5 and should be memorized because these devices are in such large-scale commercial use.

RADIATION AND THICKNESS SENSORS

Two general types of radiation sensors will be discussed here, *X-ray sensors* and (nuclear) *radioactivity* sensors. *X-ray sensing devices* are commonly used for sensing the level of liquids (as described previously) in a tank or as a sensor for measuring the thickness of moving (continuous sheets) materials. The most common X-radia-

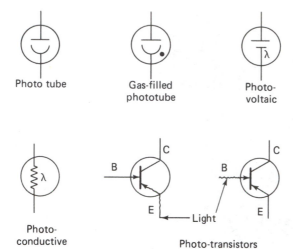

Figure 5-5 Electrical circuit symbols for light sensors.

tion sensors use either a material that will produce visible light when subjected to X-rays or a semiconductor material that changes its resistivity as a function of intensity of X-radiation. Both types are proportional (analog) sensors and require elaborate shielding and environmental protection for the safety of operators.

The intensity of radiation emitted by a source is attenuated by various materials, based primarily on type of material and thickness. Therefore, with a sensing head on one side of a material and an X-ray source on the other (Fig. 5-6) the difference in intensity sensed between the times when there is no material between the two and when the material attenuates the radiation is proportional to its thickness.

Nuclear radiation exists in several basic forms; alpha, beta, gamma, and neutron are the most commonly referred to forms. Alpha particles are relatively large (a helium nucleus), cannot travel far (a fraction of an inch in air), and are stopped by a piece of paper. Beta particles are electrons; they can travel much farther than alpha particles but are also easily stopped by a piece of paper. Therefore, neither of these two forms of radiation is often used in industry. Neutron sensing equipment is limited primarily to nuclear reactors, and therefore its applications are very specialized and will not be covered in detail here.

Gamma radiation passes through all known materials; however, it is also attenuated by all materials to some degree, depending upon the molecular structure of the material and the thickness of the material the radiation passes through (as with X-radiation). Therefore, measurement of gamma radiation is extremely important for measuring such parameters as liquid level and thickness of such materials as steel and other metals where no other adequate sensors are commercially available.

The device frequently used to sense gamma radiation is the *Geiger-Müller tube* (Fig. 5-7). It consists of a long hollow tube (the cathode) which is connected exter-

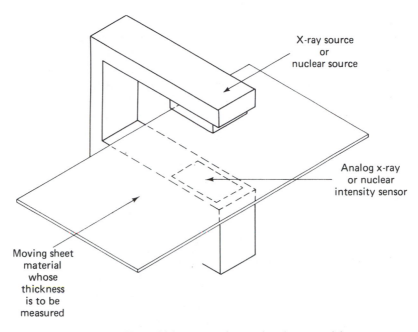

Figure 5-6 X-ray thickness gage for moving-sheet materials.

nally to the negative terminal of a source of voltage, and a wire anode which is physically located along the axis of the cathode and is connected to the positive terminal of the voltage source through a resistor. The anode and cathode are enclosed in a gas-filled tube which operates in a manner much the same as the gas-filled phototube. However, in the Geiger-Müller tube the cathode has no coating on it at all; it is not needed.

As gamma radiation passes through the walls of the tube, it (the gamma radiation itself) causes the electrons in the outer orbits of the gas to be knocked loose. These electrons, in turn, knock other electrons loose, and all the freed electrons rush to the positive anode; while the positive atoms of gas (those whch lost electrons) rush to the more negative cathode. This whole process occurs in a fraction of a second, causing a pulse of current to flow and then cease. The tube therefore "fires"

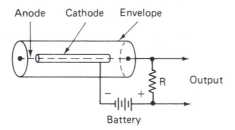

Figure 5-7 Geiger–Müller tube.

each time a gamma ray passes through the walls. These pulses can be electronically processed so that they can be heard audibly, or they can be fed to a time-based counter, which counts how many pulses (gamma rays) per second occur; either gives an indication of the intensity of the radiation at the sensing head.

There are many specific designs of radiation heads using Geiger-Müller tubes, some of which can sense the alpha and beta particles previously mentioned.

A second type of nuclear radiation sensor is the *scintillation counter*. It uses certain types of crystals which produce brief flashes of light when struck by the various types of nuclear radiation (alpha, beta, or gamma). These brief flashes of light can be sensed by some of the light sensors previously discussed for an electrical output. The scintillation counter is generally more sensitive than the Geiger-Müller tube and both can be used to sense X-rays.

HUMIDITY/MOISTURE SENSORS

Many processes require careful control over the ambient conditions under which the process is run. Temperature, air-particle content (dust)), and *humidity* or moisture content are all commonly controlled parameters. We have discussed temperature sensing, dust is controlled primarily by means of filtering and pressure control, and humidity is controlled by automation.

Humidity, or the amount of moisture present in air, can be spoken of from two different points of view. First is *absolute humidity,* normally measured in grains of water vapor per cubic foot of air. There is no sensor commercially available with the capability of directly measuring absolute humidity, but this is normally not the quantity that is to be controlled anyhow.

The ability of air to hold moisture is dependent upon both the ambient pressure and the temperature of the air. Since most processes occur under relatively constant ambient pressure conditions, we shall assume this condition. Therefore, the amount of moisture that air can contain depends upon its temperature; the hotter the air, the more total moisture it can hold; and conversely for colder air. The more useful information is then the amount of moisture that the air actually does contain under its present conditions as compared to the maximum amount of moisture it could possibly contain under the same conditions. This ratio of actual moisture content to maximum moisture content possible under the same condition is called *relative humidity*.

One relative humidity sensor takes advantage of the dependence of liquid evaporation upon the moisture-absorbing ability of the air. The moisture-absorbing ability of air is proportional to the difference between the amount of water in the air and its saturation content. The *sling psychrometer* (Fig. 5-8) consists of two identical thermometers mounted on a sling base. One of the two thermometers has a "sock" over the reservoir (sensing) end. The sock is dipped into water and the

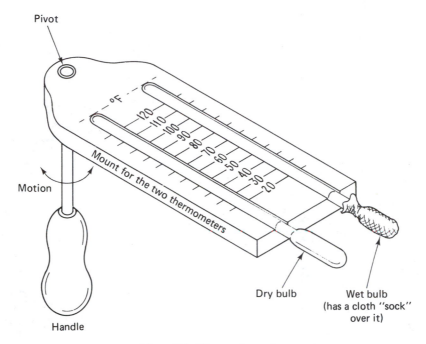

Figure 5-8 Sling psychrometer.

two are whirled around to increase the rate of evaporation from the sock to its maximum possible value.

As you may remember from physics, the process of evaporation is actually a cooling process of the surface which the liquid has evaporated from. This explains why you feel cooler after wiping your face with a cool, wet cloth on a hot summer day. Therefore, the two thermometers will register different temperatures, one being influenced by the cooling effect of evaporation.

Tables (curves) have been constructed from which the relative humidity can be directly determined when entering with these two temperatures. This is the method used by weathermen, but it is not systems-compatible in that form. Nearly any of the temperature-to-electrical transducers previously discussed can be used in place of the two thermometers to obtain the temperature difference. Also, instead of whirling the thermometers around, there could be a small fan which moves the air past stationary thermometers. Such a system is illustrated in Fig. 5-9.

Another type of humidity-sensing device relies on the moisture-absorbing qualities of certain materials, such as animal (and human) hairs. The hair changes length as its moisture content changes (with humidity changes). Figure 5-10 illustrates a mechanical system that will measure the change in length of hairs. The mechanical output could also be measured by any of the mechanical position sensors discussed earlier in Chapter 2.

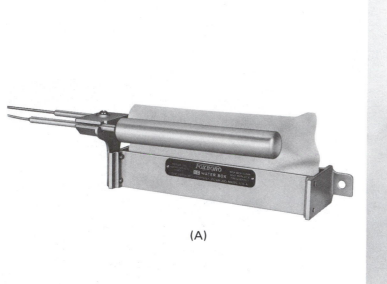

(A)

(B)

Figure 5-9 (A) Wet- and dry-bulb sensing elements for a relative humidity recorder. The dry bulb is visible; the wet bulb is covered by the wick which extends into the water box. (B) Thermal shield which also provides forced-air ventilation for relative humidity sensing heads which are located internally. [(A) Courtesy of Foxboro Company; (B) courtesy of General Eastern Corp.]

There are some chemicals which are moisture sensitive and even change their resistivity in response to moisture changes. Such devices may consist of two covers enclosing a thin layer of chemical (such as lithium chloride), which changes resistivity with moisture content. Two metal electrodes will also be held in place by the plastic cases, and these are in contact with the chemical. The electrical output will be a variable resistance which is proportional to relative humidity.

Some solids, when in powdered or granular form, can also be used as relative humidity sensors because they, too, exhibit a change in electrical resistance with a change in moisture content. Figure 5-11 illustrates an ion-exchange type relative humidity sensor which also has a variable-resistance type output. Small changes in water content of a material can produce relatively large changes in its dielectric constant. Therefore, it is also possible to use a capacitive-type relative humidity sensor.

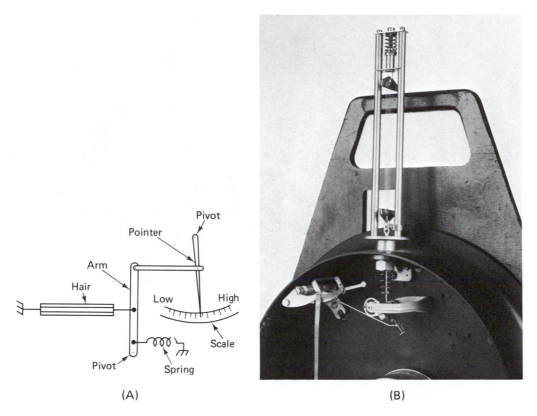

Figure 5-10 (A) Hair hygrometer; (B) cutaway view of a commercial hair hygrometer relative humidity recorder. [(B) Courtesy of Foxboro Company.]

TIME MEASUREMENT

If one were to select *the* single process control variable that requires measurement in the most systems, it would have to be *time*. Most industrial process control systems operate and exercise control based upon *real time,* that is, clock time (hours, minutes, seconds, etc.). The actual mechanism of obtaining (sensing) time can take many forms, some of which will be described in the following text. However, the requirement for time-related control forms the basis for modern automatic control system theory.

The need in most process control systems is not for time-of-day type time, but for elapsed time since the occurrence of a specific event. In many electronic systems time is internally generated by stable oscillators. If the system requires intervals of time in excess of one cycle of the oscillation frequency, various types of counters

Figure 5-11 Sulfonated polystyrene ion-exchange relative humidity sensor, which includes a platinum RTD temperature sensor. (Courtesy of General Eastern Corp.)

are used. They count the number of cycles from the oscillator commencing the instant they were started by the event that signifies the start of the process. The subsequent electronics can either continually "watch" the output from the counter or else the counter itself can have electronic circuitry associated with it which will activate a separate output after the counter has accumulated a predetermined number of cycles from the oscillator. In fact, the foregoing description even applies to the new electronic digital wristwatches which have become so popular; that is basically how they work. This technique is normally used where the time intervals are small, although it is also used in digital computer systems to keep track of time for days, weeks, and longer.

Another very common method of measuring time intervals of short duration (from *nanoseconds* to minutes) is to use the charging rate characteristics of a capacitor in an electronic circuit. A simple *RC* network is normally connected to some sort of switching device (transistor), and as the capacitor charges, it eventually reaches a level where it turns on the transistor (Fig. 5-12). The transistor, in turn, would energize a set of contacts through an electromechanical *relay* (relays are described in Chapter 16).

The contacts can be configured such that this operation either opens or closes the contacts, operates several sets of contacts simultaneously, or even causes another circuit to begin to "time-out" from the time the first one was activated. Of course,

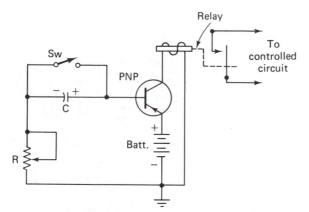

Figure 5-12 Transistorized electronic time delay. Timing starts as switch Sw is opened.

electromechanical relays operate relatively slowly (in *milliseconds* at the fastest). Therefore, for time intervals shorter than this, a transistor would replace the electrical relay. *Silicon-controlled rectifiers* (SCR) and *triacs* are also used in such applications as these.

Time in many systems is kept track of by specially designed ac motors called *synchronous motors* (operation of these devices is more fully described in Chapter 16). These motors are connected directly to power supplied by local utility companies [normally 110 to 125 volts (V), 60 hertz (Hz)]. The power utility companies maintain exceptionally close control over the frequency of the power they supply. The current practice is to maintain the number of cycles generated during a calendar day to exactly the proper number of cycles that theoretically should have been generated during that period of time.

The ac synchronous motor follows exactly the frequency of the power supplied to it; or it rotates in synchronism with the applied power, hence the name synchronous motor. Therefore, this motor provides an exceptionally accurate and stable time reference. By gearing the output from one of these motors, any desired time interval can be obtained. The common household electric clock uses a synchronous motor (operating at 3600 r/min) which has been geared down to 1 r/min to drive the second hand, then further geared down to 1 revolution per hour to drive the minute hand, and finally it is geared down again to 1 revolution per 12 hours to drive the hour hand.

There are many electromechanical devices which, using an ac synchronous motor and appropriate gearing, are designed to operate set(s) of electrical contacts based upon time since activation. Some of these devices are even designed to reset themselves and begin operation over again automatically.

One category of such devices is technically termed *interval timers*. The interval timer causes a set of electrical contacts to be closed immediately as soon as the timer is activated. After a predetermined time interval, the timer causes that set of con-

Figure 5-13 Commercial interval timers. (Courtesy of Industrial Timer Co., an Esterline Company.)

tacts to open. The interval of time that the contacts remain closed is determined by the position of a manually operated dial mounted on the timer itself (Fig. 5-13).

The *time delay timer* is a second category and is almost identical in construction to the interval timer. The only difference is that the electrical contacts are opened when it is initially activated. The contacts are then closed at the end of a preselected time interval, and in some designs are subsequently reopened at the end of a second preselected time interval. Both the interval timer and the time delay timer can use any combination of switches that operate as required for special purposes.

The *recycling timer* is similar to the interval timer except that it automatically resets itself after it "times-out" (completes its cycle) and repeats the previous cycle indefinitely until the motor that drives it is turned off.

Another category of synchronous motor-operated switches is the common *time switch* or *time-of-day switch*. This switch automatically controls the opening and closing of a switch based upon the time of the day. An example of this type of switch is the common household automatic timer, which is used, for example, to energize lights at preset times during the evening when the occupants are away (Fig. 5-14).

Some of these timers are simply connected to counters (through appropriate gear systems) such that the counter registers the total time that the motor has been running (Fig. 5-15). These are used as elapsed-time indicators, to indicate the running time of selected electrical equipment, the motors being connected to the equipment's on–off switch so that it is automatically energized as long as the equipment is in operation.

Figure 5-14 Programmable electric timer/counter, which can be field programmed to perform the functions of interval timer, time delay timer, recycling timer, or time totalizer (or as a simple counter or preset counter). (Courtesy of Eagle Signal Division, Gulf & Western Manufacturing Company.)

Another category of timing devices uses a *cam* mechanism, or a series of many cam mechanisms, attached to the geared-down output from a synchronous motor (Fig. 5-16). The cams are individually shaped to (each one) activate a switch that is mechanically associated with the cam. Using devices such as this, complicated time-based (open-loop) control systems can be designed; one very common household example is an automatic (clothes) washing machine. In industry these devices can assume very large and complicated proportions, eventually resulting in devices called "programmable controllers," which are described in Chapter 12.

Figure 5-15 Lapsed-time or running-time meters. (Courtesy of Industrial Timer Co., an Esterline Company.)

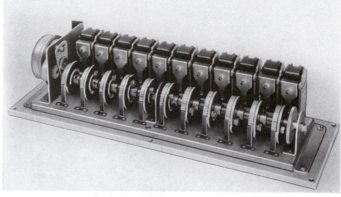

(A)   (B)

Figure 5-16 Open view of timers showing cam and switch arrangement. [(A) Courtesy of Eagle Signal Division, Gulf & Western Mfg. Co.; (B) courtesy of Industrial Timer Co., an Easterline Company.]

COUNTERS

Many situations arise in industry where there is the need for simply counting: counting the number of items passing down a production line, counting the number of feet of a material being manufactured or coated, counting the number of rejected production items, and so on. Sometimes the entire process control system consists merely of a counter that has the capability of closing a set of contacts at a specific count (e.g., loading 24 bottles into a box).

Figures 5-17 and 5-18 illustrate simple mechanical counters. By themselves they provide only a visual indication of the count; however, more exotic versions are commercially available which include electrical contacts for each of the 10 positions of each indicator wheel. These contacts can be wired either to provide remote indications of the count (Figs. 5-19 and 5-20), or else they can be wired in a manner such that an electrical circuit is completed when the wheels indicate a specific (preset) value (Fig. 5-21). The electrical continuity in the latter can be used for control: for example, counting 144 tubes of toothpaste and then shifting the production line to deliver the next 144 tubes down an alternative route.

Figures 5-17 and 5-18 illustrate counters that must establish physical contact with the material being counted. It should be obvious that photocell-type (proximity) sensors would make excellent sensing heads for counters also. In this case the output from the photodetector electronics would activate a relay that is mechanically coupled to the indicator wheels. Figure 5-22 illustrates such a solenoid-operated counting mechanism. These, too, can be constructed with the electrical

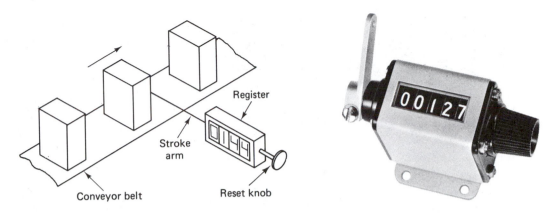

Figure 5-17 Stroke-register-type mechanical counter. (Courtesy of Veeder-Root, Inc.)

contacts on the wheels (as in Fig. 5-21), and frequently are. Devices such as these are called *preset counters,* and Fig. 5-23 illustrates two applications of such devices.

Counters take many shapes and forms, but one of the most useful ever has been the *solid-state digital counter.* Operation of these devices was explained briefly in the "time" section of this chapter, where their use to accumulate numbers of accurately timed pulses was discussed. They also find much application in accumulating counts on production lines similar to the applications of the electromechanical devices just described. The advantages of solid-state counters over electromechanical counters are numerous and they are being used in a sizable number of applications formerly reserved for electromechanical devices.

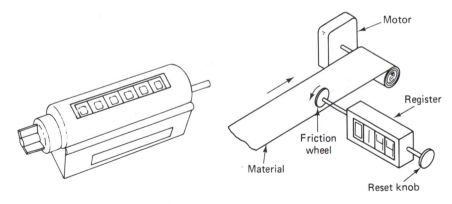

Figure 5-18 Mechanical revolution counter.

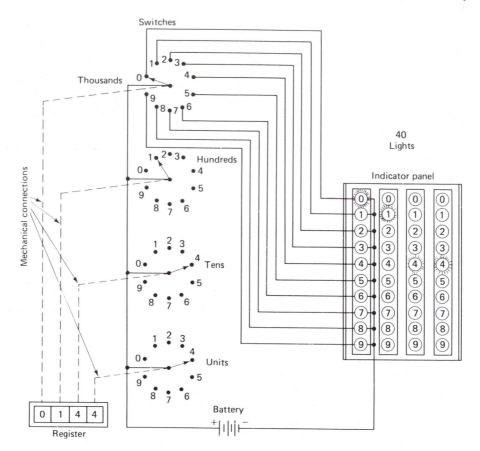

Figure 5-19 In-line indicator (only the thousands are shown wired). Count indicated is 0144.

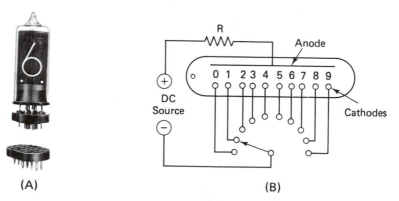

(A)   (B)

Figure 5-20 (A) Nixie tube; (B) circuit schematic showing tube wired to 10-position switch. [(A) Courtesy of Burroughs Corp.]

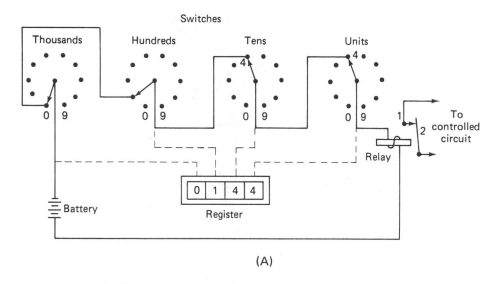

(A)

(B)

Figure 5-21 (A) Circuit of the preset counter (here preset for a count of 144); (B) commercial preset (predetermining) counter—it will count down from the manually preset number to zero, activating a set of contacts at zero. [(B) Courtesy of Veeder-Root, Inc.]

MISCELLANEOUS VARIABLES

There are numerous other variables that must be commonly measured in industrial control systems. Most of them fall into the category of rather limited application devices, and it is not appropriate to spend much time on them in this survey text, except to emphasize again that the purpose of the text is to present a survey, not a comprehensive treatment, of transducers.

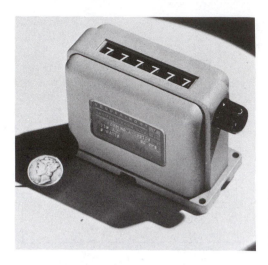

Figure 5-22 General-purpose electric counter, activated by electric contact closures. This particular counter is also electrically resetable. (Courtesy of Veeder-Root, Inc.)

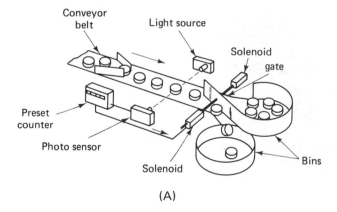

(A)

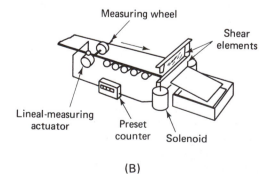

(B)

Figure 5-23 (A) Batching control employing a phototube and a preset counter; (B) use of a linear-measuring actuator and a preset counter to cut sheets of predetermined length.

Some of the process-control variables that must be measured, but are not covered specifically herein, are neutron radiation, pH, density, consistency, specific gravity, chemical content, vacuum, and viscosity.

GENERAL COMMENTS ON SELECTION OF A SENSOR

Selection of the optimum sensor must be made for each individual application. Even on the same system several different designs of sensors may be required to make the same type of measurement at different locations in that system. The following discussion is intended to point out the primary considerations that affect the selection of each sensor in a system. Many a system has required redesign because insufficient attention was paid to the proper selection of the sensors during the initial design phase of the system. Remember that the sensor sets the maximum limits that the controller can exercise as far as control system response is concerned. There is no way that the controller, no matter how sophisticated it is, can increase the system response time, sensitivity, or overall accuracy above the limits imposed by the sensor.

Of course the first consideration in selection of a sensor must be the variable to be measured. We shall assume that the nature of the variable is understood and the necessity for its measurement is established. What is the *range* and *span* of measurement absolutely required by the system? How much of an *overrange* capability must be accounted for and measured? What minimum *resolution* (sensitivity and *precision*) is required to exercise adequate control? What overall measurement *accuracy* is absolutely necessary? What is the maximum expected rate-of-change of this variable; in other words, what *frequency response* (response time) is required of the sensor? All of these terms will be discussed further in Chapter 6.

Once these questions are answered, then, before proceeding, the measurement system must be considered. The measurement system will, very possibly, reduce the resolution, frequency response, and/or accuracy of the measurement before it reaches the controller. Therefore, it may be necessary to revise the sensor's specifications so that even after the required signal processing deteriorates the critical specifications, the overall measured and processed value is within acceptable limits. Furthermore, the effect of various possible modes of failure of the sensor on the system must be considered.

Once the overall measurement system's specifications have been established and the measurement specifications of the sensor itself have been determined, the physical installation must be considered. Where will it be installed? Frequently, the measurement may be made at many possible places in the system. Are there any severe physical installation requirements that must be met, such as accessibility, vibration, radiation, or temperature? Will the sensor be in contact with corrosive or explosive substances which require special protection? How will the sensor be maintained and/or replaced if necessary? What power is available (if necessary) to supply the sensor? How will the output from the transducer get to the measurement system and the controller; does this place any extra requirements on the sensor? Once again, these criteria may well affect the choice of sensor, eliminating many otherwise possible candidates.

Finally, selection of specific sensors that can meet all the specifications and requirements heretofore determined can be made. Once all the candidate sensors are determined, it becomes a matter of comparison between them to make the final selection. Here the answers must be determined to such questions as: How reputable is the manufacturer? (What is the probability that he will be in business 10 years from now to supply repair parts; will he honor any guarantee; will he provide backup technical assistance if it is needed?). What data do the manufacturers have to prove the life expectancy, reliability, and specifications that are claimed for this transducer? What is the frequency of *calibration* and maintenance required of this sensor as compared to other candidate devices? What level of technical competence is required to install, calibrate, and maintain this sensor as compared to other candidates? Is this sensor available for timely delivery to meet installation, systems calibration, and test schedules? Will any redesign of the sensor be required for installation in this system, and can this be accomplished to meet existing schedules? And, finally, how cost-competitive is the sensor (based on the overall acquisition, installation, and projected calibration and maintenance costs over its expected operating life) as compared to other candidate sensors?

If selection of a sensor seems to be a fairly complicated and time-consuming task, just remember what is at stake—the overall success of the final operating system. The job of sensor selection is normally assigned to a project instrumentation engineer, a person whose expertise lies in this field and who from experience is likely to make sound selections. Furthermore, it is not uncommon for this process to eliminate *all* the sensors that are on the market for a specific application. In this case, redesign of existing sensors must be considered or development of an entirely new sensor must be started, or it is back to the drawing boards to reconsider the basic system design. It may be possible to redesign the system so that a related variable (for which sensors exist) may be used instead of the optimum choice variable for which none exist, and still obtain adequate overall system performance.

QUESTIONS

1. What is light?
2. List several naturally occurring phenomena that can be taken advantage of to construct light sensors; describe each phenomenon.
3. Which types of radiation are frequently used for thickness measurement in industry?
4. Describe the operation of a Geiger-Müller tube.
5. Explain the meaning of the term *relative humidity*.
6. Referring to a sling psychrometer, why will dry-bulb and wet-bulb thermometers have different readings when they are both measuring the temperature of ambient air?

7. What is the most commonly made measurement in industry?

8. Why is time measurement so critical in most process control systems?

9. Describe the physical construction and operation of a decimal mechanical counter.

10. Make a table, listing in one vertical column each of the sensors covered in this chapter. Then make four adjacent columns, labeling them Variable measured, Principle of operation, Advantage, and Disadvantage. Attempt to fill every blank space in the table.

11. List several of the critical system parameters that must be considered when selecting a specific sensor to make a particular measurement.

12. Which of the following statements *best* describes the term *radiation?*
    A. Rate at which light is radiated.
    B. Candle power.
    C. Amount of luminous flux received on a unit area.
    D. Transmission of energy through space by electromagnetic waves.
    E. One lumen per square foot.

13. Which effect refers to the property of a material that emits electrons in proportion to incident light?

14. In an industrial application use is made of nuclear radiation (gamma rays) as a primary sensing medium. What type of device is usually used to sense the quantity of radiation?

15. The following symbols, in order, represent:

    A. Photovoltaic device, photoemissive device, photoconductive device.
    B. Photoconductive device, photoemissive device, photovoltaic device.
    C. Photovoltaic device, photoconductive device, photoemissive device.
    D. Photoemissive device, photoconductive device, photovoltaic device.

16. The term pH refers to
    A. A measure of acidity of a solution.
    B. A measure of the alkalinity of a solution.
    C. A scale of 0 to 14, with the center (7) representing a neutral solution.
    D. A measure of the hydrogen-ion content of a solution.
    E. All of the answers above are correct.

17. What is relative humidity?
    A. The absolute amount of moisture in the air.
    B. The relative amount of moisture in the air compared to the maximum amount possible under the same conditions.
    C. A measure of the moisture content of some process.
    D. A measure of the atmospheric pressure at a given point.
    E. All of the answers above are correct.

18. What type(s) of process variables can be measured using counters?

19. A running-time meter or elapsed-time meter senses
    A. The shape of time-operated cams.
    B. The total number of operations of a set of contacts.
    C. The total amount of time a piece of equipment has been energized.
    D. The *RC* time constant for a time-operated relay.
    E. Any of the above.

20. Which of the following timing/counting devices provides an output which can be used directly to exercise (very simple, on–off) control?
    A. Interval timer.
    B. Time-of-day timer.
    C. Preset counter.
    D. Recycling time delay timer.
    E. All of the devices listed can actually be used directly to exercise control.

21. Which of the following phenomena is commonly used in industry to sense the thickness of solid materials?
    A. The proportionality between liquid level and pressure.
    B. The attenuation of nuclear radiation by solid materials.
    C. The variation of resistance of wire as it is deformed.
    D. The sensitivity of hair to moisture.
    E. The fact that where hydraulic flow velocity is high, the corresponding pressure will be low, and vice versa.

22. When discussing radioactive transducers, which of the emitted parameters was noted as being used primarily in industrial instrumentation systems?

23. A voltaic cell transduces _____.
    A. Mechanical motion to pressure.
    B. Hydraulic flow to differential pressure.
    C. Temperature to pressure.
    D. Light intensity to voltage.
    E. Pressure to mechanical motion.

24. Which of the following phenomena is commonly used in industry to sense relative humidity?
    A. The proportionality between liquid level and pressure.
    B. The attenuation of nuclear radiation by solid materials.
    C. The variation of resistance of wire as it is deformed.
    D. The sensitivity of hair to moisture.
    E. The fact that where hydraulic flow velocity is high, the corresponding pressure will be low, and vice versa.

25. An X-ray gage can be used to measure _____.

GLOSSARY

Accuracy   The quality of freedom from error in a measurement; a statement of the exactness of a measurement as compared to the accepted standard (or true value) for that measurement; usually expressed as a percentage of the full scale measurement capability of the instrument.

Anode   A general name for a positive electrode on an electronic device. A terminal which acts as a receiver of electrons.

Calibration   The process of applying accurate inputs to an instrument, circuit, or device and comparing the measured output to the appropriate standard for the purpose of determining (or adjusting) its accuracy.

Cam   An irregular shaped surface which, when rotated on a shaft, will cause a "follower" which is mechanically in contact with the cam to move in response to the shape of the cam.

Cathode   A general name for a negative electrode on an electronic device. The terminal which acts as a source of electrons to the device.

Frequency Response   A measure of the ability of a circuit or piece of equipment to respond to (or transmit) input signals of various frequencies which are applied to it.

Humidity   Measure of the moisture content of air.

Light (Visible)   A form of electromagnetic radiant energy in the frequency range to which the human eye is responsive.

Microsecond   One millionth of a second; one thousandth of a millisecond; 10^{-6} second (typically used in abbreviated form: μs).

Millisecond   One thousandth of a second; 10^{-3} second (typically used in abbreviated form: ms).

Nanosecond   One billionth of a second; one thousandth of a microsecond; 10^{-9} second (typically used in abbreviated form: ns). Light travels slightly less than one foot in one nanosecond.

Nuclear Energy (Atomic Energy)   Energy released in a reaction where a nucleus splits or two nuclei are joined together.

Overrange   As applied to electronic measurements, attempted measurement of an electrical quantity which has a magnitude beyond the measuring range of the device being used.

Photoconductive   That property of certain materials which results in a change in electrical resistance proportional to light intensity.

Photoemissive   An effect where light energy incident on the surface of certain materials causes the emission of electrons from that surface.

Photosensitive   Sensitive or responsive to light energy.

Photovoltaic   The conversion of light energy into electrical energy by certain semiconductor materials.

Precision   Degree of exactness of a measurement; a measure of the maximum number of possible distinguishable increments from which the sample was taken; also a measure of the readability or repeatability of a measurement when taken repeatedly under identical circumstances. This term may *imply* a degree of accuracy, but a device may be precise without being accurate (i.e., a distorted micrometer; it is still as precise as it was before being distorted, but obviously the accuracy has been affected).

Radioactivity   That natural property exhibited by certain elements by which their nuclei disintegrate into different elements, accompanied by emission of the various products of disintegration (among which are alpha particles, beta particles, and gamma rays).

Relay   An electromechanical device having two independent circuits. The activating circuit is normally an electromagnetic coil which magnetically operates set(s) of contacts which (normally) are part of other electrically isolated circuits. A binary device.

Silicon-Controlled Rectifier (SCR)   A four-layer semiconductor device which will block the flow of current in either direction through it, until "gated." Once gated, it will allow current flow in one direction (the forward direction) and this flow will continue (even though the gating signal is removed) until some external force causes the current flow to stop. It blocks all flow of current in the reverse direction regardless of the gating signal. Typically used in ac circuits where the gating signal turns on forward current and the naturally occurring current reversal cuts off the flow each cycle.

Synchronous Motor   An ac motor having a permanent magnet rotor which rotates in step
with the rotating motor (ac) field. Typically used as (low frequency) timing mechanisms.
The rotational speed depends only on the frequency of the excitation source.

Transistor   An active semiconductor device normally having three electrodes (emitter, base,
and collector) capable of amplification.

Triac   A five-layer semiconductor device which is the equivalent of two SCRs connected
together in parallel (but pointing in opposite directions) and having the gate leads connected together. Each half independently operates as an SCR, therefore current can pass
in *both* directions, flowing through one SCR in one direction and through the other SCR
in the opposite direction.

X-ray   Electromagnetic radiation similar to (visible) light but having much shorter wavelengths and capable of penetrating most opaque substances.

6

THEORY
OF MEASUREMENTS

INTRODUCTION

Before beginning a detailed discussion of process control system measurements and of the systems used to effect these measurements, it is appropriate first to discuss measurements in general. In any given process control system the overall system response characteristics (i.e., accuracy, frequency response, repeatability, etc.) are limited by the capability of the process variable measurements; therefore, the basic concepts behind making these measurements must be afforded utmost consideration.

The *measurement* system includes the transducer, which actually interfaces directly with the process variable to be measured; the signal conditioning, filtering, impedance-matching, and amplification circuits (normally either electronic or pneumatic); and the display device and/or controller for which the measurement is intended. The measurement is the value obtained by the measurement system. It is a quantity which is somehow representative of the value of the process variable that was measured.

It is the purpose of this chapter to discuss nomenclature, measurements in general, the types of errors that measurements are affected by, and the standards against which measurements are compared in order to determine the validity (accuracy) of the measurement system. The subject of standards, and the process of calibration against those standards, are considered to be so important that a special section of this chapter is devoted to discussion of those topics.

MEASUREMENT GOALS AND CONCEPTS

What are the goals of a process control measurement system? Normally, they include conversion of intelligence between energy systems while maintaining specifications such as accuracy, precision, and speed of response within acceptable limits. The intelligence referred to is the current (instantaneous) value of the process variable that is being measured, in appropriate engineering units. Of course, there must also be a good reason for wanting the measurement made, because, as you will discover, the "cost" of making a process variable measurement is considerable.

Normally, the reason for making a measurement in a process control system is either that the measured information is required in order to effect adequate control or to monitor system performance (or both). Based on the specific reason for each individual measurement, the conversion of energy systems required will be obvious, although sometimes intermediate transformations are required. For example, the speed of rotation of the wheels must be measured and converted to an electrical signal in order for an electronic speed control system in an automobile to work properly. This would require a mechanical rotational-to-electrical energy system conversion—or do you have other suggestions? Such conversions of information from one energy system to another are made by equipment called transducers. As you may remember from earlier chapters, transducers are hardware pieces of equipment which convert signals (or information, or intelligence) available in one energy system to a proportional signal in another energy system. Transducers that directly measure process control variables in order to provide that variable information for use by the process control system and/or operators are called sensors. As we will see later, there are other types of transducers besides sensors.

SCALE FACTOR

This gives rise to a concept called a conversion factor, conversion constant, or *scale factor*—different names for the same concept. The scale factor, as we will refer to it in this text, is derived by taking the span of the variable being measured and dividing it into (or by) the span of the converted value, each expressed in its own appropriate engineering units.

RANGE AND SPAN

We must now define span. Span is directly related to another term, range, and is defined in terms of range. The *range* of a variable is simply a numerical statement of the minimum and maximum values that the variable may assume. As an example, the weatherman typically reports on the range of temperatures for a given period—the minimum temperature and the maximum temperature for that period.

So much for range; *span* is simply the numerical difference between the two range values. Note that if the range includes zero as one of its values, the numerical values of range and span would be the same. For example, if the range of temperatures for a given period were 18 to 23°C, the span would be 5 Celsius degrees. If the range had been 0 to 10°C (a cold winter day), the span would be 10 Celsius degrees.

Putting this together in the automobile speed control system example referred to previously, the range of the measured variable might be 20 to 60 mph (actually measured in r/min of the drive wheels), while the speed sensor's electrical output might be 2 to 10 V dc. The scale factor in this example would then be

$$\text{scale factor} = \frac{\text{sensor output span}}{\text{sensor input span}} = \frac{10 \text{ V} - 2 \text{ V}}{60 \text{ mph} - 20 \text{ mph}} = 0.2 \frac{\text{V}}{\text{mph}}$$

or perhaps the inverse of this relationship would be more useful:

$$\text{SF} = \frac{60 \text{ mph} - 20 \text{ mph}}{10 \text{ V} - 2 \text{ V}} = 5.0 \frac{\text{mph}}{\text{V}}$$

The scale factor is used to relate any value of measured output to the actual value of the process variable which caused that measurement. Assuming a linear relationship between mph and voltage in this example, a measured voltage of 5.75 V dc would correspond to a speed of

$$(5.75 \text{ V} - 2.0 \text{ V}) (5.0 \frac{\text{mph}}{\text{V}}) = 18.75 \text{ mph plus } 20 \text{ mph} = 38.75 \text{ mph}$$

LINEARITY

As noted, this example assumed that there was a linear relationship between input and output values in the range of control desired. As you will find out in later chapters, most process control system measurements are not exactly linear relationships. Some can be "linearized" by processing them through special circuitry such as square-root extractors, others can be assumed to be linear (within acceptable accuracy limits) over certain measurement spans, while other measurements must be processed by sophisticated computational equipment prior to being used. This problem of measurement linearity will be addressed further in applicable portions of the following chapters; however, it represents a major factor in the overall conversion accuracy of a measurement, and its effects must be understood.

Under ideal conditions (which are, in fact, seldom achieved in practice) the output of a transducer would be exactly linearly related to the value of process variable being measured, which is technically referred to as the *measurand*. Figure 6-1 is a graphic plot of the characteristics for an ideal pressure-to-electrical transducer.

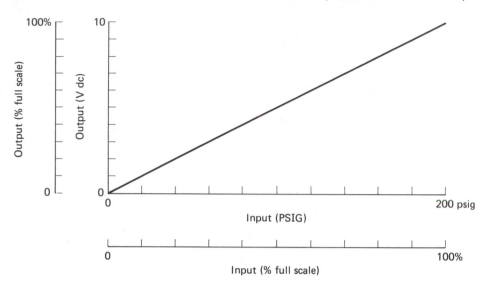

Figure 6-1 Characteristic curves for an ideal pressure-to-electrical transducer.

In fact, the output of a transducer is normally not a perfectly linear function of the measured variable; therefore, any assumption of linearity (i.e., use of a scale factor) is subject to some error. The error is also normally a nonlinear function of the actual measurement; that is, the magnitude of the error also varies as the measurand varies. Figure 6-2 illustrates this, the error (in this figure) being greater at lower and higher values of process variable and less in the middle of the range of measurement. In fact, this transducer is not really usable above approximately 95%

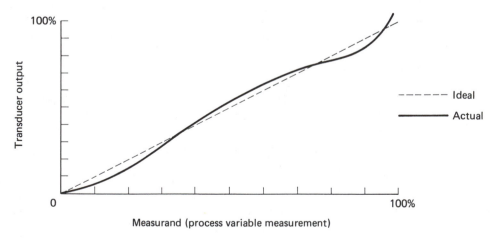

Figure 6-2 Characteristic curves for a more realistic transducer (as compared to Fig. 6-1), illustrating a nonlinear relationship.

input. For this reason most transducers are sized and selected for use near the center of their rated range of operation.

HYSTERESIS

Another significant source of error, which is somewhat similar to linearity, is hysteresis error. Basically, hysteresis error is due to the fact that the transfer characteristic curve obtained as a transducer is subjected to an increasing input signal (from 0% to 100%) may be significantly different from the same device, under the same conditions, only as the input signal (measurand) decreases (from 100% to 0%).

This effect is illustrated in Fig. 6-3. Note that the transfer characteristic curves also illustrate some nonlinearity as well (they are curved rather than being straight). The theory as to why hysteresis exists in transducers varies with the device; however, basically it is related to reactive effects in electrical circuits and such things as backlash between gears and friction in mechanical systems. Since in most transducers the hysteresis is due primarily to mechanical imperfections, the hysteresis error will change as the mechanical system wears with use, ambient temperature, and as its state of lubrication varies.

FREQUENCY RESPONSE

Frequency response is a specification which quantitatively measures the ability of a device's output signal to follow changes in the input signal. Perhaps the best explanation is through an example. If you are driving your car along a level road at, say, 30 mph, there is a position of the gas pedal corresponding to 30 mph. Similarly,

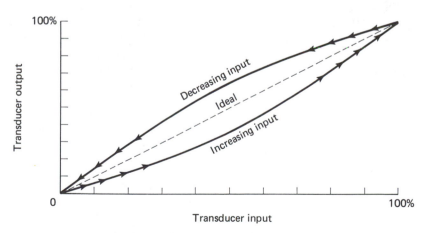

Figure 6-3 Combined effects of hysteresis and nonlinearity on transducer characteristic curves.

there is another position of the gas pedal corresponding to a speed of 55 mph. Simply moving the gas pedal from the 30-mph position to the 55-mph position does not instantaneously cause the car to change to the new speed; rather, there is a considerable delay until the car is finally up to the new speed. Note that the environment (hills, curves, bumps, etc.) also affect this rate of change.

Response time can be defined as a measure of the time interval starting from the instant that the output has changed 10% of the total change that it is going to make, until the time that the output has reached 90% of that final value, assuming that it has been disturbed with an instantaneous change in input. Frequency response is simply the reciprocal of this response time.

The primary use of this particular parameter is in indicating to perspective users the specific dynamic characteristics of a transducer (or amplifier, or piece of test equipment, etc.) to enable them to determine whether the device is compatible with the dynamic characteristics of the system with which it will be used.

As one soon learns in dealing with practical process control systems, consideration of the delays introduced into automated systems due to the frequency response characteristics of the individual components in the system itself is at least as important as the accuracy specifications. Lack of proper engineering consideration of either can result in failure of the system design.

ACCURACY

Linearity, hysteresis, and so on, are specific considerations that must be weighed when selecting and applying a sensor and a measurement system. The more general consideration is that of conversion accuracy. Accuracy will be dealt with further in the section on standards; however, it generally implies the degree of exactness to which a measurement is made as compared to its (theoretically) absolutely perfectly true value. It implies a measure of the degree of freedom from error, simultaneously referring to the maximum expected error in the reading provided by an instrument when making a measurement. A brief list of some of the more significant sources of error which combine to affect adversely the accuracy of measurements includes:

1. Friction, backlash, and inertia in mechanical systems
2. Hydraulic friction, leakage, and inertia in hydraulic systems
3. Heat loss through imperfect insulation and imperfect thermal conductivity in thermal systems
4. Distributed pneumatic capacitance and resistance (friction) in pneumatic systems
5. Offset; drift; the effects of distributed resistance, inductance, and capacitance; and the effects of electrical noise pickup in electrical systems
6. Thermal drift in all systems

In any given system it is impossible to get rid of all of these effects (and the multitude of effects not included in that list). The best we can do is to recognize and understand them individually, measure their effect on system performance, compensate for those that we are able to compensate for, and then to design the system to reduce the magnitude of the undesirable side effects to levels that will be acceptable in each measurement and control system.

These types of errors, their sources, and techniques for dealing with them will be specifically discussed where appropriate in other portions of this book. However, the topic of how we arrive at that "true" value for a process variable measurement is the subject of the section on standards and calibration.

RESOLUTION, PRECISION, AND REPEATABILITY

It is not by accident that we are going to discuss these terms after having discussed accuracy, nor is it by accident that we will discuss them simultaneously. Very commonly, these terms are used in a context that implies accuracy, when in fact this is not necessarily the case.

First let us talk about the term *precision*. Precision is a term that has more than one technical meaning, depending on the context of its usage. One definition of the term refers to the degree of exactness of a measurement. This can normally be interpreted as the number of significant digits included in the value of the measurement, but more accurately it refers to the least significant digit which contains valid information. A measurement value of 123.7 r/min is more precise than a measurement value of 124 r/min; it may not be more accurate, though, as we shall see later.

The second connotation of the term *precision* is technically interchangeable with the terms *repeatability* or *reproducibility,* which have the same technical meaning. For any series of readings taken of a static measurand, using the same piece of equipment and the same operator, under the same set of circumstances, the numerical values will vary within certain statistically determinable limits. These terms all refer to that spread in values for a particular measurement. They are normally expressed as a percentage of the full-scale readings.

Let us take an example to illustrate these definitions and the difference between them. Let us use a machinist's micrometer and make a series of readings of the same workpiece, taken at exactly the same spot and by the same operator. The micrometer may have a scale which would indicate that readings can be taken to within 0.0001 in. (the precision of the scale is 0.0001 in.). As we make our series of measurements we will note that we have a range of readings. Not all readings will be the same value; we will not be (honestly) able to reproduce the same exact reading every time.

Statistically, it can be proven that our physical readings will vary around the actual value in a random fashion. The number of significant digits in our readings is a measure of the precision; the extremes of the readings if taken by a competent

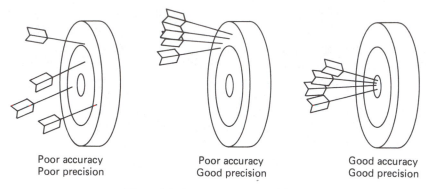

Poor accuracy   Poor accuracy   Good accuracy
Poor precision   Good precision   Good precision

Figure 6-4 Accuracy versus precision.

machinist, not us, is a measure of the repeatability or reproducibility of the measurement.

Well, what about accuracy? Isn't this also indicative of the accuracy of our measurements? Not necessarily so. Suppose that our micrometer has been mistreated and has been slightly bent. Now the accuracy of the equipment has been completely destroyed; however, the basic ability of the instrument to give readings has not been affected at all. It is still just as precise as it was before being bent, and probably it will repeat the same measurement to the same degree of repeatability as before.

In many cases a picture is truly worth a thousand words; Fig. 6-4 is a comparison of the concepts of accuracy and repeatability (precision).

Resolution is a term that reflects the ability of the output of an instrument to respond to small incremental changes at its input. It is basically a statement of the largest incremental change in the input to an instrument (or measuring device) which will produce no detectible change at its output. In a given system a determination must be made of the largest change in input measurements (resolution) which the rest of the system can effectively ignore and still respond properly.

SYSTEM ERROR ANALYSIS

The techniques and systems discussed in the following chapters are commonly used throughout the process control industry in order to interface with the (primarily) electrical outputs from various types of sensors, while introducing as little additional inaccuracy into the measurements as is economically possible. Remember, every time an additional piece of equipment is used in a measurement system, the overall accuracy of the measurement system is adversely affected to some degree.

As an example, let us assume that we use a sensor to measure a specific process variable and that the electrical output is guaranteed accurate (under certain specified conditions and then only for a specified period of time) to $\pm 0.1\%$ of the measured value. This means that the output is a 99.9% faithful reproduction of the input, introducing no more than 0.1% error in the measurement (in either the positive or negative directions). If the output is 0 to 1 V dc, the maximum expected error would be $(\pm 0.001) (1.0$ V max.$) = \pm 0.001$ V. We do not know which direction this error is in, positive or negative; whether it is constant over the useful range of the sensor; or what the nonlinearity curve actually looks like.

We then purchase the rest of the measurement system to use with this sensor, and it is also rated with a 0.1% accuracy specification. By similar reasoning, assuming that this measurement system has an output of 10 V dc, or introduces a gain of 10 into the system, the error analysis would proceed as follows.

The maximum expected voltage error at the input to this measurement system was calculated to be 0.001 V dc. This, multiplied by the gain of 10, would result in an output voltage error of $(10)(\pm 0.001$ V$) = \pm 0.01$ V dc error, even if the measurement system were perfectly accurate, which it isn't in this case and never is in actual practice. If the input to this measurement system had been perfectly accurate, we could expect a voltage error in its output of $(\pm 0.1\%)(10$ V full scale) or ± 0.01 V due to this piece of equipment only.

The overall voltage error in the output will be the sum of these two voltage errors: ± 0.01 V dc due to the sensor plus ± 0.01 V dc due to the measurement system, or ± 0.02 V dc. This gives a resultant overall system accuracy of ± 0.02 V error out of 10 V dc maximum, or a $\pm 0.2\%$ maximum expected error.

This example has illustrated the fact that the errors in a system are cumulative. The primary significance of this fact is that if an overall system accuracy of $+0.1\%$ is required, which is a common goal in practice, each piece of equipment used in the system must be significantly more accurate, such that the total effect of their inaccuracies is no greater than the overall system accuracy specification.

To state that the accuracy of hardware must be significantly better than 0.1% is relatively simple and costs nothing. However, when the time comes to purchase this equipment, the purchaser may be in for a rude awakening. If appropriate equipment with acceptable accuracy is available at all, then other factors must be considered. First, what premium in price are you willing to pay for a piece of equipment with a $\pm 0.05\%$ accuracy as compared to a piece of equipment (to do the same job) with a $\pm 0.1\%$ accuracy? The price premium typically will be significant.

Second, what other restrictions will this more accurate piece of equipment place on your system? Typically, you could expect to be required to calibrate the system more frequently, or the more accurate equipment may have more severe operating condition and environmental restrictions associated with its use. Also, typically, you could expect the longevity of this more accurate equipment to be less and the maintenance and repair costs to be increased.

Overall system accuracy is one of the most important system specifications. As noted, it significantly affects the design philosophy, cost, reliability, and maintenance of the system. There are many other specifications that have similar effects on systems. Many of these specifications will (or have been) covered in appropriate portions of this book.

STANDARDS AND CALIBRATION

Philosophy of Calibration

As human beings have developed more complex and sophisticated technologies, the requirement for more accurate measurements has also evolved. Of particular interest here are the requirements for accuracy in electrical measurements. From the early volt-ohmmeters, which boasted accuracies of better than 10%, the (art and) science of measurement has developed to the point where some electrical measurements can now reliably be made to an accuracy of better than 0.001% at the working level, and considerably higher accuracies are routinely available in controlled environments.

It is very important for technical people to understand the need for, and basically the techniques used in, calibration of test and production equipment. The advertised specifications for equipment and products offered for sale on the industrial and military market represent the performance capabilities of the advertised equipment as compared to universal standards which are maintained at various centers around the world. In the United States the standards are maintained at the National Bureau of Standards (NBS) in Washington, D.C. Other countries throughout the civilized world either maintain their own similar standards centers or have access to the standards centers of other countries.

The process of adjusting test and production equipment to conform to these standards, and verifying (certifying) their compliance to those standards, is the process of calibration.

Concerning Standards

Before proceeding further, we must discuss the standards themselves. Why are standards needed? Where do they come from? Who sets them up as standards? How are the standards distributed to the various users? Who enforces compliance with standards? The answers to these questions are discussed briefly below.

Standards are necessary in order to engage in commerce of any type. Component parts for any piece of modern machinery (or even electronics circuits, for that matter) are manufactured at many different locations (even multinational locations) and subsequently assembled into the finished product. Each individual contributor involved must manufacture, assemble, and test the component parts at each stage of development of the product. The parts must fit and work together even though they were manufactured by different people at different physical locations; in fact, many of those individual contributors may never have even met. Without common agreement as to the units of measurement and a common reference to check for instrument accuracy, such a procedure would not be possible.

Therefore, national standards have been developed in the United States and all other major industrial nations. The units of the standards might differ from nation to nation; however, conversions between them are possible as long as everyone concerned understands the standards themselves and as long as those standards selected remain constant. Since all measurements are to be made based on these standards, it really makes no difference what the particular mechanism used for each standard is; as long as it remains constant, is accessible for transfer to those who require it, and is at least as accurate as the highest accuracy required by the users. As you might guess, the mechanisms that implement the standards have changed over the years to reflect the requirement for greater accuracy; however, the standards themselves have remained relatively constant.

Dissemination of Standards and Traceability

Hardware standards are critical pieces of equipment, requiring specific environmental controls, and are not therefore transportable. Therefore, an intermediate mechanism is required to disseminate the standard to those who require it. Several techniques are used to accomplish this objective. One technique consists of the National Bureau of Standards sending to a requesting company a *transfer standard* which the NBS owns and has calibrated against *the* standard. The requesting company then calibrates their internal *primary standard* as compared to this NBS transfer standard. The transfer standard is then returned to the NBS, where they recalibrate it, tell the company how accurate it was while in their possession, and the process is repeated with another company. It is interesting to note that the NBS standard and transfer standard are passive devices, devices that require no external power source and/or contain no internal elements which cause amplification of any input signals. These passive devices are used at the higher levels of accuracy, but they require special handling and qualified operators.

A second common technique for establishing primary standards at any particular company is for the company to send their own transfer standard (normally, their internal primary standard) to the NBS for calibration by them against *the* standard. It is subsequently returned to the company, which maintains it until it is again due for recalibration.

A third technique is also possible with certain types of measurement standards. These standards are based upon some locally reproducible phenomenon. A good example is length measurement, where that standard is based upon the wavelength of a certain type of light. It is possible, though expensive, to reproduce this standard accurately in the field; therefore, certain facilities are able to maintain their own standards without resorting to the processes described above.

In any case, the procedure is very expensive and time consuming; calibration of a primary standard can require a period of several weeks. Therefore, very few companies actually maintain standards which have been directly calibrated against

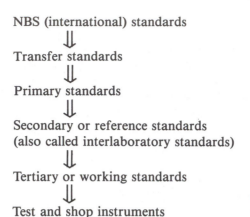

NBS (international) standards
⇓
Transfer standards
⇓
Primary standards
⇓
Secondary or reference standards
(also called interlaboratory standards)
⇓
Tertiary or working standards
⇓
Test and shop instruments

Figure 6-5 Traceability chart for calibration of instruments.

the NBS standards. Most companies send their own *secondary* or *reference standards,* which may be either passive devices or very high quality active devices, to one of the calibration companies which maintain their own primary standards. These secondary standards are calibrated against those primary standards and are returned.

All standards referred to thus far are so critical that they require environmental controls and handling by specially trained persons. Therefore, they are not useful to operating personnel for everyday measurements. The equipment used routinely in the field is frequently active (not passive) and at the *tertiary,* or *working, level* of calibration.

The tertiary level of calibration is normally accomplished directly by larger companies which maintain their own primary or secondary standards; and many specialized companies exist to accomplish this function for companies not maintaining their own standards. These standards should never be used directly for measurements, only for calibration of routine shop test instruments and measuring apparatus.

Therefore, there is a "tree of traceability" for the calibration of the instruments used in routine measurements by working technicians. When an instrument is referred to as having its calibration *traceable,* it has been subjected to the calibration process we have been discussing. Figure 6-5 is a tabular summary of the calibration process.

Specific Standards

To go into detail concerning the specifics of standards used at the various levels of accuracy (refer to Figure 6-5) is beyond the scope of this text. However, some general comments will be made concerning selected (more common) electrical measurement standards as examples.

Resistance standards are quite common. They are normally passive at all levels of calibration accuracy and are commercially available in sizes from 0.0001 to 10,000 Ω, with accuracies in the range 0.1 to 0.01%. They are relatively stable devices, but their values do vary somewhat with time, temperature, and humidity. Figure 6-6A schematically illustrates a resistance standard.

Dc voltage standards are also quite common and at the lower levels of calibration accuracy are normally provided by *standard cells*. Standard cells are electrochemical systems consisting of two dissimilar electrodes immersed in an electrolytic solution (Fig. 6-6B). Their materials and design are selected such that they provide a very reproducible and stable source of emf for long periods of time. The accuracy of these devices is within $\pm 0.005\%$ and everything about their construction and use is critical. They must be maintained at a constant temperature and are sensitive to being moved; therefore, they must be maintained at a fixed location in a controlled environment. For the technician it is important to note that they may never be used to pass current, and even instantaneous accidental shorting of the terminals can destroy their usefulness.

The standard for ac voltages is the thermal transfer standard. Here the heating effect of ac current (rms measurement) is used. It consists basically of a thermocouple and a series impedance (resistance) element. The ac to be measured is applied across that impedance element (Fig. 6-6C). The emf output from the thermocouple is a measure of the heating effect of the applied ac, therefore, its rms value. Implementation of this technique is considerably more complex than this explanation may indicate. They are capable of making measurements to within 0.02% accuracy.

Standards exist for all basic measurements; however, this brief discussion should suffice for the purposes of this text. More comprehensive and detailed descriptions are available directly from the manufacturers of this type of equipment.

Calibration

Calibration of various instruments must be accomplished at periodic intervals due to internal drift of components with age and change in ambient conditions (predominantly temperature and humidity), accumulation of dirt, and so on, which causes shunt paths for currents, and, of course, equipment abuse (accidental or use of equipment outside its specifications).

Instrument manufacturers will recommend maximum recalibration intervals (as long as the equipment is maintained within specified limits on temperature, humidity, etc.) for their equipment. Based on past experience or particular local conditions under which their instruments are used, companies frequently set up their own recalibration schedules for all of their equipment. Normally, specially trained persons are assigned the task of recalibration and maintenance of all local standards; equipment is sent to them and they certify recalibration by applying a dated and signed calibration sticker to each piece. Typical recalibration periods run from 90 days for most instruments up to one year for certain instruments. Critical instruments may have specified recalibration intervals scheduled in terms of weeks.

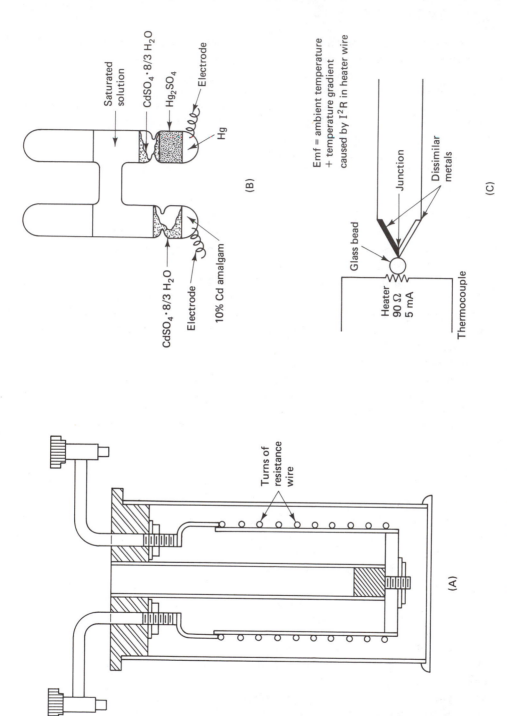

Figure 6-6 Selected electrical measurement standards: (A) resistance; (B) dc voltage; (C) ac voltage/current.

Remember, no matter how well a piece of equipment has been calibrated, the measurements may still be corrupted by all types of "noise," including ground loops, common-mode effects, electrostatic and electromagnetic pickup, power supply variations (if appropriate), thermal emf's generated by dissimilar metals at junctions, distributed capacitance, resistance and inductance of test leads, and so on.

SUMMARY

This chapter has generally discussed the goals and philosophy of making measurements in industrial process control systems. Choice of measurements to be made are controlled on the one end by available access points in the process (i.e., you may not be able to stop a chemical reaction in midreaction in order to measure some critical variable) and on the other end by available transducers and by the requirements of the rest of the system.

Once the variables to be sensed and those to be controlled have been selected, the problem becomes one of working with available transducers and subsystems while maintaining the overall system control within acceptable performance limits and cost. Many of the different types of errors have been discussed, along with their effects on systems. System error analysis was presented as a technique that effectively combines the effects of accuracies (or inaccuracies if you prefer) of the various system elements, helping to predict overall system performance based upon accuracy specifications of each of those subsystems.

The chapter included many definitions of the sources of (in)accuracy in systems and their individual elements. The term *accuracy* has so many constituent parts and connotations that considerable time was spent discussing it. The presentation culminated with a section on standards and calibration. There the philosophy of calibration, traceability of calibration, and standards used to calibrate equipment were discussed in general terms.

This chapter is a lead-in for the next chapter, which discusses the techniques and hardware used to interface directly with sensors in order to maintain the systems within acceptable response limits.

QUESTIONS

1. Why are sensors such critical components in process control systems?
2. In your own words, explain the term *accuracy.*
3. Differentiate between accuracy and precision.
4. What are industrial standards, and how are they used?
5. What is the general goal of a calibration procedure?
6. Give your own definition of measurement system.

7. Referring to the discussion of an automobile cruise control system, discuss why the following variables would or would not be adequate quantities to measure instead of the speed of the rear wheels.
   A. Airspeed past the car.
   B. Engine r/min.
   C. Engine vacuum.

What is the range and span of a measurement device (or system) required to make the following measurements?

8. The speedometer of an American automobile.

9. The odometer of an automobile.

10. The air-speed indicator of a commercial jet airliner.

Determine the scale factor for the following measurement systems.

11. The speedometer of a car that goes from 5 mph to 85 mph on a scale 8 in. long.

12. A circular thermometer having a scale with a 6-in. mean diameter and an indicating range of −30 to +110°F.

13. How do the terms *range, span, linearity,* and *scale factor* relate to the plot in Fig. 6-2?

14. Referring to the tree of traceability (Fig. 6-5), where would the following instruments be classified?
   A. A VOM.
   B. A standard cell.
   C. Each of the four pieces of equipment shown in Fig. 6-7.

15. Standard cells are useful tools to the technician as
   A. Constant-current sources (voltage varies).
   B. Constant-voltage sources (current varies).
   C. Voltage standards.
   D. Current standards.

16. An instrument that was calibrated against instruments which themselves were ultimately calibrated against an NBS standard is said to have a _____ calibration.
   A. Valid.
   B. Accurate.
   C. Expensive.
   D. Traceable.
   E. Precise.

GLOSSARY

Accuracy   The error present in a measurement as compared to the true value of that quantity. Typically expressed as a percentage of the full-scale value of that measuring system.

Characteristic Curve   A graph plotted to show the relationship between the input variable measurement of a process and the output measurement of a transducer or measurement system.

Hysteresis   A measure of the difference in response of a device or system as the input signal increases from minimum to maximum and subsequently decreases from maximum to minimum over the same range.

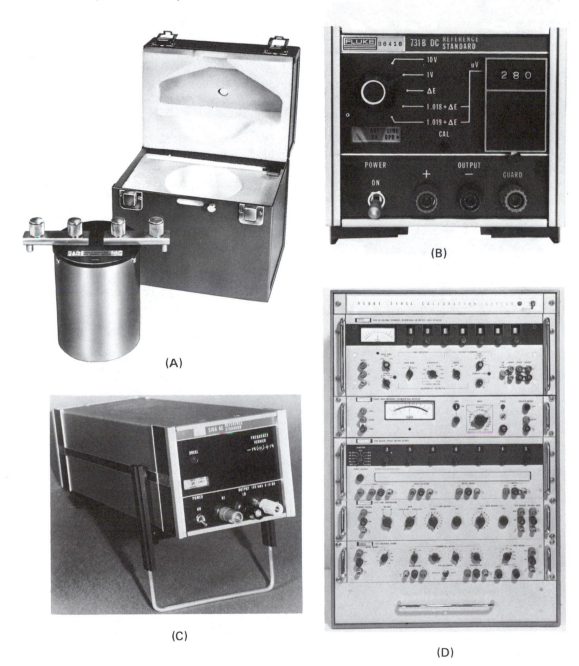

(A)

(B)

(C)

(D)

Figure 6-7 Standards whose calibration are traceable to the National Bureau of Standards: (A) resistance standard; (B) dc reference standard; (C) ac reference standard; (D) calibration system. [(A) Courtesy of Leeds & Northrup Co.; (B), (C), and (D) courtesy of John Fluke Mfg. Co.]

Linearity   A measure of the deviation of an actual process variable measurement from an ideal straight-line response of that same measurement over the same range.

Measurand   The physical variable being measured by a sensor or measuring instrument.

Measurement   The value obtained by a measurement system. It is representative of the physical quantity being measured.

Measurement System   The integrated combination of the sensor, signal conditioning, filtering, impedance matching, amplification, and display device(s) which makes process variable measurements.

Scale Factor   Also called conversion factor and conversion constant. A numerical ratio between the span of the input and the span of the output; of a transducer or measurement system.

Standard   The value accepted by scientists as being the exact or true value for each particular variable. It is this value that is used in the determination of accuracy for other pieces of equipment.

MEASUREMENT TECHNIQUES AND HARDWARE

INTRODUCTION

Probably the most critical components in modern process control systems are the system's sensors. These devices are the primary elements for measuring the system variables and converting the variable information to an energy system convenient for use by the controller.

Obviously, then, the maximum possible system accuracy is going to be controlled by the accuracy of the sensors. There is no way to subsequently improve upon the absolute accuracy of the initial variable measurements. Furthermore, every time the output signal from a sensor is processed in any way, a portion of the initial measurement accuracy is sacrificed (as explained in Chapter 6).

This chapter is dedicated to discussion of the techniques and hardware commonly used throughout industry in order to interface directly with many commonly used sensors to provide the necessary impedance matching. The topics of signal conversion to standard levels, and power amplification, are treated elsewhere in this book, although the type of equipment discussed in this chapter typically is included within those other types of circuits.

TYPICAL SENSOR (ELECTRICAL) OUTPUTS

Taking a broad look at the most common available sensors which transduce from any energy system to electrical outputs, it will be noticed that these electrical outputs can generally be categorized into just a few general types.

Many of these sensors have an electrical output in the form of a variable resistance. A few typical examples would include strain gages, thermistors, potentiometers used for measuring position, photosensitive semiconductor devices, resistance temperature bulbs, and humidity indicators.

Many other sensors develop an electrical *potential* (voltage) at their output, the magnitude of which is proportional to the parameter being measured. Some of the more common examples of this type of transducer would include thermocouples, photosensitive semiconductors, piezoelectric devices, some magnetic devices such as those which operate using the Hall effect, magnetic proximity sensors, and linear variable differential transformers (LVDTs and RVDTs).

A third type of electrical output from sensors would be an ac or frequency-type output. This type would differ from the voltage type in that the intelligence is contained in the instantaneous frequency rather than the value of the voltage. Rotary digital shaft encoders, shaft r/min counters, relays, proximity sensors, microswitches, photoelectric devices, and radiation counters would be a few common examples of sensors having this type of output.

The rest of this chapter will be dedicated to discussion of the electrical circuits which interface to these typical sensor electrical outputs.

BRIDGE MEASUREMENTS

The topic of bridge measurements in the field of process control systems is all important. So very many measurements are made using some type of bridge technique to interface to the electrical outputs from sensors that considerable effort will be made herein to give the reader a basic background in bridge theory, operation, hardware, and applications. There are numerous different types of bridges and also numerous different types of measurements are made using bridge techniques; we must, therefore, find some starting point from which to develop. The starting point selected for this text will be dc measurement of small values of resistance. We will then discuss dc measurement of small values of voltage using the theory and techniques developed for resistance measurements.

NULL BALANCE

Basic to any discussion of measurement techniques is a knowledge of the concept of "null balance." The null balance measurement technique is one in which an unknown quantity is compared to a known quantity; then the value of the known quantity is adjusted until the values of both the known and unknown are equal (i.e., the difference between them is null). Thereby, since they are equal and one is known, the value of the unknown quantity has been determined. The physical technique that is used to indicate when the two quantities are equal is termed a null balance technique.

As an example, consider a pan-balance scale for measuring weights. The unknown weight is placed on one pan, tipping the scale out of balance. Gradually, known weights are placed on the opposite pan of the scale until the two weights balance each other—or the difference between the two weights (both the known and the unknown) is minimized (nulled). When the scale is balanced, it is at its null. This is a physical example of a technique which is much used throughout industry in its electrical form.

VARIABLE-RESISTANCE-TYPE MEASUREMENTS

VOM Technique

Probably the simplest and most basic circuit used to measure dc resistance is the volt-ohmmeter-type circuit, illustrated in Fig. 7-1. This circuit works by connecting a known voltage source across the resistance element to be measured, and measuring the resulting current flowing through the element. The meter must initially be adjusted such that the voltage applied to the resistance is accurately known. Since a known (and constant) voltage is connected across an unknown resistance element, then by Ohm's law ($R = V/I$), we can measure the current passing through the resistance element and it will be directly proportional to the value of unknown dc resistance.

This circuit is economical, practical, and for many manual test and troubleshooting purposes, it is sufficiently accurate. However, for continuous, automated, high-accuracy, industrial process control applications, it suffers from several fatal shortcomings.

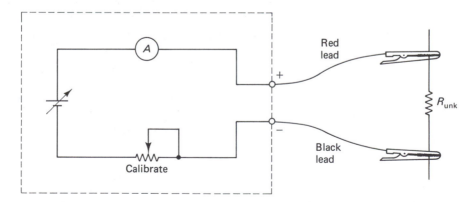

Figure 7-1 Schematic of typical VOM resistance measurement circuit (range switch capability not illustrated).

The first drawback is the dependency of the accuracy of the final measurement upon the accuracy of the voltage source. To compensate for voltage variations in commercial VOMs, an OHMS-ADJUST manual adjustment is included. This adjustment has basically the same effect as the "calibrate" potentiometer illustrated in Fig. 7-1, even though it may, in fact, be connected differently. It requires that the operator manually measure and adjust the system to compensate for voltage variations prior to each measurement. This is an unacceptable requirement for an automated system, where the adjustment would have to be made automatically and the overall measurement accuracy would be further reduced by inaccuracies in the automatic adjustment hardware.

The second disadvantage is that fairly heavy and somewhat variable currents flow through the measuring circuit. This causes variable heating of the variable-resistance element, which introduces another source of error.

The third fatal flaw is that all measurements are made on scales which start referenced to zero ohms, regardless of the range scale used. As an illustration of why this is a problem, let us assume the need for measurement of the resistance of a bonded wire strain gage,* having a no-strain resistance of 95 Ω and a full strain resistance of 100 Ω. We would use a VOM resistance scale which would have this measurement read about in the middle half of the meter scale for best overall measurement accuracy. If we assume that we have a 1-kΩ scale on the meter, using the divide-by-2 switch, we would have a meter with a range of 0 to 500 Ω. On such a scale, the 5-Ω variation expected from our strain gage would produce a 1% change in the reading. It should now be obvious to the reader that it will not be possible using this VOM technique to interpret the variations in strain gage readings with any useful accuracy or precision.

The final reason (for our discussion here) for not being able to use this circuit (if the others were not enough) is the inaccuracy introduced (in the circuit of Fig. 7-1) by the calibration resistor. Any variations in the value of resistance measured will cause proportional variations in the value of current flowing not only through the resistance element to be measured but also through the calibration resistor, thereby changing the voltage across the resistance element being measured. If you will remember back to the beginning of this discussion, a constant and known value of voltage across the measured resistance element was assumed.

Some of these undesirable variables can be compensated for in the markings of the scale on the meter; however, basically, this technique is totally unacceptable for the large majority of automated industrial resistance measurement systems.

Now that we have discussed (in detail) why the VOM-type circuit cannot be used in automated measurement systems, where do we go from here? The answer is to bridge-type circuits; the "why" will be explained in comparison to the previous discussion.

---

*What strain gages are, how they work, and how they are used was covered in a previous chapter; suffice it to say here that they are commonly used in industrial systems to measure distortion and they are very precise devices.

RESISTANCE-BALANCED WHEATSTONE BRIDGE

Basic Wheatstone Bridge

A very simple circuit that overcomes many of the disadvantages of the VOM-type circuit is the Wheatstone bridge circuit. The basic balanced Wheatstone bridge circuit is illustrated in Fig. 7-2. It consists of a voltage source (V), two precision fixed-resistance elements (R_1 and R_2), a galvanometer (G), and a precision variable-resistance element (R_4) having a dial calibrated directly in resistance units. Resistance R_3 represents the resistance element whose value is to be determined using this system. It is typically an electrical sensor such as the bonded wire strain gage previously mentioned. Each resistance element is included between nodes (labeled *a, b, c,* and *d* in Fig. 7-2), where it connects with other elements. The circuit between each of these nodes is commonly referred to as a "leg" of the bridge.

Those unfamiliar with electrical concepts should note that the voltage at all points connected together by wires (only) is the same. Therefore, the voltage at the positive terminal of the battery V and the "tops" of resistors R_1 and R_2 will be identical; however, the currents through those resistors are not necessarily the same. In fact, the current flowing in the wire connected to the positive terminal of the battery will be the algebraic sum of the currents flowing in the wires connected to resistors R_1 and R_2. These elementary electrical concepts, including Ohm's law and Kirchhoff's laws, will be assumed knowledge in the material that follows.

The various electrical elements of the bridge are sized based upon the intended applications. The following discussion of the theory of operation of the basic balanced Wheatstone bridge circuit, as illustrated in Fig. 7-2, will use values chosen for this example only, sized to permit ease of calculations.

To begin our discussion, let us ignore the presence of the galvanometer. This

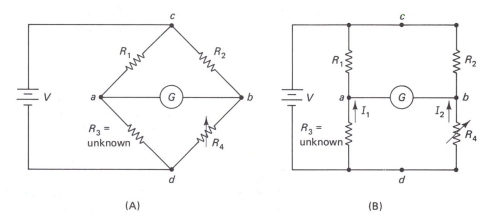

(A)   (B)

Figure 7-2 Two equivalent schematic representations of the basic balanced Wheatstone bridge resistance measurement system.

leaves the battery (V) connected to a resistive network consisting of two parallel–series paths which can be investigated individually. The first path consists of the series combination of R_1 and R_3, which is in parallel with a second path that consists of the series combination of R_2 and R_4.

Let us pick resistor values such that when the bridge is balanced the total values of resistance in both legs of the bridge will be equal. In mathematical form this means that $R_1 + R_3 = R_2 + R_4$ when the bridge is balanced. Since both legs of the bridge are of equal resistance and they are connected in parallel across the battery, the currents flowing through both legs will be equal ($I_1 = I_2$). This relationship will prove very useful in the examples that follow; however, it is not necessary that all bridges have these currents equal; they are mathematically more difficult to solve, though.

For the voltage at point a to be equal to the voltage at point b, so that no current will flow through the galvanometer (null) when we reconnect it in, the voltage dropped across R_1 must equal the voltage dropped across R_2, and the voltage dropped across R_3 must equal the voltage dropped across R_4. In equation form, this is expressed as

$$V_{R_1} = I_1R_1 \quad \text{which must equal} \quad V_{R_2} = I_2R_2$$

and

$$V_{R_3} = I_1R_3 \quad \text{which must equal} \quad V_{R_4} = I_2R_4$$

By waving our algebraic magic wand at these equations, first setting $V_{R_1} = V_{R_2}$ and then $V_{R_3} = V_{R_4}$ and then solving for the ratio of currents, I_2 divided by I_1, we arrive at the following relationship:

$$\frac{I_2}{I_1} = \frac{R_1}{R_2} \quad \text{and} \quad \frac{I_2}{I_1} = \frac{R_3}{R_4}$$

Setting these two quantities equal to each other and solving for the value of R_{unknown} (R_3 in the equation), we arrive at the basic Wheatstone bridge equation:

$$R_3 = \frac{R_1}{R_2} R_4$$

Notice that R_3, the unknown value of resistance, is related to R_4, the calibrated potentiometer, by the ratio of resistors R_1 and R_2; therefore, these resistors are commonly referred to as the ratio arms or ratio legs of the bridge. Note also that this relationship does not require that either $R_1 + R_3 = R_2 + R_4$ or that $I_1 = I_2$. These choices were made only for example purposes; the basic relationship is valid even when these conditions are not met.

We now have an algebraic relationship that we can use to calculate the value of the unknown resistance (R_3), whose value will be measured in terms of known resistance values. Notice that the value of the battery voltage does not affect this relationship at all; therefore, the calculated value of resistance is essentially independent of the battery voltage. This is one outstanding advantage of the Wheatstone bridge resistance measurement technique over the VOM technique previously discussed.

If we now select values for R_1 and R_2 which are equal in this example ($R_1 = R_2$), then (when the bridge is balanced) our equation reduces to

$$R_3 = R_4$$

Therefore, if a calibrated dial is mounted on R_4, when the bridge is balanced, the reading on the dial (of R_4) will indicate a value equal to the value of the unknown resistance.

Bridge Balancing

Now we had better discuss the concept of balancing the bridge. Up to this point, we have ignored the presence of the galvanometer in the bridge circuit. The galvanometer is a very sensitive, zero-center, indicating ammeter (current meter). Any ammeter, in order to indicate the electrical current flowing through it, must be connected in series with the circuit; and it should have very low resistance to the flow of current through it, or be very nearly an electrical short circuit from one of its connections to the other. Therefore, the galvanometer in the bridge circuit of Fig. 7-2 literally shorts points *a* and *b* together; and this is proper for the bridge circuit to operate correctly.

By definition, the bridge will be balanced when the current flowing through the galvanometer leg of the bridge circuit is zero. Referring to Fig. 7-2, the only way that the current through the galvanometer can be zero is for the voltage at point *a* to be *identical* to the voltage at point *b*. Since we already chose the series combinations of $R_1 + R_3$ to equal $R_2 + R_4$, we have constrained the current flowing through $R_1 + R_3$ to be equal to the current flowing through $R_2 + R_4$. Therefore, I_1 must equal I_2 when the bridge is balanced.

By careful application of Ohm's law to R_3 and R_4, we can determine that the only way for the voltage at point *a* (V_a) to be equal to the voltage at point *b* (V_b) is for R_3 to equal R_4. Therefore, when the value of R_4 (which is variable and has a calibrated dial on it) is equal to the value of R_3 (the resistance we are trying to measure), then V_a will equal V_b and no current will flow through the galvanometer leg of the bridge.

Turning this procedure around, we can state that to balance the bridge (with a value of R_3 connected in), the procedure is to adjust the calibrated knob on R_4 until the galvanometer reads zero; at that point the value of R_4 will equal the value of R_3, and the calibrated knob on R_4 will indicate their value.

The foregoing discussion outlines the theory and operation of balanced Wheatstone bridges; however, there are a few more considerations to be made before leaving this topic. Our original example was measurement of the value of resistance of a bonded wire strain gage, which varied in resistance from 95 Ω to 100 Ω as it measured the process variable to which it was connected.

These devices are very heat sensitive, especially to heat generated internally due to current being passed through them by the rest of the measurement system (I_1 in Fig. 7-2). The value of I_1 is not important to the accuracy of the bridge circuit as long as it is equal to the value of I_2 (in our example). Therefore, if we select R_1 and R_2 to be very large values of resistance, the values of I_1 and I_2 will still be equal (since R_1 still equals R_2 in value); however, they will be very small in magnitude. Since a very small value of current will be flowing through the bonded wire strain gage (R_3 in Fig. 7-2), the heat generated by this current, I_1, will also be relatively low (remember that heat dissipated in a resistance can be calculated by the equation: heating power $= I^2R$; $P = I_1^2R_3$ in this example) and the self-heating effect of the current through the strain gage can be minimized to acceptable levels. Also, since the value of I_1 is effectively a constant value, the effects of self-heating will be constant and can be calibrated out. This is the second major advantage of the Wheatstone bridge circuit in industrial applications.

Bridge Sensitivity

The third major advantage lies in the fact that when the bridge is balanced, the "zero" is at the current value of R_3 (the unknown resistance). What this means is that once the bridge is balanced, any variation at all (even a small fraction of an ohm) in the value of R_3 will throw the bridge slightly out of balance. This out-of-balance condition will be noted by a deflection of the galvanometer needle, indicating that current is flowing through it (i.e., the voltages at points a and b are no longer equal). Simply readjusting the value of R_4 will bring the bridge back into balance and the new value of R_3 can be read from the dial. The major point to note is that the circuit does not respond to the dc value (95 to 100 Ω in our example); rather, it is sensitive to *changes* in value. Therefore, it is theoretically just as sensitive and accurate regardless of the dc value of the unknown resistance element. This is quite an advantage over the VOM circuit previously discussed.

Extended Range Measurements

The manually balanced Wheatstone bridge circuit is extensively used throughout industry in portable test equipment applications. The usefulness of the test equipment is normally extended by including a selection of ranges.

To understand the operation of the variable range selector, refer to Fig. 7-3. Note that the only change from the resistance balanced Wheatstone bridge (in Fig. 7-2) is resistor R_2, which has been replaced by a combination of 11 elements.

For example, let $R_1 = 100$ kΩ and R_2 be comprised of a 99.9-kΩ resistor plus ten 10-Ω resistors (R_2 totals 100 kΩ). Therefore, $R_1 = R_2$; also R_1 and R_2 are chosen to be so large (compared to R_3 and R_4) that variations in R_3 and R_4 will not significantly affect I_1 and I_2 (as discussed earlier).

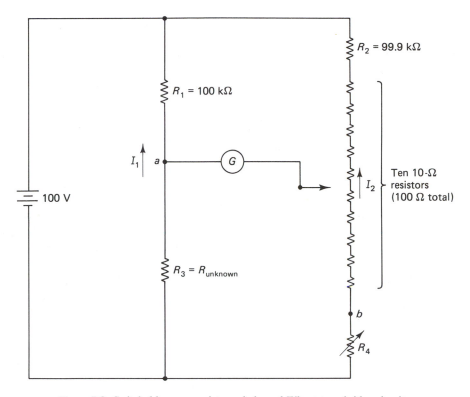

Figure 7-3 Switchable-range resistance-balanced Wheatstone bridge circuit.

I_1 and I_2 can be calculated by Ohm's law as follows:

$$I_1 = \frac{100 \text{ V}}{R_1 + R_3} \quad \text{or} \quad I_2 = \frac{100 \text{ V}}{R_2 + R_4}$$

However, R_1 and R_2 are so much greater in value than either R_3 or R_4 that the latter can be ignored. Therefore,

$$I_1 = I_2 \approx \frac{100 \text{ V}}{100 \text{k}\Omega} \approx 1.0 \text{ mA}$$

The fact that the original single resistance, R_2, has been physically broken into 11 parts does not change the electrical properties of the bridge, except to permit access to additional points which can be used instead of the original point b (in Fig. 7-3). Since 1.0 mA flows through R_2 (all 11 parts), the voltage dropped across each 10-Ω resistor will be 10 mV. By algebra, this translates into increases in the range of resistance (R_{unk}) which can be measured in increments of 10 Ω.

If R_4 is a 10-Ω precision potentiometer, then, with the galvanometer connected to point $b,$ the range of the bridge is from 0 to 10 Ω. With the galvanometer connected to the top of the first 10-Ω portion of R_2, the measurement range is changed. With R_4 set to zero, the voltage at the right side of the galvanometer will be such as to require a minimum value of R_{unk} of 10 Ω to balance the bridge. Therefore, the range of the bridge will now be from 10 to 20 Ω.

Each time the galvanometer leg tap is moved up or down, the range of the bridge is changed by a factor determined by the values of the resistors. Notice, however, that no matter where the tap is, the potentiometer (R_4) always adjusts the bridge for the last 10 Ω to actually balance the bridge. Therefore, the precision with which the final balance (and, therefore, the measurement) is made is the same throughout its measurement range.

Another common technique is to alter the ratio of R_1 to R_2, instead of having them be equal in value as we have done previously in our examples. If you will refer back to the basic equation for a resistance-balanced Wheatstone bridge, the solution equation was

$$R_3 = \frac{R_1}{R_2} R_4$$

If the ratio of R_1 to R_2 is altered, we can change the ranges of resistance values the bridge can measure. Of course, the value read off the dial for R_4 would also have to be modified. It is very common for manufacturers to select resistance values for R_1 and R_2 which result in ratios that are in powers of 10 (i.e., 10:1, 100:1, 1000:1, etc.). Figure 7-4 schematically illustrates this type of circuit, and most commercial Wheatstone bridges have this capability.

In any of these bridge circuits, use of a more sensitive galvanometer and a more precise (and accurate) resistance element for R_4 will result in corresponding greater precision and accuracy in the measurement of the unknown resistance element. These devices are in common industrial use where manual adjustment of R_4 is acceptable. Some can be purchased which will provide accurate resistance readings to six significant digits. Figure 7-4B illustrates a commercially available Wheatstone bridge.

THREE-WIRE RESISTANCE MEASUREMENTS

When making remote resistance measurements using a Wheatstone bridge (as is frequently necessary in a process control system), the unpredictable resistance of the two leads connecting the unknown variable resistance sensing element to the bridge (which can be very long) can introduce intolerable errors into the measurement. As illustrated in Figure 7-5A, both of the leads add their resistance to the same leg of the bridge. Therefore, to compensate for the lead resistance, the three-wire resistance-measuring system was developed.

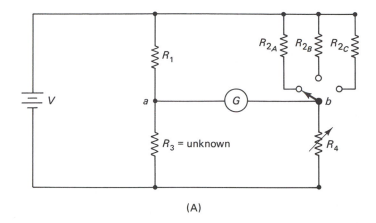

(A)

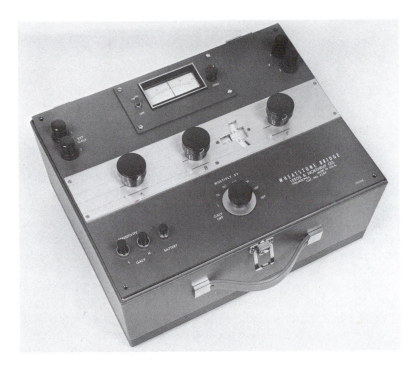

Figure 7-4 (A) Basic Wheatstone bridge modified for range switching; (B) commercial Wheatstone bridge. Measures resistance from 1 Ω through 11.01 MΩ with ±0.05% maximum error. [(B) Courtesy of Leeds & Northrup Co.]

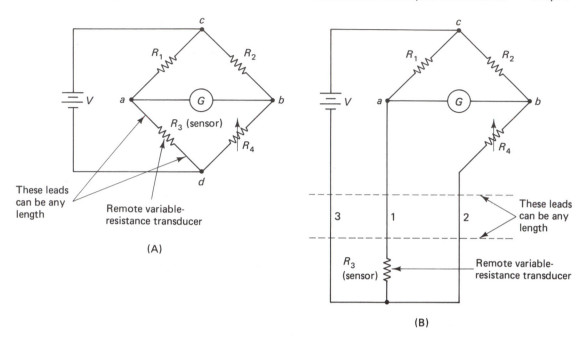

Figure 7-5 Modifications to Wheatstone bridges for three-wire measurements: (A) basic Wheatstone bridge circuit; (B) modified to make a three-wire resistance measurement.

The three-wire technique effectively moves point d in Figure 7-5A to the remote site. Therefore, three very long leads are required to connect the variable-resistance transducer and point d back into the (now remotely located) Wheatstone bridge. It would seem as though there were no real advantage to this technique until one realizes that of the three long measurement leads, one is connected into the leg between points a and d, the second is connected into the opposite leg between points b and d, and the third between point d and the battery.

Referring to Fig. 7-5B, any resistance effects due to lead 1 will affect the left leg of the bridge, and any resistance effects due to lead 2 will affect the right leg. If leads 1 and 2 are both the same length, of the same material and size, and are run together, any effects on the measurement due to lead length will affect both legs of the bridge identically; therefore, the changes will not affect the output from the bridge since they will balance out. Lead 3 will affect both legs the same; therefore, it, too, will have no effect on the operation of the bridge circuit, and lead resistance is effectively compensated for. This type of measurement is so common in industry that most resistance temperature detectors come from the manufacturer having three leads attached.

Note that this technique, as with all the others discussed, also makes a measurement of some unknown (and normally variable) resistance as compared to other known values of resistance. Therefore, the overall accuracy of the measurement is always limited by the precision of the other resistors in the system. Measurements to an accuracy of 0.01% are about the maximum practical using these techniques.

FOUR-WIRE RESISTANCE MEASUREMENTS

As mentioned previously, resistance bridge measurement techniques are limited in the accuracy of the final measurement by the accuracy of the other components in the circuit. For extremely accurate measurements, these techniques are not adequate.

When extreme accuracy is necessary (i.e., for calibration purposes), the resistance value is not measured as a comparison to other resistors; rather, it is derived by passing an accurately known current through the resistance element (whose value is to be determined) and simultaneously measuring the voltage across it.

Figure 7-6 schematically represents the circuit for this type of measurement. Commercially available equipment which uses this technique has the capability of making resistance measurements with an accuracy of 0.001%. Needless to say, this is specialized equipment and has limited application; however, its use is necessary for very accurate measurements.

UNBALANCED BRIDGE MEASUREMENTS

Often an unbalanced bridge is used, wherein no attempt is made to null the bridge. This technique is especially popular in systems where the dynamic characteristics of the process and the sensor would require frequent and rapid rebalancing. In this

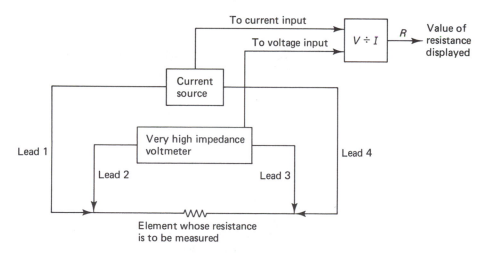

Figure 7-6 Four-wire resistance measurement.

application, the basic Wheatstone bridge circuit can be used; however, the only varying element(s) is/are the sensor(s). Figure 7-7 schematically illustrates several unbalanced bridge circuits, using one, two, and even four sensors in the same bridge. The theory behind use of more than one active sensing element in the same bridge is beyond the scope of a survey text; however, the multiple sensors are connected into the circuit such that the effects of changing resistance are additive, thereby increasing the sensitivity of the bridge to variations.

In the unbalanced bridge application (as with the balanced bridges), the output from the bridge is the voltage difference between points *a* and *b* (Fig. 7-7). Use of this type of circuit creates several problems which must be understood and compensated for. The first problem is that when using a dc power supply, a change in resistance of either arm (by itself) will cause a change in the current flowing in that

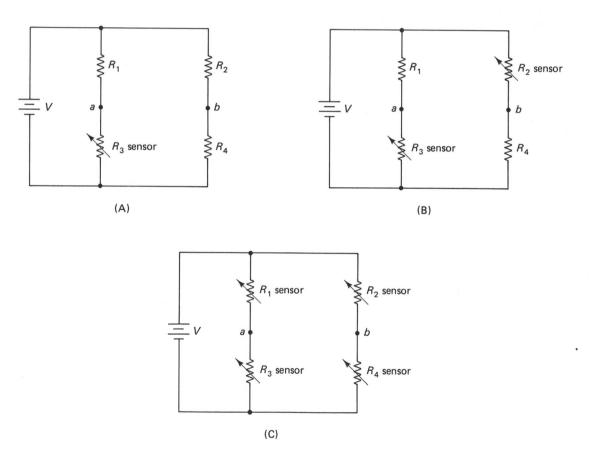

Figure 7-7 Unbalanced Wheatstone bridge schematics. *Note*: In part (B), R_3 and R_2 both change resistance in the same direction. In part (C), R_3 and R_4 both change resistance in the same direction, also R_1 and R_2 both change resistance in the same direction; however, R_3 and R_4 change direction in opposite polarity to the change in R_1 and R_2.

arm and, therefore, a changing voltage drop across the fixed element in that side of the bridge. This particular problem can be somewhat overcome by using multiple sensors and putting them into the bridge circuit such that the changes in voltage (between points a and b) due to changes in current through the fixed circuit components will affect both sides similarly, thereby somewhat canceling out this effect.

The changing current in a single-sensor bridge circuit (Fig. 7-7A) results in a nonlinear relationship between sensor variations and the output voltage. The equations describing the nonlinearity are complex; however, several things can be done to minimize its effects. First, the output will be sensitive to changes in the dc supply voltage; therefore, it must be well regulated. Second, the nonlinearity varies considerably as the unknown resistance varies above or below a certain design value; therefore, the bridge is designed so that the variation is always in the "best" direction throughout its measuring range. Third, maintain the ratio of change in transducer resistance to its static value as low as possible.

There are so many variables to be considered in unbalanced bridge measurements that specific values must be used to develop this topic further. Therefore, further discussion will be left to appropriate outside sources.

CURRENT-BALANCED WHEATSTONE BRIDGE

As useful and wonderful as manually balanced Wheatstone bridges are, they suffer from one fatal disadvantage when we try to apply them to automated industrial processes. The manually turned precision resistance value (R_4 in Fig. 7-2) must be automated by some type of servomechanism. If we try to do this, the final accuracy of our measurement will be limited by the precision with which the servomechanism can adjust R_4. Furthermore, R_4 is a precision (and very expensive) element and will not withstand the constant usage that an automated system would require of it. Therefore, the manually balanced Wheatstone bridge is not acceptable for use in an automated process; rather, it is a test instrument circuit.

It certainly seems a shame to have to disqualify a system with so very many outstanding characteristics due to one major fault, so let us try to design around the requirement for a precision variable-resistance element in the circuit. If you think back to the beginning of this discussion, you will remember that in order to balance the bridge circuit (Fig. 7-2) it was necessary to make the voltages at points a and b identical to each other. This could theoretically be accomplished by replacing R_4 with a precision variable-voltage source. If this was done, however, we would still be faced with the task of automatically varying some element, the voltage source in this case; besides, precision variable-voltage sources are quite expensive. Well, back to the drawing board for another try.

Another technique would be to make R_4 a fixed value and to include it as a common element in a second circuit also, such that another current also flows through it. By varying this second current we could vary the voltage across that common resistor, thereby balancing the bridge. We would now be faced with the

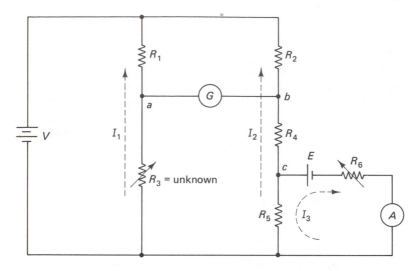

Figure 7-8 Basic current-balanced Wheatstone bridge circuit.

requirement for a precision current-controlling device in order to balance the bridge. Fortunately, such a device currently exists, is inexpensive, and is reliable: the ordinary transistor.

Figure 7-8 illustrates the basic Wheatstone bridge circuit as modified to be current balanced; we will refer to this circuit as the *current-balanced Wheatstone bridge circuit*.

The major modification to the basic resistance-balanced Wheatstone bridge (Fig. 7-2) is replacement of the precision variable-resistance balancing potentiometer with two fixed-value precision resistors (R_4 and R_5 in Fig. 7-8). Then a second circuit is added to the measurement system, the current-balancing circuit. The current-balancing circuit consists of resistor R_5, potentiometer R_6, ammeter A, and battery E. Note that R_5 is a common element in both the basic bridge circuit and the current-balancing circuit.

Currents I_1 and I_2 will flow in the basic bridge circuit exactly as described for the resistance-balanced bridge circuit. However, an additional current will flow in resistor R_5; this second (independent) current (I_3) is generated and controlled by the current balancing circuit.

To understand the operation of this circuit, we must rely on a basic electrical (dc) circuit analysis of the circuit illustrated in Fig. 7-8. To balance the bridge, the voltages at points a and b (with respect to circuit common) must be identical within the indicating capability of the galvanometer. The voltage at point a will be $V_a = I_1 R_{unk}$. The voltage at point b will be the sum of the voltages across resistors R_4 and R_5. Therefore, $V_b = V_{R_4} + V_{R_5} = I_2 R_4 + I_2 R_5$, *plus* the voltage across R_5 due to the current I_3. The complete equation now looks like this:

$$V_b = I_2 R_4 + I_2 R_5 + I_3 R_5$$

Refer to the schematic for the basic current-balanced Wheatstone bridge in Fig. 7-8.

1. V_a must $= V_b$ when the bridge is balanced.
2. Since $R_1 = R_2$ and they are so large in value that reasonable variations in the value of R_3 will have negligible effect on the value of I_1, then I_1 very nearly equals I_2, and they are essentially constant values.
3. Therefore, $I_1R_1 = I_2R_2$ within the measurement accuracy of the galvanometer, and any differences in the voltages at points a and b will not be affected by R_1 and R_2.
4. Therefore, since I_1R_2 must equal $I_2R_4 + I_2R_5 + I_3R_5$ and $I_1 = I_2$, this equation reduces to

$$R_3 = \frac{I_2(R_4 + R_5)}{I_2} + \frac{I_3R_5}{I_2}$$

$$= R_4 + R_5 + \frac{I_3}{I_2}R_5$$

$$= (1 + \frac{I_3}{I_2})R_5 + R_4$$

Figure 7-9 Derivation of the solution equation for the value of the unknown (measured) resistance element in a current-balanced Wheatstone bridge.

By waving our magic algebraic wand at these relationships, we can arrive at the basic equation to solve for the value of the unknown resistance element (R_3) in terms of the value of the current flowing in the current-balancing portion of the bridge measurement system and the values of two fixed-value precision resistors. For those of you who are so inclined, Fig. 7-9 summarizes the major steps in deriving this equation; however, it is not necessary to understand the derivation to continue.

$$R_3 = R_{unk} = (1 + \frac{I_3}{I_2})R_5 + R_4$$

Note that the only variable in this equation is the value of I_3, the current flowing in the current-balancing circuit; all other values are essentially constants.

Common sense now says, "so what? We are no better off than we were with the resistance-balanced bridge; in fact, we are worse off, because we have introduced more components and still have to vary a potentiometer." This is true—but!

The precision variable-resistance element in the resistance-balanced Wheatstone bridge circuit has been replaced by two fixed-value precision resistance elements. The combination of voltage source (E) and potentiometer (R_6) in the current-balancing circuit is required only to produce a precise value of current; it is not necessary that they be precision elements themselves. As an example of what I mean, if the value of the voltage source E varies, the current I_3 will proportionately vary. This will affect the voltage at point b on the bridge, throwing it out of balance. The out-of-balance condition will be noted on the galvanometer and the potentiometer R_6 will be adjusted to return the current I_3 to the proper value to balance the bridge. At the time of balance, the actual precise values of E and R_6 are unimportant as long as the combination generates the proper value of I_3 to cause balancing of the bridge.

Since neither the voltage source E nor potentiometer R_6 need be precision elements, the logical choice for R_6 would be a solid-state variable-resistance element. This description fits the common transistor. The transistor can be loosely (but accurately) described as an electronically controlled variable-resistance element. The resistance between the collector and emitter terminals is controlled by the value of current flowing through the base terminal.

With this information we can now proceed to develop an automatically balanced bridge circuit, having *no* manual operations required to operate it and *no* moving parts.

Refer to Fig. 7-10 for the following discussion of the operation of the automatically balanced Wheatstone bridge. The balanced condition is represented by the fact that no current (I_4) flows in the galvanometer leg of the bridge. Another precision resistance element, a sampling resistor, can be included in this leg of the bridge so that when current is flowing a voltage will be produced across it (by Ohm's law). The galvanometer is illustrated as being included in the circuit, but it serves no useful purpose in an automatically balanced bridge except as an indication of proper operation; therefore, it could be, and normally is, omitted.

The voltage across the sampling resistor (R_S) is connected to a special electronic circuit called a null-balance amplifier. This amplifier produces a current at its output which is of the proper magnitude and polarity to control the transistor (which replaced potentiometer R_6), such as to balance the bridge automatically.

Now that we have developed an automatically balanced bridge circuit, and no

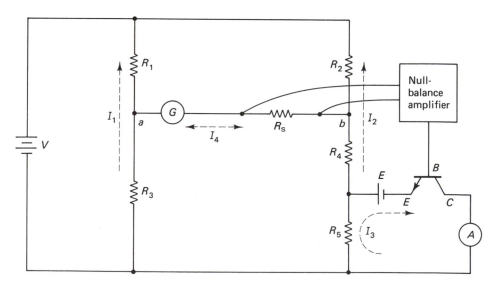

Figure 7-10 Automatically balanced Wheatstone bridge circuit.

longer have a dial on a variable-resistance element to read the value of the unknown resistance element, how do we know what its value is? Refer back to our equation (developed in Fig. 7-9) and, knowing the values of the "fixed" parameters in the circuit, the value of the unknown resistance element can be computed by measuring the value of current flowing in the current-balancing circuit and solving the equation. Another possibility is to calibrate the scale of the ammeter directly in terms of resistance.

As will be discussed later in this book, a current-type variable signal is a very desirable feature in process control systems where the variable information frequently must be sent (or "transmitted") over long wires to the control system which will use that information. Therefore, this basic current-balanced Wheatstone bridge is very nearly an ideal circuit for use in automated industrial process control systems.

Let us now take an example of how the bridge works, using numbers. Refer to Fig. 7-11 for the circuit and values to be used and the following procedure.

1. Calculate values of I_1 and I_2. Remember, we chose R_1 and R_2 to be large so that variations in R_{unk} will not affect I_1.

$$I_1 = \frac{100 \text{ V}}{100 \text{ k}\Omega + R_{unk}} \simeq \frac{100 \text{ V}}{100 \text{ k}\Omega} = 1.0 \text{ mA}$$

$$I_2 = \frac{100 \text{ V}}{100 \text{ k}\Omega + 10 \text{ }\Omega + 20 \text{ }\Omega} \simeq \frac{100 \text{ V}}{100 \text{ k}\Omega} = 1.0 \text{ mA}$$

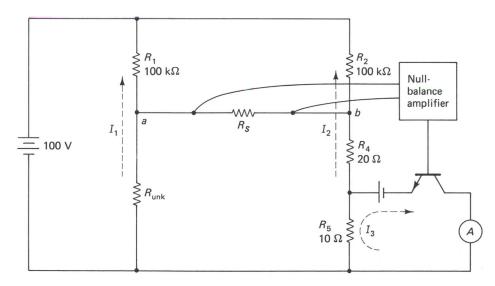

Figure 7-11 Automatic balanced Wheatstone bridge circuit for example problem.

2. Using the equation derived in Fig. 7-9 and the value of I_3 for the particular measurement, calculate the value of R_{unk}.

   For this example, assume that when the bridge automatically balances, the ammeter reads 6.25 mA. What is the value of R_{unk}?

$$R_{unk} = (1 + \frac{I_3}{I_2})R_5 + R_4 = \left(1 + \frac{6.25 \text{ mA}}{1.0 \text{ mA}}\right) 10 \text{ } \Omega + 20 \text{ } \Omega$$

$$= (7.25)(10 \text{ } \Omega) + 20 \text{ } \Omega = 92.5 \text{ } \Omega$$

There is another procedure by which the value of R_{unk} can be calculated once we know the values for I_1, I_2, and I_3, besides remembering the equation. The procedure is to calculate the voltage at point b and then determine the value of R_{unk} necessary to produce exactly the same voltage at point a.

In the preceding example, the voltage at point b will be equal to the sum of the voltages across R_4 and R_5.

$$V_{R_4} = I_2 R_4 = (1.0 \text{ mA})(20 \text{ } \Omega) = 20 \text{ mV}$$
$$V_{R_5} \text{ will be } I_2 R_5 + I_3 R_5$$
$$V_{R_5} = (1.0 \text{ mA})(10 \text{ } \Omega) + (6.25 \text{ mA})(10 \text{ } \Omega) = 72.5 \text{ mV}$$

Therefore,

$$V_b = V_{R_4} + V_{R_5} = 20 \text{ mV} + 72.5 \text{ mV} = 92.5 \text{ mV}$$

Next, calculate the value of R_{unk} necessary to cause the voltage at point a to be exactly the same as the voltage at point b.

$$R_{unk} = \frac{V_a}{I_1}$$

Since $V_a = V_b$ and $I_1 = 1.0 \text{ mA}$ in this example,

$$R_{unk} = \frac{V_b}{1.0 \text{ mA}} = \frac{92.5 \text{ mV}}{1.0 \text{ mA}} = 92.5 \text{ } \Omega$$

Both procedures are just as accurate as each other; simply select the one that is easier for you to remember and use.

Summary

In summary, the Wheatstone resistance bridge measurement systems are commonly used throughout industry to measure resistance values from around 1 Ω up to the megohm range. The overall measurement accuracy of the bridge itself is limited primarily by the accuracy of the fixed bridge elements and by the sensitivity of the galvanometer (null detector). Other sources of error include changes in all resistance values due to the heating effect of current flowing through them, thermal emfs generated internally by contact between dissimilar metals, and of course by lead resistance, as has already been discussed.

As noted previously for extremely accurate resistance measurements, one would use the three- or four-wire resistance measurement technique. Normally, the accuracy of basic resistance bridge circuits is adequate for automated process control systems.

VARIABLE-VOLTAGE-TYPE MEASUREMENTS

A large percentage of the sensors used throughout industry have as their output a dc voltage. Typically, this voltage is in the millivolt (and milliwatt) range, and the purpose of the following text is to explain the techniques for interfacing to these sensors. Undoubtedly, the largest number of this type of sensors are devices called thermocouples, which are discussed in Chapter 4; however, they have at their output typically a small millivolt dc potential and must interface with circuitry that requires (draws) virtually no power from the sensor. These requirements are readily handled by null-balance techniques.

POTENTIOMETRIC MEASUREMENTS

The technique used is called the potentiometric method. A potentiometer is a device that measures an unknown (and variable) voltage difference (electrical potential) by balancing that potential difference, wholly or partially, against an accurately known potential. Typically, the known potential is a voltage derived from a calibrated voltage source and divided down by a rheostat.

To explain the theory of operation of the common potentiometer circuit, let us take it one part at a time. Figure 7-12 schematically illustrates the calibration or standardizing portion of the circuit. The objective of the calibration procedure,

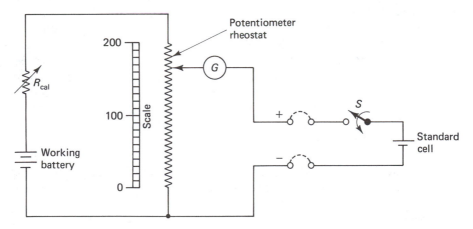

Figure 7-12 Standardizing circuit for a potentiometric measuring system.

which from now on will be called "standardizing" the potentiometer, is to ensure that a very precise value of working voltage is applied across the potentiometer rheostat.

This precise value is supplied by a standard cell. The subjects of standards and standard cells are covered in Chapter 6; for our purposes here, the standard cell is a very fancy chemical battery whose voltage is known to within a tolerance in excess of the accuracy of the overall potentiometer measurements. It maintains its voltage very stable over long periods of time, however, it cannot be used to provide any appreciable quantity of power and still retain its accuracy. Therefore, it is used only to calibrate a second voltage source, which can then be used to provide the power necessary for routine measurements. This working source, however, will have a voltage that varies appreciably with respect to time and use; therefore, the measuring system must be frequently recalibrated (or restandardized).

The objective of standardization will be to force the working battery to provide all the current required by the rheostat and to calibrate the scale on the rheostat such that the markings reflect exactly the voltage at each marked position. The rheostat itself is a very precise and linear variable-resistance element.

To standardize the circuit, the rheostat is first set (using its scale) to exactly the value of voltage listed on the standard cell. Then the switch (S in Fig. 7-12) is depressed; if the measuring system is properly calibrated, the voltage at the rheostat wiper (as read on the scale) will be exactly the same as the voltage from the standard cell. Any inaccuracy will result in a current flowing through the galvanometer. Simply adjust the calibration resistor (R_{cal}) until the galvanometer reads zero. At that point, the working cell is providing all the current required by the measuring system *and* the voltage at that spot on the rheostat wiper is exactly equal to the standard cell voltage. Since we have used a precision linear rheostat, the markings on the scale from zero to maximum will accurately reflect the voltage at the wiper when it is at each marking.

Figure 7-13 illustrates the measuring system we have just standardized connected to an unknown voltage source. The voltage source might be a temperature sensor (i.e., a thermocouple), a light sensor (i.e., a photovoltaic cell), or the output from any of a wide variety of other sensors. Regardless of the source of the voltage, it represents some process variable measurement whose value must be accurately determined. To operate the potentiometer, the Det switch is depressed and the rheostat wiper is moved until the galvanometer reads zero. At that point the voltage at the wiper is exactly (within the sensitivity of the measuring system) the same value as the voltage from the sensor (V_{unk}). Furthermore, since the galvanometer reads zero, *no* power is being taken from the sensor.

The latter characteristic of potentiometric measurements is extremely important since most of the sensors we would be using with such a measurement system are incapable of supplying much power to the measuring system without significant sacrifice in accuracy, linearity, and/or sensitivity. The potentiometric measuring system is therefore essentially an infinite-impedance voltmeter, in that it requires negligible power from the unknown voltage source.

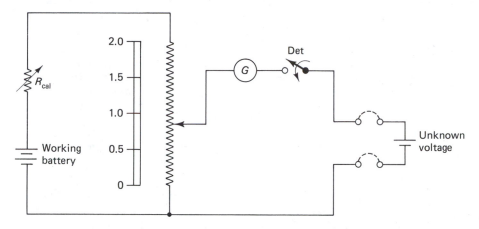

Figure 7-13 Measuring an unknown voltage with a standardized potentiometer system.

To extend the usefulness of this technique, the rheostat is normally provided as a circular element rather than a linear one; this reduces the space requirements and makes it more convenient to package and use. Figure 7-14 schematically illustrates a circular potentiometer. It also illustrates an economical feature where fixed values of resistance are switched in increments for course readings and then a vernier rheostat is used to obtain the null balance precisely.

Another of the possible refinements to the circuit in Fig. 7-13 is schematically illustrated in Fig. 7-15. To eliminate the need for adjustment of the measuring rheostat in order to standardize, a separate (additional) standardizing resistor is included in the circuit. Here, the standard cell is connected across the standardizing resistor only (not the rheostat) and R_{cal} is adjusted as noted previously until the bridge is standardized.

Since a known voltage has been connected across a known resistance element, R_{cal} is adjusted until exactly the proper value of current is flowing through R_{STD} (and, therefore, through the measuring rheostat since they are electrically in series with each other); at that point, the galvanometer will read zero and the circuit is standardized.

CURRENT-BALANCED MILLIVOLT POTENTIOMETER

As with the manually balanced Wheatstone bridges, the presence of the manually adjusted rheostat renders the potentiometer circuits heretofore discussed unsuitable for continuous operation in an automated process control system. In our search for an automatically (and electronically) operated circuit, we need look no further than the basic Wheatstone bridge circuit. Remember the basic theory concerning the balancing of a basic Wheatstone bridge? It required that the voltages at points *a* and

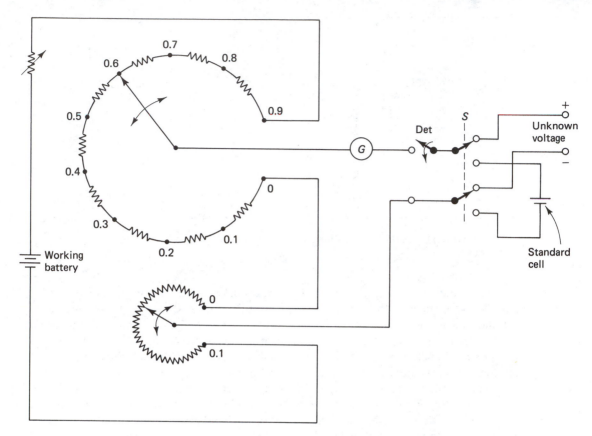

Figure 7-14 Simple potentiometric measuring system.

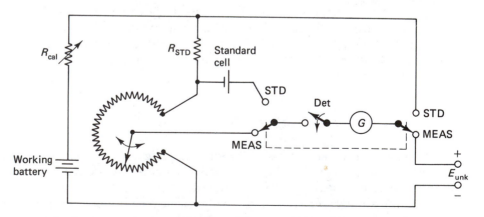

Figure 7-15 Using a standardizing resistor in a potentiometer circuit. Also, illustrates the switch connections for STD, MEAS, and DET switches.

b in Fig. 7-2 (reproduced in Fig. 7-16A) be exactly equal, within the measuring sensitivity of the circuit, at the point the circuit was balanced.

By using a precision fixed resistor for R_3 (in Fig. 7-16A) and inserting a small voltage source into the galvanometer leg, we have effectively modified the bridge so that it will measure small voltages. The modified bridge circuit is illustrated in Fig. 7-16B and this is how it works.

Rheostat R_4 is varied, as with the basic Wheatstone bridge, to cause the galvanometer to read zero (balance the bridge). However, when this bridge is balanced the voltages at points *a'* (not *a*) and *b* will be equal. Therefore, the setting of the dial on the rheostat (R_4) will be proportional to the value of V_{unk}. With V_{unk} equal to zero, the dial on R_4 should read zero at balance; as V_{unk} increases from zero to its maximum value, R_4 must be changed proportionately, and its dial can be calibrated directly in terms of the value of V_{unk}.

When the bridge is balanced, the current flowing in the galvanometer leg is zero; then (at balance) effectively zero current is flowing through V_{unk}, and therefore no power is being supplied to this bridge by V_{unk}. Since most of the sensing elements that would use this measurement technique have essentially no power to give, this is an almost ideal (very high impedance) voltage measurement circuit.

When we try to automate this bridge, we run into similar problems as discussed for the resistance measurements using a manually balanced Wheatstone bridge. The solution to the automation problem is also similar; use the current-balance technique.

Figure 7-17 is the schematic of a current-balanced voltage-measuring Wheatstone bridge. If you refer back to Fig. 7-11, you will see that the only alterations to the resistance-measuring bridge to convert it to measuring small voltages (millivolt potentials) are (1) replacement of the resistive sensor (R_3) by a precision fixed resistor, and (2) inclusion of the voltage-type sensor in the galvanometer leg of the bridge.

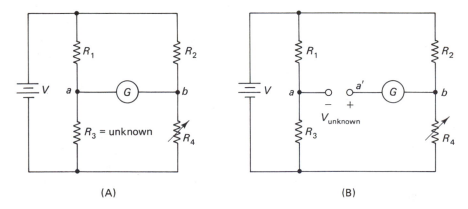

(A)   (B)

Figure 7-16 (A) Basic Wheatstone bridge; (B) Wheatstone bridge modified to measure small (millivolt) potentials.

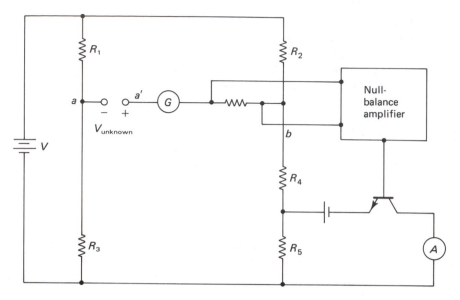

Figure 7-17 Current-balanced, voltage-measuring Wheatstone bridge.

The theory of operation of this bridge circuit is identical to that of the resistance-measuring bridge and will not be repeated here. For this measuring system, the current flowing in the current-balancing leg of the bridge will be proportional to the value of V_{unk}.

FREQUENCY-TYPE MEASUREMENTS

The final category of sensory outputs to be dealt with are those which provide process variable measurement information in the form of a frequency (or pulse)-type electrical output. The actual electrical outputs can vary from sinusoidal-type outputs to pulse-type output waveforms; and obviously the circuitry to interface with these waveforms must be different.

As is frequently the case in systems design and circuit design, the interface design problem may be more easily understood if we start with its output and then work back to its input (the output of the sensor). Our following system circuitry normally will require either a pulse that is of a uniformly square shape, or it will require a dc voltage level that is proportional to the frequency.

First, let us consider a circuit that is designed to interface with a frequency-type sensor and provide a "calibrated" output *pulse train* which has exactly the same frequency as the pulses from the sensor. Sometimes there is only the problem of interfacing with a square-wave pulse train arriving from the sensor and increasing its power capability for subsequent use in a digital controller. This can be accomplished by a simple (high-frequency) transistor amplifier stage. This type of circuit is commonly referred to as a *single-shot* or *one-shot amplifier.*

Frequently, the output from the sensor has an impulse, spike, sawtooth, or triangular waveform. The single-shot amplifier (for example) can have a passive *RC* network connected to its input which will result in a square-wave-type output. This passive input network can also be designed to accept a sinusoidal-type waveform at its input and result in a square-wave-type output from the single-shot.

Now let us turn our attention to the interface circuitry necessary to provide a dc voltage level output which is proportional to the signal frequency from the sensor. Probably the simplest circuit would consist of a passive *RC* network following the single-shot circuits previously described. This network smooths out the pulses and results in a varying dc voltage output which is proportional to the number of pulses (per unit time) that charge the filter capacitor. This circuit requires a calibrated pulse from the one-shot amplifier, which means that each pulse must have exactly the same height and width regardless of the waveform of the pulse train arriving from the sensor. This leaves the frequency of the pulses as the only variable at the output of the single-shot. Therefore, the rate of charge of the filter capacitor (and therefore its dc voltage level) is directly proportional to the frequency of the pulses that are used to charge it. Figure 7-18 illustrates this in block-diagram form.

The entire circuit is commonly referred as a frequency-to-dc converter and is an elementary form of digital-to-analog converter (which will be discussed in detail in a later chapter). Furthermore, the process of smoothing the pulse train from the single-shot circuitry into a dc voltage is called *integrating* the pulse train. The term "integrating" is exactly the same term, with the same definition, as when used in a calculus course. The smoothing, or integrating, network can also be viewed as a (very) low-pass filter.

Certainly, other types of interface circuits can be used with these frequency-type sensors, but the ones discussed here are typical and effective. Any other circuits that you may run into in industry will operate using basically the same principles.

One further point must be understood when dealing with frequency-type measurements. The problem is one of obtaining more than one output pulse from the interface circuitry for each input "pulse" (whatever its waveform may be).

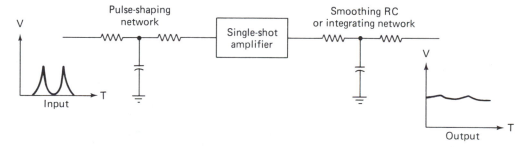

Figure 7-18 Variable frequency-to-dc voltage conversion.

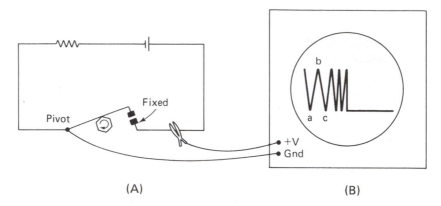

(A)   (B)

Figure 7-19 (A) Mechanically operated high-frequency switching device; (B) voltage waveform across the set of closing points.

To illustrate the source of this problem, let us assume that the interface circuitry must interface with a set of automotive-type points. These points are simply a mechanically operated (relatively) high-frequency switching device (illustrated in Fig. 7-19A). If you observed the voltage waveform across the set of points as they closed, you would observe the waveform illustrated in Fig. 7-19B (on an oscilloscope).

Theoretically, you would expect a nice square wave when the points operated; however, this is normally not the case when dealing with mechanical contacts of any sort. Actually, the contacts are brought together by spring pressure (normally) and the shock of collision when they close (point *a* on Fig. 7-19B) causes them to bounce open (point *b*) for a fraction of a second before the spring pressure causes them to reclose (point *c*). This process can go on for many cycles until the contacts finally rest against each other solidly.

When dealing with digital circuitry, the circuitry is fast enough to respond to *each* bounce, thereby indicating many pulses, when in fact the contacts closed only once. Therefore, when dealing with any mechanical set of contacts, this bouncing must be considered and appropriate filtering networks must be used to filter out the bounces.

ELECTRICAL GROUNDING CONSIDERATIONS

Before we can close the discussion of measurement theory I feel it necessary to discuss another very common source of measurement value "corruption." It really has nothing to do with the specific type of transducer or interface equipment, nor the type of measurement being made, nor the calibration of the equipment used;

however, it has a potentially devastating effect on overall measurement accuracy. This source of inaccuracy and error is due to improper grounding of equipment.

Typically, in a process control system the electrical "common" (or "ground") potentials at various points in the system may vary by values of from several millivolts to values of several hundred volts. Needless to say, these differences must be successfully dealt with in order for the system to work properly. In the following text we investigate the sources of grounding problems, the effects of improper grounding in systems, present some proper grounding techniques, and discuss some symptoms to look for to detect the existence of a grounding problem.

Origin of Grounding Problems

Some grounding problems result from improper use of equipment; however, most grounding problems are a direct result of imperfections in electrical equipment. When the power transformer for a piece of electrical equipment is wound, trade-offs are made between quality of materials used and product cost. The (primary and secondary) windings are electrically insulated from each other and from the chassis to an extent based upon the quality of insulating materials used.

Figure 7-20A is a schematic of a typical (and very basic) dc power supply and consists of only a transformer and a regulator. Of primary interest to us is the transformer in this circuit. Due to imperfect insulation there will be some finite value of leakage current directly between the primary and secondary windings and between both sets of windings and the equipment chassis. Therein lies the primary source of the type of grounding problem we are interested in here. A secondary path for electrical (ac) current is through the stray capacitance between primary and secondary windings and between the windings and the chassis. The direct leakage is illustrated by resistance elements on the schematics, and the capacitance is illustrated by capacitor symbols.

These two effects combine to provide both ac and dc current paths directly between the "low side" of the power source (typically 110 V ac or 220 V ac at 50 or 60 cycles) and the negative (ground or common) terminal of the dc power supply output. To compound matters, most electrical codes require that electronics equipment be fitted with a three-prong plug for personnel safety, where the third wire is connected between power source ground and the equipment chassis. Therefore, the situation as it normally exists is one where there are current paths between power source (third wire) ground and the negative terminal of the electronics equipment (power supply in our illustration). Note that the problem exists because of the power transformer, and therefore this situation will exist to some extent in all electronic equipment that uses ac power as its primary source (as compared to battery-powered equipment).

Refer now to Fig. 7-20B, which is the same circuit as Fig. 7-20A except that the distributed resistance of the third-wire ground lead and the negative output lead

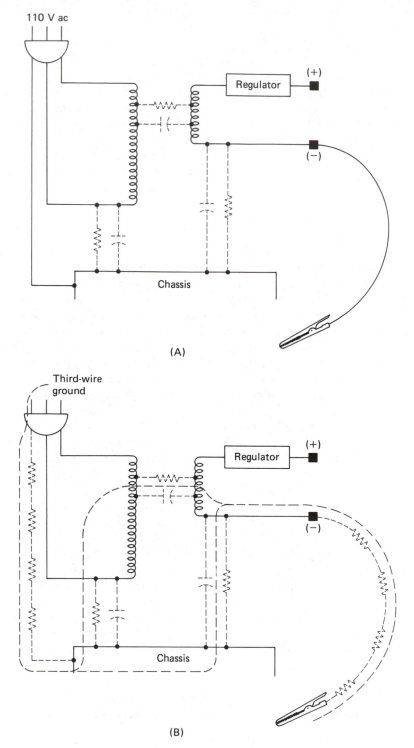

110 V ac

Regulator

(+)

(−)

Chassis

(A)

Third-wire
ground

Regulator

(+)

(−)

Chassis

(B)

Figure 7-20 (A) Schematic of a power supply showing leakage paths; (B) schematic showing lead and wire distributed resistance and potential ground current path.

is illustrated. The ground lead current path, which is what we are worried about here, is illustrated by dashed lines.

How Grounding Problems Affect Circuits

To illustrate the effects of (ground) leakage currents, let us now construct a very simple circuit and instrument that circuit for testing. Figure 7-21 illustrates a common-emitter transistor dc amplifier circuit, instrumented to observe input and output parameters. It is not necessary to understand this circuit in order to understand the grounding problem it is intended to illustrate.

Figure 7-22 illustrates the same circuit but includes two of the dc ground paths (dashed and dotted lines), both of which include at least one of the test circuit elements. There are other possible leakage paths but they do not include any test circuit elements and are therefore not of interest here.

If all three pieces of electrical test equipment (the DMM, the power supply, and the oscilloscope) are of the plug-in (ac) type, and therefore have the leakage paths between third-wire ac common and their negative leads, it is obvious (from Fig. 7-22) that any leakage currents that flow will also flow through test circuit elements, the emitter circuit of the transistor in this example. Therefore, the readings of input and output parameters will be corrupted by the algebraically additive effects due to those leakage currents, or "ground loops," as they are technically referred to. Any emf's occurring in the part of the test circuit that includes this ground loop will potentially cause current to flow; this is explained further in the next section.

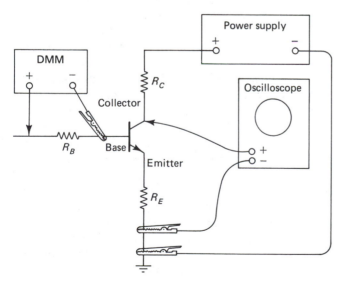

Figure 7-21 Typical test circuit with instrumentation (improperly grounded).

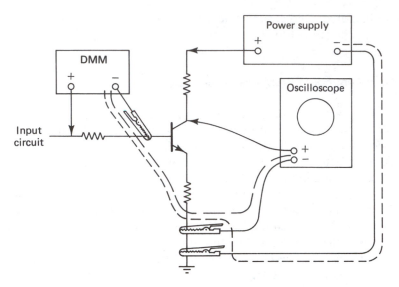

Figure 7-22   Circuit of Fig. 7-21 showing potential ground leakage paths.

Typical Test Circuit Loops

To illustrate other sources (beside the transformers) of the emf's which cause ground currents to flow in this circuit, refer to Fig. 7-23. This figure illustrates a more complex test setup, with only the negative (ground) leads from each of the pieces of test instrumentation shown as being connected (of course, the other leads must also be connected for the circuit to function properly). The chassis ground (third wire) for each of the pieces of equipment which are to be connected to ac power are also illustrated (note that the current loop is closed through these third-wire grounds on the power plugs). Remember, each of these wires has a finite value of distributed resistance.

Putting this circuit into a realistic environment, each of the leads illustrated is effectively an antenna for electromagnetic radiation from any source (radio stations, nearby heavy power equipment, fluorescent lighting, utility wires, etc.), including crosstalk between wires in the same circuit. There is also some degree of electrostatic coupling from any available nearby source. Finally, as noted previously, there are the emf's due to voltage differences between various windings on the transformers and each chassis ground.

Figure 7-24 illustrates several of the potential ground loops for the circuit of Fig. 7-23. Several considerations are worthy of note here; first, in several instances the same lead is actually connected into several ground loops simultaneously; second, if either of the oscilloscope leads is disconnected from its illustrated position and reconnected elsewhere in the system, several of the ground loops will be affected and will carry different values of current than they previously did; third, even though

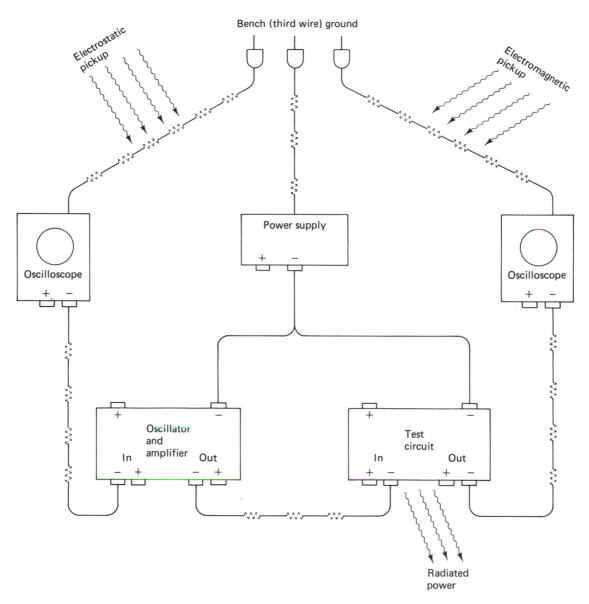

Figure 7-23 Bench test setup example, illustrating wires which will potentially conduct ground loop currents and potential sources of emf to cause leakage currents to flow.

no actual circuit components are included in the ground loops (as they were in the circuit of Fig. 7-22), voltage drops across all of the circuit common leads will affect the indicated measured values in the circuit (i.e., the circuit values may be correct; however, the test instruments will indicate incorrect results).

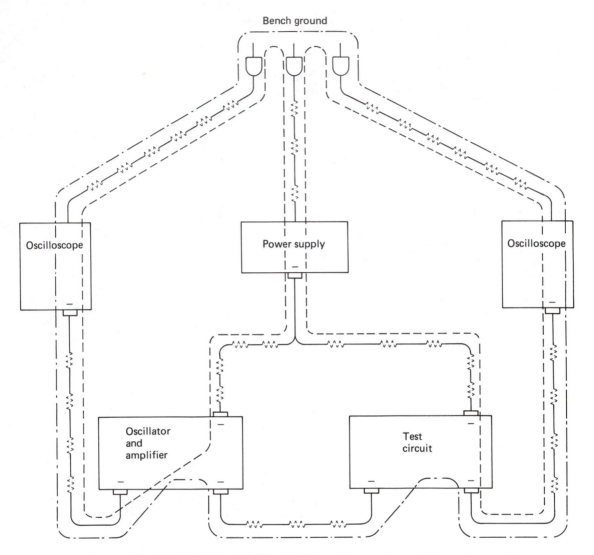

Figure 7-24 Test circuit of Fig. 7-23 illustrating several potential ground current loops.

Coping with Ground Loops

As we have seen, there are a multitude of possible emf sources which could (and normally do) cause currents to flow in "ground" systems. There is no practical technique for predicting the magnitudes of the currents that will flow or for completely eliminating them; however, some commonsense techniques can be used to minimize their undesirable effects.

Some commonly used techniques for reducing the effect of these ground loop currents on circuits are:

1. Use battery-powered test equipment.
2. Screen the equipment from electromagnetic and electrostatic fields.
3. Reroute leads away from high-noise areas.
4. Use heavy-gage ground wires (buses) to reduce resistance.
5. Shield test equipment signal leads.
6. Use a single common ground reference point for all leads.
7. "Float" the equipment above third wire-ground (a potentially dangerous and possibly illegal solution because there will be no ground-fault protection for personnel).

Since there is no practical way to eliminate the ground loops and we must therefore learn to live with them, let us investigate implementation of some of these techniques in our example test circuit. Figure 7-25 illustrates one very simple, effective, and very commonly used technique. Here, leads *A, B,* and *C* are short lengths of relatively heavy gage wire. Interconnecting the system in this manner effectively removes the signal current-carrying leads from the ground loop paths, thereby reducing the magnitude of the ground loop effects on the measurements. Note, however, that the signal-carrying leads must still be protected from other sources of interference.

A second test system is illustrated in Fig. 7-26, with both proper and improper grounding illustrated. The indication of a grounding problem in this setup (to an experimenter) lies in the oscilloscope waveform (Fig. 7-26A), which is nearly identical to the waveform from the signal generator. Here the capacitor in the circuit is effectively shorted out by the oscilloscope ground lead. Figure 7-26B illustrates proper grounding practices and the oscilloscope pattern is clearly more accurate. Once again, however, the test leads may require shielding to reduce electromagnetic and electrostatic pickup.

Process Control System Grounding Problem

Sometimes the ground loop paths are not as obvious as those previously illustrated; however, their effects on circuits can still be every bit as disastrous. Let us take a common industrial process control measurement circuit as an example. Figure 7-27 illustrates a general sensor having an electrical output which is connected via a long shielded cable to an instrumentation amplifier, which, in turn, is connected to a controller.

Either the lead between the electrical transducer's output and the amplifier or the lead between the amplifier and the controller is commonly quite long, even hundreds of feet. For our example, let us assume that the lead between the transducer and the amplifier is the long one, and that it is a two-conductor shielded cable.

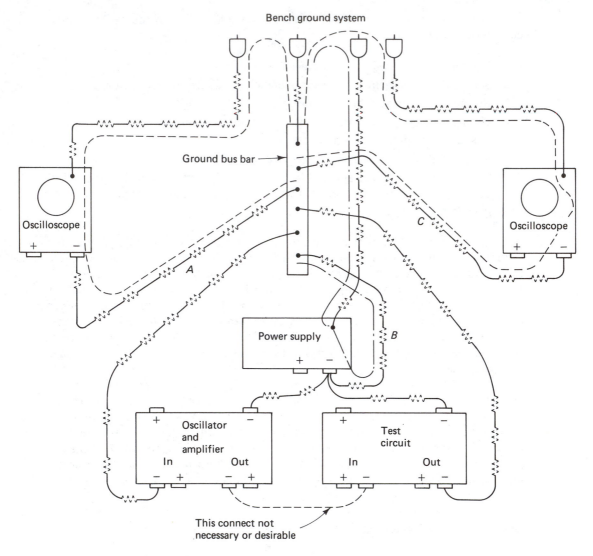

Figure 7-25 One technique for reducing the ground loop problems associated with the circuits illustrated in Figs. 7-23 and 7-24.

Figure 7-27 also illustrates the stray capacitances that will exist. C_{ls} represents the total effective capacitance between the common lead and the shield, C_{sg} represents the effective capacitance between the shield and earth ground; G_p is the earth ground at the process and G_a is the earth ground at the amplifier's location. There will be an effective voltage between these two grounds in any practical system, even if they are physically quite close. Therefore, since the electrical transducer is normally grounded at the process, and with no ground connection for the shield, there is an

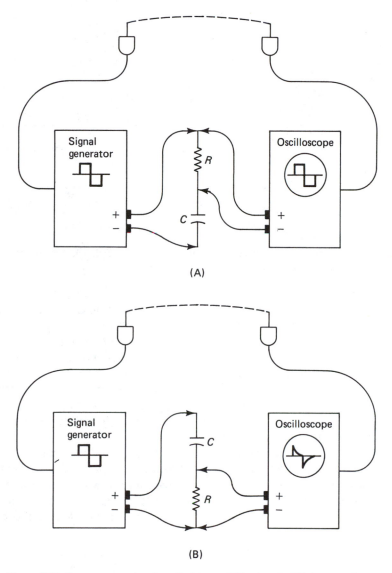

Figure 7-26 Square-wave signal applied to an *RC* network: (A) improperly connected oscilloscope ground effectively shorts out the capacitor; (B) correct grounding practice.

ac ground loop which is completed through capacitors C_{ls} and C_{sg}. The negative signal lead forms part of this ground loop and any group loop currents will flow through the distributed resistance (not illustrated) of that lead, thereby corrupting the electrical signal present at the input to the amplifier.

The most common mistakes normally made in connecting such a circuit in-

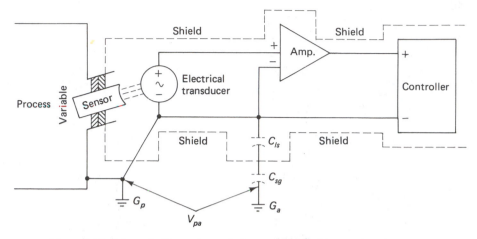

Figure 7-27 Schematic illustrating typical sensor electrical connections in process control systems and potential ground loop paths.

clude grounding the shield at the amplifier (shorting C_{sg}) and/or grounding the negative signal lead to the shield at the amplifier (shorting C_{ls}). Shorting either C_{sg} or C_{ls} will still leave the other one in a series circuit (which includes the negative signal lead) as a conductor for those ground loop currents. Shorting out both C_{ls} and C_{sg} is the worst possible solution since the resulting ground loop path (which still includes the negative signal lead) provides a low-impedance dc path also. Actually, the latter solution is probably the most commonly made mistake, where a conscientious technician (or engineer) decides (s)he will "ground everything" in order to get rid of ground loop problems.

The recommended technique involves grounding the shield to the same earth ground point as the transducer is grounded, and *only* to that point. Therefore, the undesirable ground loop currents are forced to flow in the shield, not in the negative signal lead. As a general rule, ground the system in only *one* place and make all ground connections to that point *only,* using the heaviest applicable gage wire.

How to Recognize a Grounding Problem

I honestly wish there were a straightforward, commonsense technique for determining when a system is being adversely affected by a grounding problem; but there isn't! The key to recognizing that system integrity is being violated by ground loop problems lies in an understanding of the types and sources of the ground currents that may be present.

Very commonly, the desired circuit signal will be modulated by a 60 (or 50)-cycle ac waveform where it should not be present, and this could (but not necessarily) be a clue. More commonly, the ground loop problem shows up as unexpected values measured in the system, while individually all subsystems appear to be working correctly. These errors are (normally) unpredictable; they may either be fairly constant in value or may consistently change their value, making a valid measurement nearly impossible. They may exhibit different effects on the system as temperature or humidity varies, at different times of the day, or after some modification has been made to the system. Sometimes simply connecting a piece of test equipment into the system will have an unexpected deteriorating effect on the system's performance. In extreme (but not that uncommon) situations the ground loop problems will cause repeated unexpected component failures.

DC-to-DC Coupling Networks

Sometimes the grounding problems in a system are so severe that complete electrical isolation must be provided between the transducer, which may be grounded at the process, and the controller, which has its own ground reference. Devices called dc-to-dc converters are available for this purpose.

Figure 7-28 diagrammatically illustrates two of the most common techniques used to provide this isolation. Note that each half of the circuit (input half and

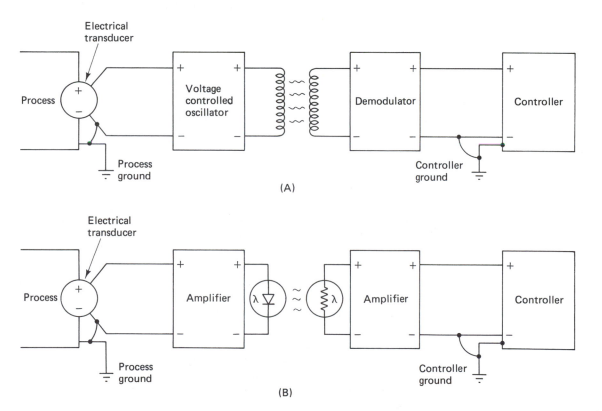

Figure 7-28 Dc-to-dc coupling techniques: (A) magnetic coupling; (B) optical coupling.

output half) must have its own power supply; and those supplies must not themselves share a common ground reference. Therefore, use of these devices in a system is quite expensive, and they are used where no other solution will work, or where electrical isolation is extremely important.

SUMMARY

The purpose of this chapter has been to point out the need for careful consideration of the circuitry which is connected directly to sensor outputs in order not to affect seriously the system's measurement accuracy. It has further gone into discussion of the importance of properly grounding equipment to maintain the accuracy of measurements already made.

Subsequent chapters will develop the types of circuits and devices that use the circuitry discussed in this chapter as their input. Therefore, the circuitry discussed herein has been primarily for the purpose of impedance matching between sensor and subsequent signal-processing equipment, and the maintenance of accuracy and precision of the original measurement by this interface equipment.

QUESTIONS

1. What is meant by the term *impedance matching,* and what are its systems implications?
2. Into what general categories can the electrical outputs from most sensors be classified?
3. List the disadvantages of using VOM-type resistance-measuring techniques in an automated system.
4. Explain in your own words the theory of operation of a basic Wheatstone bridge circuit for measuring variable resistances.
5. Why is the basic Wheatstone bridge unsuitable for use in automated systems?
6. Explain in your own words the operation and effect of the current balancing circuit when added to the basic Wheatstone bridge.
7. List several systems advantages of the current-operated (current-balanced) Wheatstone bridge circuit over the basic Wheatstone bridge circuit.
8. Describe how a current-balanced resistance-measuring Wheatstone bridge circuit can be modified to measure small voltages.
9. What is the most outstanding advantage of any Wheatstone bridge circuit in measuring small voltages?
10. Make a sketch of a current-balanced Wheatstone bridge circuit which can be used to measure either resistance or voltages, showing the appropriate switches.
11. Explain in your own words the three-wire resistance measurement technique. List its advantages and disadvantages compared to two-wire Wheatstone bridge techniques.
12. Repeat Question 11, but referring to the four-wire resistance measurement technique.
13. What are the advantages and applications of unbalanced bridges?

14. Describe the null-balance principle commonly used in measurement equipment.

The next eight questions refer to the following circuit diagram:

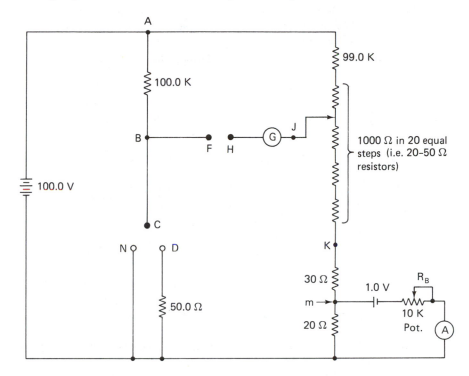

15. Which points are at equal electrical potentials (voltages) when the bridge is balanced?

16. Of what importance is the precise value of the 100-V battery?

17. To measure millivolt potentials, what connections are required?

18. What connections must be made to measure small values of resistance?

19. What is the purpose of breaking the 100-kΩ total resistance in the upper right leg into 99 kΩ plus 20 equal 50-Ω resistors (which total 100 kΩ)?

20. What is the range of voltage measurements possible using this bridge circuit?

21. What is the range of resistance measurements possible using this bridge circuit?

22. Referring to small voltage measurements, is this a high-impedance or a low-impedance measuring technique?

23. Describe at least one method of converting a frequency-type electrical output from a sensor into a variable-amplitude dc voltage. (Do *not* refer to the diagram above.)

24. Referring to balanced Wheatstone bridge measurements, what basic circuit parameter is used to indicate a balanced bridge?

25. What is the greatest advantage of using a Wheatstone bridge to measure small voltages in an industrial automatic control system?

26. What justification would lead you to specify a three-wire resistance Wheatstone bridge measuring system?

27. List several methods of handling pulse-type signals most commonly used in process control.

28. A high-impedance measurement system _____.

29. The only practical measurement technique for use with bonded strain gages is the Wheatstone bridge. Why?

The next three questions refer to the following circuit diagram:

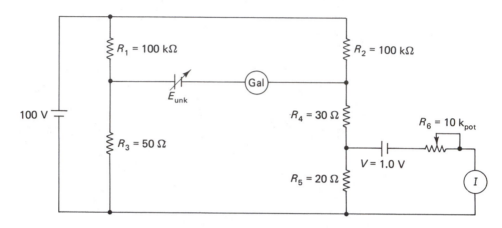

30. What is the maximum value of E_{unk} that can be measured by this bridge using the current-balancing technique?

31. What is the minimum value of E_{unk} that can be measured by this bridge using the current-balancing technique?

32. Which of the following statements is true concerning minimum and/or maximum values of voltage that a current-balanced Wheatstone bridge can handle?
    A. The measurable values of E_{unk} depend upon the relative values of R_1 and R_5.
    B. The measurable values of E_{unk} depend upon the relative sizes of R_4 and R_5.
    C. The maximum value of E_{unk} that can be measured depends upon the value of V.
    D. The minimum value of E_{unk} that can be measured depends upon the value of R_6.
    E. All of the statements above are true.

33. Explain in your own words the origin of electrical ground loop problems.

34. How would you detect the presence of ground loops?

35. How do you correct ground loop problems?

Referring to the automatically balanced Wheatstone bridge circuit in Fig. 7-11, answer the following four questions.

36. If the ammeter reads 14.75 mA when the bridge is balanced, what is the value of R_{unk}?

37. If the value of I_3 is constrained to range from 4 to 20 mA, what will be the range of resistances that this bridge will be able to measure?

38. Select a value of resistor to replace R_5 in order to extend the measuring range (of R_{unk}) to an upper limit of 1000 Ω for a bridge output current (I_3) of 20 mA.

39. Referring to Question 38, what will be the minimum value of resistance that can be measured by the modified bridge?

40. Explain the step-by-step procedure for standardizing the potentiometer in Fig. 7-12.
41. The most practical technique to keep from having grounding problems adversely affect a circuit is to
    A. Make all ground connections in the system to the same point.
    B. Not make any ground connections in the system.
    C. Make each ground connection in the system to a different point.
    D. Insist on using only battery-operated test instruments.
    E. All of the techniques above are equally practical and effective.
42. In general, the best solution to grounding problems in a system is
    A. To anticipate them.
    B. To correct them.
    C. To understand their effects.
    D. To plan the ground loop paths.
    E. All of the above are equally good solutions.
43. Normally, the origin of electrical grounding problems comes from
    A. Power transformer stray leakage and capacitance.
    B. The use of battery-operated equipment.
    C. The use of equipment having grounded chassis.
    D. The way in which grounded equipment is used by technicians.
    E. The fact that technicians make electrical measurements using two leads.
44. Some of the symptoms of a grounding problem in a system *may* include
    A. Drifting voltages.
    B. 60-cycle ac in unexpected places.
    C. Unexpected changes in circuit values when a new piece of equipment is connected in.
    D. Malfunction of a circuit in which you can find no other cause.
    E. Any or all of the symptoms above may be noted.

GLOSSARY

Galvanometer   A very sensitive electrical-current measuring and indicating device.

Ground Loop   An undesirable feedback condition in an electrical system in which currents flow in ground (or common) wires, causing undesirable electrical noise in the system, with detrimental-to-disastrous results on system performance.

Impedance   As used herein, a measure of the opposition to current flow offered by a circuit, device, or piece of measurement equipment.

Input Impedance   A measurement of the impedance "seen by" one circuit or device when "driving" or providing an input signal to another. A device having high input impedance draws very little power from any circuit driving its input; this is normally desirable for measurement equipment.

Null Balance   A technique where the measurement of an unknown quantity is taken by comparison to a known quantity.

Output Impedance   A measurement of the impedance "presented by" one circuit or device when "driving" or providing an output signal to another. A device having low output impedance will not have its output significantly affected by "loading" from the circuit it is driving; this is normally desirable for power supplies and amplifiers.

Potential   As used herein, a voltage; source of emf.

Pulse Train   A series of momentary changes in voltage which occur at a single point in a circuit as time varies.

Volt-ohmmeter (VOM)   A piece of electrical test equipment normally having the capability of directly measuring and indicating ac voltage, dc voltage, resistance, and current over various ranges. It is probably the most common piece of electrical test equipment in use and is frequently portable.

Wheatstone Bridge   A common electronic circuit typically used to measure either resistance or small dc voltages in process control systems. It can be operated on any of several basic principles and may be used in either balanced or unbalanced configurations.

THE OPERATIONAL AMPLIFIER: A SYSTEM BLACK-BOX COMPONENT

INTRODUCTION

The field of electronics was born in 1906 when Lee De Forest added the third element to Fleming's diode detector, thereby inventing the first electronic device capable of amplification. Subsequently, numerous improvements were made in the design of vacuum tubes, and then came the era of solid-state electronic devices, which have almost completely replaced the vacuum tube as electronic amplifiers.

The vacuum tube or transistor is the primary component capable of *amplification;* however, it is useless until connected into a circuit along with other electronic components. Together, they perform the function of electronic amplification. There is essentially an infinite variety of electronic amplifier designs, and until the advent of solid-state electronics and integrated circuits, a very large number of electronic engineers were employed specifically to design amplifiers.

Eventually several of these amplifier designs emerged as superior in performance, superior in general-purpose usefulness, and as being exceptionally well suited for a multitude of applications. There was only one slight hangup to their general acceptance, and that was their cost. The first universal amplifier designs were, of course, made using vacuum tubes and they were so large, so inefficient (as far as power consumption is concerned), and so expensive that they found application mainly as elements in electronic computers, where cost was not a primary consideration.

These computers consisted, primarily, of racks of electronic amplifiers which were in no way dependent upon each other. The inputs and outputs of these amplifiers were brought to connectors on the front of the computer and it was programmed by running jumper leads between the amplifiers in order to connect them to serve the desired purpose.

The desired purpose in all cases was the electronic *simulation* and solution of mathematical equations. Each of the amplifiers was connected with appropriate networks of external electronic (and electromechanical) components in order to perform a particular mathematical function on the input electronic signals. In other words, the voltage appearing at the output of the amplifier was related to the input voltage by a specific mathematical relationship. Therefore, the amplifier was said to perform mathematical operations and was dubbed the *operational amplifier* or *op amp.*

Operational amplifiers had some very specific electronic capabilities which made possible their use by people who were not amplifier design specialists. These capabilities, or design specifications, made it possible to simply jumper the output of one amplifier to the input of any other amplifier(s), called *dc coupling.* The specific mathematical relationship between the input and output voltages of any specific op amp was determined (normally) by relatively simple combinations of *passive electronic components,* the configurations and values of which were available in tables. Therefore, these operational amplifiers (which together were termed an *analog computer*) could be used by other than electronic engineers; the user did not need to be very familiar with electronics in order to program the analog computer.

Owing to their ease of use, *analog computers* (Fig. 8-1) found widespread application. With time the design of operational amplifiers improved, and with the advent of solid-state electronic components, the size and cost of operational amplifiers decreased. As performance improved and cost decreased, use of operational amplifiers increased, and this was accompanied by an increase in the numbers of companies producing these devices. As usage of op amps increased, cost decreased further, designs were further improved, and reliability was increased. They were eventually encapsulated and advertised (and sold) as "black boxes" that could be used by almost anyone as ordinary electronic amplifiers.

Integrated-circuit technology broke down all remaining cost barriers and op amps found their way into nearly every type of electronic equipment made. It has become impossible to economically justify the discrete component design and fabrication of an electronic amplifier for a large majority of applications when a temperature- and frequency-compensated operational amplifier (consisting of four to five stages of amplification and 20 to 30 active electronic elements) can be purchased for only a few cents and occupies not much more space than a single small-signal transistor.

This brief history may give the reader some insight into the use and potential of the *active electronic circuit component* called the operational amplifier. It is encountered in nearly every conceivable type of electronic equipment, and its operation, application, and use must be thoroughly understood by the modern electronic technician.

Figure 8-1 "Desktop" analog computer. (Courtesy of Electronic Associates, Inc.)

BASIC OPERATIONAL AMPLIFIER

Physically an operational amplifier is a multistage dc-coupled electronic amplifier having very high *gain;* whose overall response characteristics (in any given circuit) are determined by external components. It is a circuit itself in that it is a complete amplifier, consisting of possibly 20 to 30 transistors and all the other components necessary for it to perform as an amplifier.

Frequently, these *discrete components* are *encapsulated* into a single enclosure with only power supply and input–output terminals available for the user. With the advent of solid-state microelectronics, many integrated-circuit operational amplifiers have also become available. When you use these devices it is easy to lose sight of the fact that they are complicated electronic circuits themselves, mainly because they are packaged so that they may be used as a single component that performs a function, like a transistor, and there is no need to worry about the individual internal components. Figure 8-2 is an illustration of high voltage and high power op amps.

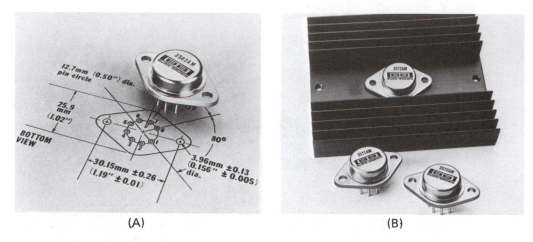

Figure 8-2 (A) High voltage–high current op amp; (B) power op amp used to drive PM dc motors. [(A) and (B) Courtesy of Burr-Brown Research Corp.]

Open-Loop Amplifier Characteristics

Since op amps basically are amplifiers, they must exhibit the same types of characteristics as any other electronic amplifier, such as gain, input and output impedances, power capabilities, *drift, offset,* and so on. When the amplifier itself is considered as a complete and independent entity, with no external components added to the input or output, it is being considered in its *open-loop* configuration. Remember, the whole amplifier is being considered as a black box, with input and output terminals only (don't worry about power supplies yet). This does not imply that there are no feedback paths wired into the internal circuitry inside this black box. By referring to the open-loop configuration, it is simply meant that no external components have been added by the user.

The primary purpose for most amplifiers is gain, so let us investigate the gain of this black box called the op amp. Gain itself is simply the relationship between the output and the input *under any given set of conditions*. The significance of the phrase "under any given set of conditions" will become apparent as more knowledge is gained of the use and capabilities of op amps. To begin with, let us restrict the operation of this amplifier to essentially dc signals. From your prior knowledge and experience with amplifiers, you should be aware that the gain characteristics of all electronic amplifiers are frequency-dependent; operational amplifiers also suffer to varying degrees with frequency-dependent gain.

The operational amplifier itself is an exceptionally good voltage amplifier. Typically, the open-loop gain (at low frequencies) of an operational amplifier is from several hundred thousand to several hundred million (10^5 to 10^9). This seemingly fantastic gain is necessary to simplify the application of these devices but is never used in its entirety. Typically, the output of an operational amplifier is ±10 V maximum. For an open-loop gain of 1 million, the full-scale input voltage would be 10 V divided by 1 million, or 10 microvolts (μV).

Ten microvolts is such a small electronic signal voltage that in most circuit applications it would be completely masked by electronic noise, *unless* the signal possessed sufficient power. In this instance that would mean that the amplifier would have to be requiring a significant amount of current. However, another important characteristic of the op amp is its extremely high input impedance. Input impedances range from several megohms up to higher than thousands of megohms, which means they require "negligible" current to their inputs.

Therefore, on these bases, several important assumptions can be made. The open-loop dc gain is so high that for all practical purposes it can be considered to be infinite—the same with input impedance. Furthermore, the voltage appearing at the input terminal to the op amp is so small that it is negligible; in other words, the input terminal to the op amp is *virtually* at zero volts with respect to system ground. The voltage potential appearing at the input terminal is termed *virtual ground*.

Open-loop gain and input impedance are very sensitive to input signal frequency. The exceptionally high values quoted previously refer to essentially dc signals, signals up to 10-kHz range. For signals with much higher-frequency components, the gain decreases and so does the input impedance; the actual rate of deterioration varies widely with different op amps.

One other important open-loop consideration that must be covered is the amplifier's output impedance. This refers to the deterioration of the output voltage as more and more current is drawn from (or sunk by) the amplifier. For all operational amplifiers a maximum output current is specified by the manufacturer and, as long as this value is not exceeded, there will be no appreciable deterioration in the output voltage as a function of current provided by the amplifier. Therefore, over this range of output currents, the output impedance appears to be so low that it can be neglected; or the output impedance can be considered to be zero.

Op-Amp Symbols

Figure 8-3 illustrates the symbol for an operational amplifier, which is simply a triangular shape having two inputs (normally) and a single output. Note that the two inputs are different. The one labeled noninverting input is normally designated (only) by a small plus sign next to the point where the terminal intersects the symbol. A positive voltage applied to this input will result in a positive voltage at the amplifier output. Frequently, this input is not used and some op amps do not even have this terminal available for use. When the noninverting input is not provided, it is omitted from the schematic diagram and the only terminal shown is the inverting input—even though it may not be marked as such.

The inverting input causes a reversal in the polarity of the input signal at the amplifier output and is normally illustrated as that input with a small negative sign

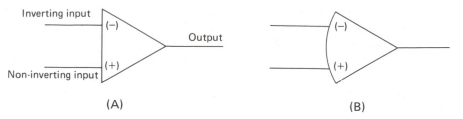

Figure 8-3 (A) Open loop op amp; (B) alternate, equivalent symbol.

close to the point it intersects with the triangular schematic. Remember, when only one input is shown, it is the inverting input.

There is only one output terminal. The other terminals the technician must be concerned with are the power supply, ground, and compensation terminals. These terminals are normally not shown on the schematic.

The symbol, as illustrated in Fig. 8-3A, is for an op amp by itself (open-loop). Frequently, the symbol in Fig. 8-3B is shown; normally (and for our purposes here), there is no difference in their meaning.

Closed-Loop Amplifier Characteristics

As was previously mentioned, the op amp is seldom (if ever) used in its open-loop configuration. Connecting external components in series with the amplifier input terminal(s) and connecting external components between the input and output terminals completely changes all of the amplifier's characteristics. These added components essentially create a new input terminal (the same output terminal is used), and use of the operational amplifier modified with these (externally) added components is termed *closed-loop* operation.

Let us now investigate the effects of these external components upon the electronic behavior of the op amp. The inverting input is the one normally used as the closed-loop amplifier input terminal, so its use will be discussed first.

USE OF THE OP-AMP INVERTING INPUT

Now let us connect the op amp into a practical circuit in which it will perform some useful function. First consider a circuit as shown in Fig. 8-4. This circuit is constructed using the inverting input; therefore, the voltage appearing on the output will be opposite in polarity to that supplied to the input. The noninverting input is not used and therefore need not be shown. In a practical circuit, it would ordinarily be connected to system ground (if it is available but not used).

Somehow the problem of presenting the inverting input terminal with an elec-

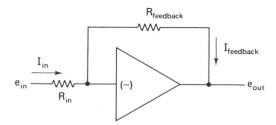

Figure 8-4 Single input to an op amp.

trical signal with a maximum value in the microvolt region, but with enough power that noise will not affect the signal, must be solved. The only way that this can happen is if this microvolt (input) signal is developed using relatively heavy currents, giving it significantly more power than would normally be associated with the anticipated electrical noise. This can be accomplished by using a resistor voltage-divider network, illustrated in Fig. 8-4 as $R_{feedback}$ in series with R_{in}.

The only current flowing through this network is directly between e_{in} (the source voltage) and e_{out} (the amplifier output voltage). There will be essentially *no* current through the inverting input terminal of the op amp due to its exceptionally high input impedance. Since e_{in} and e_{out} are of opposite polarities, and the same current flows through both resistors, it can be shown that the voltage drop across R_{in} will be equal to e_{in}, and the voltage drop across $R_{feedback}$ will be equal to e_{out}. Therefore, the voltage at the junction of the two resistors will be effectively zero volts, or ground potential. In actuality, this voltage will be exactly maintained at the proper microvolt level by the op amp itself, since it is establishing the value of e_{out} and therefore the value of current flowing through the two resistors.

Further analysis would prove that if it were not for the presence of R_{in}, the op amp would attempt to increase the magnitude of the current flowing through $R_{feedback}$ until the voltage dropped across it was nearly equal to e_{out}. This means that the voltage at the inverting input terminal would effectively be at ground potential (virtual ground), which, in turn, would mean that whatever circuit was providing the voltage e_{in} would be almost shorted to ground by the op amp. Therefore, R_{in} really is protecting the source of input voltage to the op amp.

To see how this circuit works, let us take an example. Assume an operational amplifier with a maximum permissible output voltage of 10 V dc, and an open-loop gain of 1 million (10^6). Therefore, the maximum voltage appearing at the amplifier inverting input must be limited to 10 μV. Also assume that a 1 V dc (maximum) input signal is to be amplified to 10 V maximum at the amplifier output; or the final (closed-loop) gain of this amplifier circuit is to be 10.

First, the input voltage must be reduced from 1 V dc down to 10 μV. This requires an input resistor that will be equal to (1 V–10μV)/I_{input}. The I_{input} is unknown at this point and is therefore equal to

$$\frac{1 \text{ V} - 10 \text{ } \mu\text{V}}{R_{input}}$$

Next let us look at the output. The feedback resistor must drop 10 V down to 10 μV. Therefore, its value must be equal to $(10\ V - 10\ \mu V)/I_{feedback}$. The $I_{feedback}$ is also unknown and must be equal to

$$\frac{10\ V - 10\ \mu V}{R_{feedback}}$$

Since the input impedance is so high that it could be considered to be infinite, the amount of current drawn by this amplifier (with infinite input impedance) would be negligible, or zero.

Referring back to Fig. 8-4, then, I_{input} must be equal to $I_{feedback}$. Therefore, from the foregoing discussion,

$$\frac{1\ V - 10\ \mu V}{R_{input}} \quad \text{must equal} \quad \frac{10\ V - 10\ \mu V}{R_{feedback}}$$

Since 10 μV is so small compared to 1 V or 10 V, then neglect it and the equation reduces to $1\ V/R_{input} = 10\ V/R_{feedback}$. Algebraically, rearranging this expression the relationship between the two resistors is

$$R_{feedback} = (10)R_{input}$$

or, in another form,

$$\frac{R_{feedback}}{R_{input}} = 10 = \text{gain}_{closed\text{-}loop}$$

Remember that we wanted an overall amplifier gain of 10. This brief discussion has shown that the overall (closed-loop) gain of an op-amp circuit is determined simply by the ratio of the input and output resistors—what could be more convenient to work with? Note also that the negative feedback (negative feedback will be discussed further in Chapter 11) has degenerated the overall system gain from 1 million (open-loop) down to 10 (closed-loop). Another example: to construct a circuit with a gain of 30, the ratio of feedback to input resistors would be 30:1. See if you can verify this by reasoning similar to the foregoing presentation.

The closed-loop gain of practical operational amplifier circuits is normally limited to less than 1000, the reason being that the input resistor is usually limited to no less than 10 kΩ, which means that (for a gain of 1000) the feedback resistor must be 10 MΩ. Very large resistors such as this tend to introduce excessive noise into the system, due to the resulting voltage drop caused by very small changes in the current flowing through them. The output of the op-amp system would contain this noise amplified by a factor of 1000. Therefore, as a *practical limit,* the closed-loop gain of an op amp is normally no greater than 500.

The fact that the negative feedback reduces the systems gain has already been noted. However, what has happened to the other op-amp characteristics? The input circuit now electrically appears to be simply the input resistor connected to (virtual) ground. Therefore, the closed-loop input impedance has been reduced from a nearly infinite value down to the value of the input resistor. For a 10-V system and a 10-kΩ input resistor, the input current requirement would be a maximum of 1 mA. The amplifier output appears as the parallel combination of the amplifier output (nearly zero ohms) and the feedback resistor (many kilohms), which is referred to virtual ground; therefore, the output impedance is relatively unaffected. The feedback resistor, however, has introduced the requirement that the op amp provide (or sink) the same amount of current as supplied to (or by) the amplifier input circuit (since I_{input} must equal $I_{feedback}$). Therefore, the feedback has reduced the drive (fan-out) capability of the op-amp circuit.

A more rigorous and theoretical investigation would reveal that some of the other effects of closed-loop operation include the fact that the *bandwidth* of the circuit is increased as the closed-loop gain is decreased and stability of the amplifier circuit is also increased as gain is decreased. Other effects will not be covered in this presentation.

The theory applied to the single input resistor can be extended to each of several inputs to an op-amp circuit; and all can be connected to the same op-amp input terminal (see Fig. 8-5). Since this terminal is at virtual ground, there will be no interaction between the different circuits (A, B, and C) supplying the input resistors. Each input resistor is sized so that the particular signal is amplified by the proper gain. For example, the gain for input A is R_F divided by R_{in-A}; the gain for input B is R_F/R_{in-B} and is R_F/R_{in-C} for input C. Note that all inputs share the same feedback resistor but are completely independent of each other, including the specific gain accorded each input.

Furthermore, the frequency and relative polarity of each of the input signals are completely independent of each other. The instantaneous voltage appearing at the output of this amplifier will be the algebraic sum of the voltages appearing at the input at that same instant—each multiplied by its own gain.

Obviously, the input voltages must be algebraically summed at the op-amp (inverting) input terminal; therefore, this terminal is given the name *summing junction* (SJ).

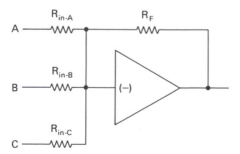

Figure 8-5 Multiple inputs to an op amp.

A couple of quick examples should help to clarify the operation of summing several input voltages by an op-amp (closed-loop) circuit.

Example 1

Referring to Fig. 8-6A, the output due to input A would be (-1 V) $(100 \text{ k}\Omega/10 \text{ k}\Omega)$ or $+10$ V; due to input B, it would be $(+1 \text{ V})$ $(100 \text{ k}/20 \text{ k})$ or -5 V; and due to input C it would be $(+3 \text{ V})$ $(100 \text{ k}\Omega/50 \text{ k})$ or -6 V. There is only one output and it cannot have three different values simultaneously, so it is the algebraic sum of the three outputs, or $+10$ V -5 V -6 V, which is -1 V.

Example 2

Referring to Fig. 8-6B, the output would be a 2 V peak-to-peak sine wave riding on a -6-V dc level.

USE OF THE OP-AMP NONINVERTING INPUT

Remember that a voltage appearing on the noninverting input will cause a voltage of the same polarity to appear at the amplifier output. Therefore, if some of this voltage was fed back to the noninverting input terminal, it would add to the voltage already there and would cause the amplifier to saturate immediately. This is typical of the unstable operation which occurs in the presence of positive feedback and is a useless type of operation. Therefore, the noninverting terminal of the op amp is almost never used with feedback from the op-amp output.

The general use of the op-amp noninverting input is with a voltage source connected directly to it plus some feedback to the inverting input, with the inverting input resistor connected to system ground. This circuit configuration is illustrated in Fig. 8-7.

Basically this op amp (having both inverting and noninverting inputs) amplifies the voltage difference between the two inputs. In the instance, where only the inverting input was used, the noninverting input was connected to system ground. Therefore, the amplifier amplified the difference between the voltage on the inverting input and ground.

Now back to Fig. 8-7: assume that current is flowing through the (inverting terminal) input and feedback resistors. The voltage at the summing junction (SJ) is equal to $R_{in}/(R_{in} + R_F)$ times the output voltage. The output voltage must also be equal to the open-loop gain times the difference between the voltage at the summing junction and the voltage applied to the noninverting input terminal. By algebraically manipulating these two relationships it can be shown that the output voltage is equal to the voltage applied to the noninverting input times

$$\frac{R_{in} + R_F}{R_{in}}$$

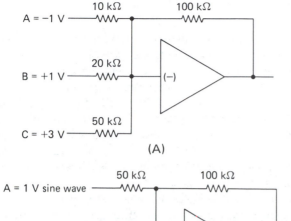

(A)

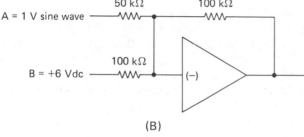

(B)

Figure 8-6 Op-amp example problems.

Therefore, the closed-loop gain is again determined by the ratio of resistors. Note that if the input resistor (R_{in}) is removed from the circuit (R_{in} becomes infinite), and if the feedback resistor is shorted out, the gain becomes exactly equal to 1. In this condition it is called a *unity gain voltage follower.*

The primary advantage of using the op-amp noninverting input is that the circuit input impedance is the input impedance of the amplifier itself (hundreds of megohms) rather than the impedance presented by an amplifier input resistor [less than 1 megohm (MΩ)]. This particular configuration is often used to buffer high impedance signal sources (extremely low current drive capability) for further signal processing. It is also perfectly feasible to use both the inverting and the noninverting inputs (simultaneously) to perform mathematical operations.

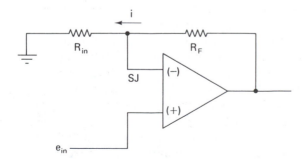

Figure 8-7 Use of the op amp in its non-inverting mode.

VOLTAGE-TO-CURRENT CONVERSION

It is frequently necessary in process control systems to convert from a variable-voltage type of signal (i.e., from a sensor) to a standard variable-current type of signal for transmission over long distances. The voltage-to-current conversion circuit is a very basic application of operational-amplifier circuit theory.

Refer to Fig. 8-8. Here we have an inverting op-amp circuit with a single input. If, as in Fig. 8-8, we connect a varying source of emf (i.e., as from a sensor having a voltage-type output) to the input of the op amp, the currents that will flow through R_{in} will vary from 4 mA (when the input is 0.25 V) up to 16 mA (when the input is 1.0 V). These are two standard ranges of electrical quantities which are commonly used in process control applications.

The op amp would not be necessary to make this conversion if there was no load placed in the circuit; however, if the output of the op amp is used to drive a load, the op amp will automatically adjust its output to compensate for the loading effects (up to its maximum ratings) so that the value of the output current is independent of the effects of loading. Without the op amp the output current would be directly affected by any value of output load; even worse, it would vary as the load dynamically varied in the system.

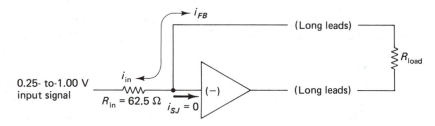

Figure 8-8 Voltage-to-current conversion operational-amplifier circuit.

OP AMPS AS DIFFERENTIAL AMPLIFIERS

As was previously noted, op amps having both inverting and noninverting inputs actually amplify the difference between the voltages present on the two inputs. This is a description of *differential amplification*. A differential amplifier ideally has an output that is a function *only* of the difference between the signals applied to its two inputs, both of which are referenced to the same electrical common.

In order for an amplifier to act as an ideal differential amplifier, the internal gain from each of the two input terminals to the output would have to be of exactly the same magnitude but of opposite polarity. When the gains of the two inputs (with respect to system ground) are not identical, primarily as a result of the imperfections in the symmetry of the input circuitry of the amplifier itself, then an error (called a *common-mode error*) will be present in the amplifier output signal. The net effect of *common-mode error* (CME) can also be thought of as a small offset (error) voltage apparently appearing between the input terminals of a differential amplifier.

Common-mode error arises because, although the differential amplifier is (theoretically) responding only to the difference between two signals, each of these input signals has a voltage potential above system ground. For example, consider two dc signals, one of 9.300 V and one of 9.100 V. These two signals applied to a differential amplifier should produce an output of exactly 0.200 V times the gain of the amplifier. However, the amplifier inputs are also sensitive to the fact that each input is actually more than 9 V above ground, and this sensitivity produces an output which is not dependent upon the difference between the signals, and which is a common-mode error.

Common-mode error is most often described by a factor called *common-mode rejection ratio* (CMRR). It expresses the degree to which common-mode voltage is rejected by the amplifier circuitry. CMRR is defined by the ratio of common-mode voltage to common-mode error (in differential amplifier circuits). CMRR can also be defined as the ratio of open-loop (differential) gain to common-mode gain. CMRRs of several hundred thousand to 1 are attainable, which means that the differential amplifier gain is several hundred thousand times the common-mode gain.

Note that common-mode error is a problem only when using the noninverting input to an op-amp.

BRIDGE AMPLIFIERS

At this point we can finally correlate the use of the op amp in connection with the Wheatstone bridges discussed in Chapter 7.

Figure 8-9 illustrates several op-amp bridge amplifier circuits. It can be shown that even with the very simple circuit of Fig. 8-9A, the output voltage will be a linear function of the variations in the amount of resistance of the sensor; if the restriction is placed on the circuit, the variations are small compared to the value of R. This restriction is necessary because the voltage difference is multiplied by the (very large) open-loop gain of the op amp (in that circuit) and it will quickly saturate for larger values.

Figure 8-9B and C illustrate other very basic op-amp bridge amplifier circuits. Each has its own applications, advantages and disadvantages. A more comprehensive text should be consulted before attempting to use those circuits in practical applications.

INSTRUMENTATION AMPLIFIERS

The instrumentation amplifier must be noted as one of the most versatile electronic circuits used in modern instrumentation systems. It is also used extensively in the modern digital computer-controlled systems. Basically, the instrumentation ampli-

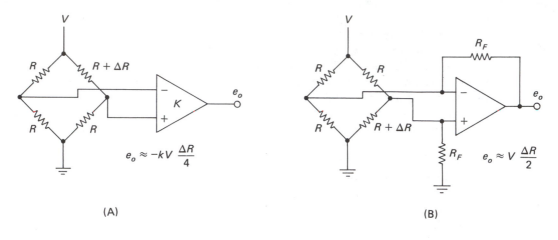

(A)   (B)

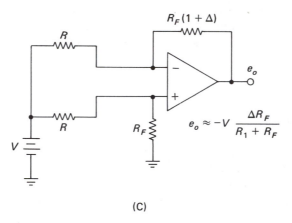

(C)

Figure 8-9 Bridge circuit applications of operational amplifiers: (A) basic bridge amplifier; (B) wider-range bridge amplifier; (C) wide-range bridge amplifier.

fier is an electronic differential input amplifier with very high input impedance and common-mode rejection. In addition, it (normally) provides adjustable voltage gain and low output impedance, and can be modified to include level shifting of the output signal, among other possibilities. They can be designed for ac or dc signal amplification, or both.

These amplifiers find application in modern instrumentation systems which commonly require an amplifier to be connected to transducers in such a way as to respond to the difference between two signals, as compared to responding to the value of a single signal with respect to its own common reference level. The two signals are referenced to the same circuit common, and normally we are amplifying small differences between relatively larger values; therefore, the amplifier circuit must have a high common-mode rejection. Furthermore, the transducers frequently have a high source impedance, which generates the requirement that the amplifier have a very high input impedance for optimum system performance.

No single simple circuit can simultaneously meet all of these requirements; therefore, the instrumentation amplifier (which was originally specifically designed to meet them all) is a highly complex, very sophisticated, multistage, modern electronic amplifier. The development of the operational amplifier and concurrent developments in integrated-circuit technology have considerably reduced the component count and apparent complexity of the modern instrumentation amplifier to the point where it can also be used as a system black-box component; however, the actual sophistication and complexity of the device itself must be appreciated in order to apply it properly.

In order to develop an understanding of its capabilities and potential uses, let us now develop the basic circuits which, when combined together, comprise the modern instrumentation amplifier.

Differential Amplification

A differential amplifier is designed basically to accept two independent voltage inputs with the same electrical polarity and circuit common and provide a single-output voltage referenced to that same circuit common. Internal amplifier circuitry inverts the polarity of one of the two inputs and then additional circuitry "adds" the two voltages together. The rest of the amplifier therefore responds to the difference between the two input voltages, providing an output voltage equal to this difference multiplied by the amplifier's gain. As an applications example, if the two input voltages each originated from pressure sensors, the output voltage would be proportional to the differential pressure. Figure 8-10 is a block diagram functionally illustrating the differential amplifier.

The real value of differential amplifiers lies in their ability to reference one variable signal (or parameter) to a second, rather than being restricted to only amplification of one signal with respect to its own common. This results in two outstanding advantages. First, the differential amplifier can be used like a battery-operated volt-ohmmeter, placing either of its inputs at any desired point in a circuit (or system) without the restriction that one of the two leads be connected to circuit common. Second, very small differences in relatively large values can be amplified directly. For example, the difference between 100.0 and 101.2 V can be amplified directly (1.2 V) rather than having to amplify the full 100.0-V common-mode value and then subsequently subtracting one voltage from the other. Figure 8-11 illustrates a common circuit schematic for a practical differential amplifier which requires use of only one operational amplifier.

Ideally, the differential output will be strictly a function of the difference between its input voltages; however, in practice, the output will also have a component that is due to the common-mode voltages at the inputs (100.0 V in the last example). Modern differential amplifiers do an excellent job of rejecting this common-mode component at the output; a 100-dB common-mode rejection ratio is common in

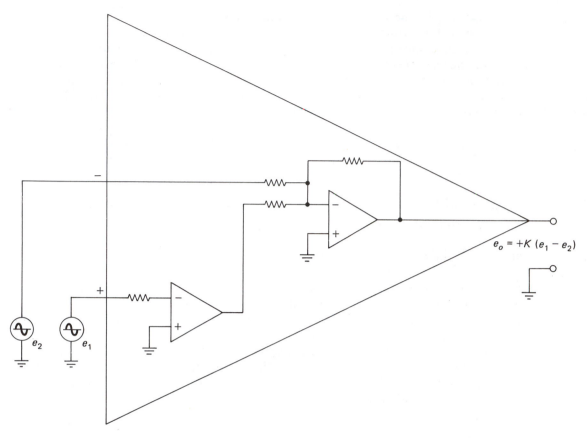

Figure 8-10 Basic differential amplifier functional block diagram.

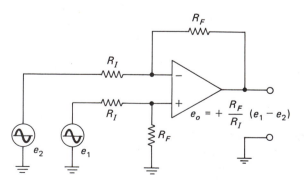

Figure 8-11 Practical differential amplifier using a single operational amplifier.

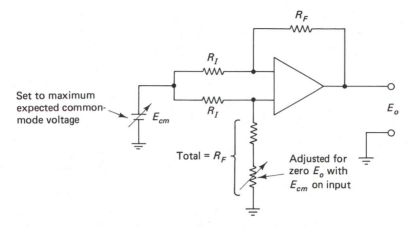

Figure 8-12 Technique for reducing the common-mode output voltage.

high-quality differential (and therefore instrumentation) amplifiers. If common mode still presents a problem, a circuit similar to that illustrated in Fig. 8-12 can be used to reduce further the common-mode component of the output. Note that the voltage simultaneously applied to both inputs (in Fig. 8-12) is for test and calibration purposes only, and represents the expected common-mode input voltage.

The differential amplifier, as discussed thus far, comprises the basic instrumentation amplifier. Other circuits and components are added to increase the versatility and usefulness of this basic differential amplifier, and to meet the performance characteristics ascribed to the instrumentation amplifier. Some of these refinements are discussed below.

Buffered Inputs

The differential amplifier, as discussed thus far, is very useful; however, it has a serious drawback in that the input impedance to the amplifier is much too low for practical measurement system application. The solution to this input impedance problem is quite easy and relatively inexpensive. By adding unity-gain voltage followers (specific application of the noninverting op amp) to each input, the signal source electrically "sees" an input impedance equal to the input impedance of the op amp itself, a very acceptable value. Figure 8-13 illustrates the buffered input circuitry for the differential amplifier. Note that although each input is amplified with respect to common, the output (across R_L) is a true differential voltage, neither side of which is referenced to circuit common.

Variable Gain

This circuit, useful though it may be, would be much more versatile if it had an easily adjustable gain. This additional feature is relatively easy to add by simply

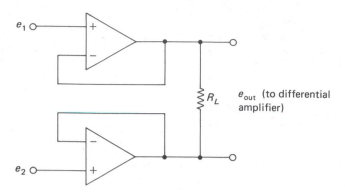

Figure 8-13 Buffered input circuit for differential amplifier.

replacing R_L (in Fig. 8-13) with three resistive elements connected as illustrated in Fig. 8-14; they provide that adjustable gain feature.

Referring to Fig. 8-13, even if we were able to make R_L a potentiometer we could not change the value of e_{out}, only the value of current through R_L, at least until the op amp's output current limitations were exceeded. By redistributing R_L into two fixed resistors and a potentiometer, as illustrated in Fig. 8-14, a variable output (across the voltage-follower outputs) can be realized.

As discussed previously, the voltages appearing at both the inverting and non-inverting inputs to each unity-gain voltage follower op-amp circuit will be equal (for sufficiently high open-loop gains). Therefore, the voltage across the potentiometer will be equal to $(e_1 - e_2)$, with polarity depending upon which input voltage is larger in magnitude. The current through this potentiometer will be $= (e_1 - e_2)/gR$. This current will flow from the output of one op amp through the three resistors to the output of the other op amp (since negligible current will flow through the inverting

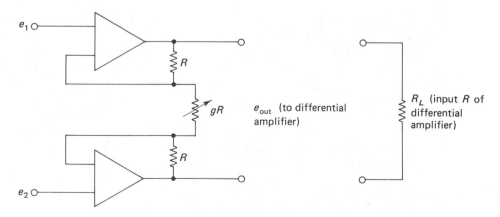

Figure 8-14 Buffered input with gain-adjust feature for differential amplifier circuit.

inputs of the op amps themselves), thereby creating voltage drops across the potentiometer (gR) and both other resistors.

Therefore, the voltage drop across the entire string of resistors (e_{out}) will be

$$e_{out} = I\,(R + R + gR) = \frac{e_1 - e_2}{gR}\,(2R + gR)$$

$$= (e_1 - e_2)\,(\frac{2R}{gR} + \frac{gR}{gR}) = (e_1 - e_2)\,(\frac{2}{g} + 1)$$

This output will be maintained by the buffer op amps themselves; therefore, the effects of loading by R_L (which represents the low input impedance of the differential amplifier) will be negligible. The two resistors in the feedback paths to the inverting inputs of the buffer op amps will now allow the voltages from the output of one buffer op amp to the output of the other to vary not only as a function of the difference between the input signals (e_1 and e_2) but also as a function of the gain setting of the potentiometer (gR). The lower the value of potentiometer resistance (value of gR), the higher the gain and therefore the higher the value of e_{out}.

This configuration gives a gain range from a low of approximately 3 (with g = 1.0 and the potentiometer resistance equal in value to the feedback resistors) to an upper value determined by the maximum output current of the op amps and the maximum desired gain (assume that the differential amplifier will have unity gain). For a maximum output of 10 mA (one op amp providing the 10 mA and the other sinking it) and 10 V maximum e_{out},

$$R_{total} \text{ must } = \frac{10 \text{ V}}{10 \text{ mA}} = 1 \text{ k}\Omega$$

If a 10-mV differential input voltage is to be amplified to 10 V at the output, there will be 10 mV across the potentiometer with 10 mA of current through it. Therefore, the potentiometer must have a minimum resistance of (10 mV/10 mA) = 1.0 Ω, and the other two resistors would evenly split the remaining 999 Ω, or each would be approximately 500 Ω. The potentiometer required would therefore be a 500-Ω potentiometer with reliable resolution down to 1.0 Ω, and the resulting gain range would be from approximately 3 to 1000. These are reasonable and practical values.

Complete Instrumentation Amplifier

Figure 8-15 schematically illustrates the complete (basic) instrumentation amplifier circuit. It includes the differential amplifier with a common-mode adjustment, the unity-gain buffer amplifiers, and the variable-gain circuitry. The industrial schematic symbol for the complete instrumentation amplifier is illustrated in Fig. 8-16.

Further Refinements and Features

Other features are commonly incorporated into the basic instrumentation amplifier circuit, some of which will be presented without detailed analysis of their operation.

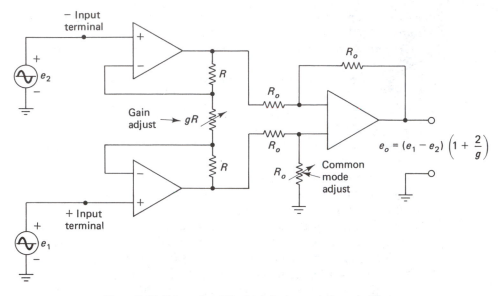

Figure 8-15 Schematic of the basic instrumentation amplifier.

Normally, in process control system applications, the output from the instrumentation amplifier is desired in an offset format. For example, an output of 1 to 5 V rather than a 0 to 5-V form. Figure 8-17 illustrates one simple technique for shifting the level at the output of the differential amplifier to realize an output voltage other than zero with a zero differential input voltage. The reference voltage illustrated as a battery in Fig. 8-17 would, in practice, be derived electrically from a reference voltage source.

Also, the output from the instrumentation amplifier can be converted to a current output rather than a voltage output. This can be accomplished by breaking the feedback path of the differential amplifier and connecting those broken leads into the remote wiring (not practical in electrically noisy environments) or by following the basic instrumentation amplifier with a voltage-to-current circuit, which is basically illustrated in Fig. 8-18. One outstanding advantage of the circuit in Fig. 8-18 is that the load drive current is provided by the op amp; the instrumentation (differential) amplifier "sees" only the high input impedance of the voltage-to-current converter op amp.

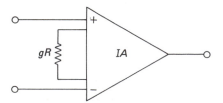

Figure 8-16 Basic instrumentation amplifier symbol.

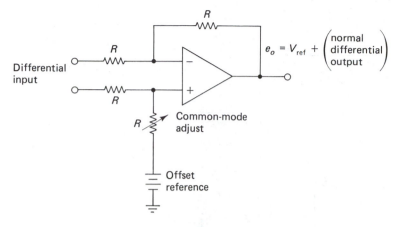

Figure 8-17 Offsetting the output from a differential amplifier.

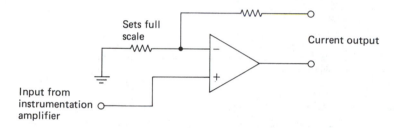

Figure 8-18 Current output from an instrumentation amplifier.

To permit these, and other, variations in the applications of instrumentation amplifiers, they are commercially supplied with several of the terminals available to the user. These are illustrated in Fig. 8-19. Further development of the circuits and uses for these terminals are beyond the scope of this text.

OTHER OP-AMP PARAMETERS

There are numerous other extremely important parameters which must be understood in order to properly apply op amps. A thorough investigation of all of them is well beyond the scope of this text; however, some of the more important ones will be briefly discussed here. Most of these parameters apply to *any* amplifier, not only op amps.

One assumption normally made when using an amplifier is that for zero volts on its input, the amplifier will produce zero volts at its output. This is nice in theory; however, in a normal amplifier, it will be true only if the amplifier *offset* has been adjusted. The offset adjustment (sometimes called *zero adjust*) compensates for the

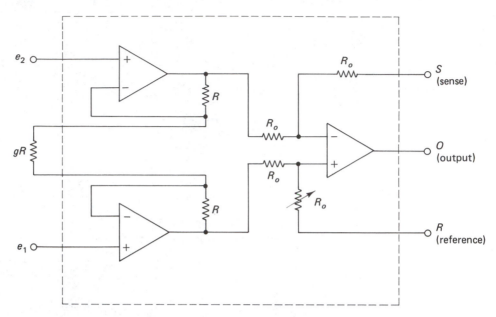

Figure 8-19 General-purpose instrumentation amplifier with sense and reference terminals.

fact that zero volts on the input may not result in exactly zero volts at the amplifier output. This adjustment provides an appropriate signal to the amplifier to drive its output to zero when the amplifier input is connected to ground. Once adjusted, the amplifier will perform properly at that time and under those specific ambient conditions. The true output zero (for zero input) will "drift" (change value) as the temperature of the amplifier changes, as the value of power supply voltage varies, and as the amplifier components change their characteristics with use. Therefore, it will be necessary to periodically rezero the amplifier in service. Many op amps are specially designed such that they are effectively "temperature-compensated" to reduce the magnitude of the drift as the temperature changes; however, it cannot be eliminated.

Another group of problems with amplifiers has to do with how fast the amplifier can react to a changing condition at its input terminal. Again, theory would like to have the amplifier respond to provide exactly the proper output at the same instant that the input changes. In fact, all amplifiers introduce some time delay in responding to changing input signals. Furthermore, owing to the physical limitations of the material the amplifier components are made of, the output of the amplifier is definitely limited in its maximum rate of change, a condition technically referred to as *slew rate*.

These limitations team up to restrict the maximum frequencies at which a particular amplifier can be used. What happens at frequencies above the amplifier's maximum frequency-response capabilities is that the output starts changing, responding to a change in the input; however, the output lags behind the changing input. When the input reaches its maximum value, the output has not yet responded to that maximum value (i.e., the output is not yet equal to the input times the amplifier's gain). If the input then starts to decrease, the output never reaches the maximum it was headed for, and it starts to respond to the decreasing input value. Therefore, the output signal never reached the value it should have attained, and the maximum peak-to-peak amplitude of the output begins to decrease simply because the amplifier can no longer keep up with the changes, *not* because of a decrease in amplitude of the input signal. This is an example of the amplifier's gain variation with frequencies above its maximum rating.

Under certain conditions, it is possible to design additional circuits which can compensate somewhat for the fact that the magnitude of the output no longer reaches the values it should at higher frequencies. This is technically termed *frequency compensation* or *phase compensation*.

Still another problem with amplifiers is that the gain is not only dependent on the ratio of fixed components, as previously explained; it is also somewhat dependent upon the temperature and aging characteristics of the individual components that make up the amplifier (even if it is an integrated circuit). Therefore, even the gain of the amplifier will vary with time. A measure of how much the gain will vary is given by its gain stability specification.

There are so many other parameters that should be understood for proper application of op amps that entire handbooks exist simply to explain them and give illustrative examples and applications. Obviously, no attempt can be made here to include them, but the interested reader is urged to refer to these other resources for more thorough coverage of this subject.

USE OF THE OP AMP AS AN INDEPENDENT CIRCUIT ELEMENT

By considering the op amp as a black-box circuit element, it can be used in much the same way as a transistor, resistor, or any other electronic circuit element. The procedure is to accept the inputs, outputs, and *transfer function* as specified by the manufacturer, not worrying about what is internal to the op amp.

Using this philosophy, the op amp can be used for nearly any purpose that any other electronic amplifier can (with the same power, accuracy, and frequency specifications). The user can go to any of literally hundreds of references that contain tables of schematics illustrating op amps in specific circuits. Selecting the circuits from these references and interconnecting them appropriately, very complicated electronic circuits can be quickly and easily fabricated.

The fabrication of circuits using op amps is much simplified because each portion can be constructed, tested, and adjusted independently. Then, after each portion of the circuit works, they can be interconnected with reasonable assurance that the complete circuit will work (i.e., that interactions among the various portions will be negligible).

A partial listing of the functions that are easily performed using op amps would include:

Voltage or current amplification	Zero-crossing detectors
Integration	Root generators
Comparison	Summation of signals
Impedance matching and circuit isolation	Differentiation
Signal generation (sine wave, sawtooth, square wave, other nonlinear waveforms)	Limiting (voltage or current)
	Discrimination
	Current/voltage sources (reference sources)
Current/voltage conversion	Differential signal amplification
Modulation of signals	Ac-to-dc conversion
Instrumentation amplifier	Peak detection circuits
Bridge amplifier	Filters
Absolute-value circuits	Multiplication of signals
Sample and hold	

Having gone over this partial listing of the applications of op amps, it should not be difficult to believe that they are presently in widespread use and that sooner or later every technician will be faced with circuits that contain these devices.

QUESTIONS

1. In your own words, just what is an operational amplifier?
2. For what application were op amps originally designed?
3. What applications are op amps primarily used for in modern times?
4. Why *must* systems-oriented students become familiar with the use of op amps?
5. Describe what is meant by the open-loop configuration of an op amp.
6. What is meant by the term *virtual ground*?
7. List several of the outstanding characteristics that differentiate op amps from ordinary amplifiers.
8. What is the significance of high input impedance for an amplifier?
9. What is the significance of low output impedance for an amplifier?
10. Referring to the symbol for an op amp, what electrical connections are typically provided in the device but not shown in the diagram?
11. What is the difference in use between the inverting and noninverting op-amp input terminals?
12. What are the most significant differences between the op amp as used in its open-loop versus its closed-loop configurations?

13. Referring to the op amp when using its inverting input only:
    A. What is the purpose of the feedback resistor ($R_{feedback}$)?
    B. What is the purpose of the input resistor (R_{in})?
    C. What is the (voltage) gain of this circuit?
    D. What would be the ratio of resistors necessary to obtain an op-amp circuit with a gain of 100?
    E. What are some of the overall circuit effects when using negative feedback around an op amp?
    F. Is the voltage appearing at the op-amp inverting input terminal actually zero volts (ground)? What would happen if it was zero volts?

14. Referring to the op-amp circuit when using its noninverting input only:
    A. Why can't the noninverting input terminal be connected to the op-amp output through a feedback resistor in a manner similar to that used for the inverting input terminal?
    B. What will be the difference in voltages appearing at the two op-amp input terminals at all times?
    C. What is the closed-loop gain of this circuit?
    D. What is the closed-loop circuit input impedance when using this configuration?

15. When using the op amp in a closed-loop configuration (using either input terminal, or both input terminals), what happens to the value of the open-loop gain?

16. In your own words, describe where common-mode effects come from and their effect on closed-loop system performance.

17. Repeat Question 16 using drift instead of common mode.

18. Repeat Question 16 using frequency response instead of common mode.

19. Make a sketch of an op amp in its inverting mode, illustrating the basic symbol input resistor, feedback resistor, input voltage, summing junction voltage, output voltage, and both open- and closed-loop gains (both in terms of voltages and in terms of resistances).

20. Repeat Question 19 for an op amp in its noninverting mode.

21. Referring to the op-amp symbol, what electrical connections are not illustrated but are assumed to be present in the amplifier?

For Questions 22 through 25, refer to the following diagram:

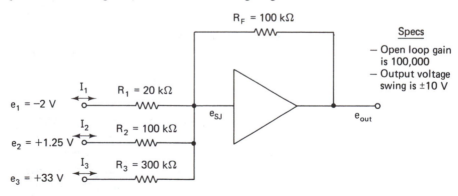

22. What is the output voltage?
23. What is the actual voltage on the summing junction?

24. What is the circuit input impedance for each of the signal sources e_1, e_2, and e_3?

25. What value of current must each of the three signal-source circuits be capable of handling as a result of their being connected to this op-amp circuit?

For Questions 26 through 30, refer to the following diagram:

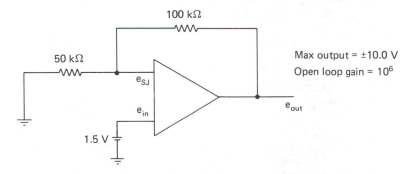

26. What is the output voltage?

27. What is the *approximate* voltage appearing at the summing junction?

28. Define common-mode error in terms of e_{sj}, e_{in}, and e_{out}.

29. What is the input impedance as seen by the 1.5-V battery (e_{in})?

30. If the 50-kΩ resistor were simply removed from the circuit and the 100-kΩ resistor were shorted by a wire across its ends, what would the closed-loop gain of the circuit be?

The following seven questions refer to the following diagram:

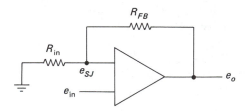

31. What is the open-loop gain of this op amp?

32. What e_{sj} voltage would you expect to measure at the op-amp inverting input terminal under normal operating conditions?

33. What is the common-mode error?

34. What is the closed-loop gain of this circuit?

35. What is the (closed-loop) input impedance of this circuit?

36. If the input resistor (R_{in}) were removed from the circuit and the feedback resistor (R_{FB}) were shorted across its ends, what would the output voltage (e_o) be?

37. If you were trying to connect this circuit in lab, what would happen if you connected the input terminals up reversed (i.e., e_{in} to the top input terminal and R_{FB} and R_{in} to the lower input terminal)?

38. Referring to the symbol for an operational amplifier, what is *not* shown but is assumed to be present in the amplifier?

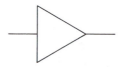

    A. Power supply terminals.
    B. Offset compensation and adjustment.
    C. Any frequency compensation that is necessary.
    D. A noninverting input terminal that is not being used.
    E. All of the answers above are correct.

39. What are some of the characteristics of negative feedback around an electronic amplifier?

40. Which of the following characteristics would be specifically attributed to operational amplifiers?
    A. Extremely high gain.
    B. Very low input impedance.
    C. Very high output impedance.
    D. Low output power capabilities.
    E. All of the above apply specifically to op amps.

The following four questions refer to the following figures.

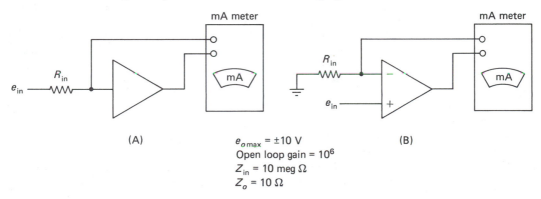

$$e_{o\,max} = \pm 10 \text{ V}$$
$$\text{Open loop gain} = 10^6$$
$$Z_{in} = 10 \text{ meg } \Omega$$
$$Z_o = 10 \ \Omega$$

41. Calculate the value of R_{in} necessary to convert a 1- to 5-V dc input voltage to the circuit in Fig. A to a 4- to 20-mA output (as measured on the milliammeter illustrated).

42. Calculate the power (maximum) delivered by the source (e_{in}) in the circuit of Fig. A to the circuit. (Do not use the value of R_{in} in this calculation.)

43. Calculate the value of R_{in} necessary to convert a 1- to 5-V dc input voltage to the circuit in Fig. B to a 4- to 20-mA output (as measured on the milliammeter illustrated).

44. Calculate the power (maximum) delivered by the source (e_{in}) in the circuit of Fig. B to the circuit. (Do not use the value of R_{in} in this calculation.)

45. Why is positive feedback seldom used with op-amp circuits?

46. Which of the following characteristics would you normally expect when using op amps?
   A. Extremely high gain.
   B. Extremely high bandwidth.
   C. Extremely low input impedance.
   D. Very high output impedance.
   E. All of the above are characteristic of op amps.

47. What is *the most* important characteristic of an op amp (when used with feedback) that makes the equation $e_o = -e_{in} (R_{FB}/R_{in})$ valid?
   A. Very low output impedance.
   B. Very low input impedance.
   C. Extremely large closed-loop gain.
   D. Extremely large open-loop gain.
   E. Very small bandwidth.

48. Referring to Fig. 8-7, explain what happens to the gain as the input resistor (R_{in}) gets very large while simultaneously the feedback resistor (R_{FB}) gets very small. In the limit $R_{in} \to \infty$ and $R_{FB} \to 0$, what is the gain approach and of what use would such a circuit be?

GLOSSARY

Active Electronic Components   Electronic devices (components) capable of controlling voltages and/or currents in such a manner as to cause amplification. Transistors and vacuum tubes are typical active electronic components.

Amplification   A statement of the degree to which any particular electronic parameter is increased in magnitude as the result of electronic circuit manipulation. Typically expressed as a ratio of output-to-input.

Analog Computer   A collection of analog electronic circuits (primarily operational amplifiers) which can be conveniently "programmed" for the solution of mathematical equations by letting electrical quantities (voltages and currents) represent mathematical quantities. They also are used to simulate other, non-mathematical quantities (such as temperature, flow, light intensity, etc.) and can be programmed either to simulate entire processes or to control these processes.

Bandwidth   The range of input signal frequencies over which a specified circuit or device is capable of being operated while continuously providing a valid output within its stated specifications.

Common-Mode Error   The output response of a differential amplifier due to the absolute values of the two inputs (referenced to their common grounds), as compared to the output response due to the difference between them. Commonly expressed in decibels as common-mode rejection ratio.

Common-Mode Rejection   A measurement of how well a differential amplifier rejects or does not respond to identical, varying signals simultaneously applied to both inputs (ideally the output response should be zero to all identical signals simultaneously applied to both inputs).

Differential (Input) Amplifier   An amplifier so designed that it (ideally) provides an output which is proportional *only* to the difference between the two inputs (both inputs being referenced to a common ground).

Discrete Components   Individual electronic devices which are capable of only one function, are packaged separately, and are used individually.

Drift   An undesired change in the output of a circuit device or measurement system (as a function of time) with no corresponding change in any of the input variables (such as power supply, etc.).

Encapsulated   Electronic components and their interconnections which have been enclosed in a protective coating of plastic.

Gain (of an electronic amplifier)   A statement of the difference in voltage, current, or power of an electrical circuit measured at its output as compared to that measured at its input. Typically expressed as a ratio of output divided by input, and expressed either in decimal or decibel form.

Gain Stability   A measure of the variation of the sensitivity (or gain) as a function of time.

Integrated Circuit (IC)   A functional electronic device in which any number of active and passive electronic components have been inseparably fabricated and interconnected on a single substrate.

Offset   A measurable (constant) output from a circuit or device when a zero signal is applied to its input. Also, the difference between what the output should (theoretically) be and its actual measured value for any valid input. Commonly compensated for by a "zero-adjust" device.

Operational Amplifier (Op Amp)   A category of electronic amplifier that originally was designed to perform mathematical functions as integral parts of Analog Computers. Modern usage applies this term to any amplifier having characteristics similar to the original op amps (such as extremely high open loop gain), regardless of its use; very few modern op amps are actually used in analog computers.

Passive Electronic Components   Components not capable of amplification; electronic devices which typically use up (dissipate) energy rather than regulating it. Resistors and capacitors are passive electronic components.

Simulation   A model of a device or process which reacts to various stimuli in a manner statistically analogous to the device or process being simulated. Typically simulations are mathematical models which can be studied more conveniently and/or economically than the process they represent (simulate).

Slew Rate   The maximum rate of change of the input signal from a circuit or device which will still result in an output which is within stated specifications (accuracy, linearity, etc).

Transfer Function   The mathematical relationship between two variables which are correlated.

9

CALCULUS, OPERATIONAL AMPLIFIERS, AND ANALOG COMPUTERS

Now that the elementary uses of operational amplifiers have been discussed, it is necessary to realize that closed-loop use of op amps with (only) external resistors constitutes a relatively small percentage of the total applications of op amps. The majority of op-amp applications make use of capacitors in the input and/or feedback circuits. Therefore, it is absolutely necessary to become familiar with the operating characteristics of the op-amp circuits in which the external component networks include capacitors.

CAPACITORS AND INTEGRATION

The easiest way to understand the effect of inserting capacitors into the feedback circuits of op amps is to review the relationship between the current/voltage characteristics (of a charging capacitor) and mathematics. First, let us go back to technical mathematics and review, with an attempt to understand, the concept of computing the areas under curves and making a plot of the numerical values of the area at different points.

Refer to Fig. 9-1A, which is the plot of a constant (voltage, current, etc.) versus time. This plot could be obtained by simply measuring the voltage of a dc battery at specific points in time. The problem will be to compute the total area under this curve at each point in time and plot the results.

At point A in time the total area under the curve will be $0.0 \times 1.0 = 0.0$. Now plot this value on the graph in Fig. 9-1B and label the data point a. Next, compute the area under the curve between points A and B. This increment of area

210

will be $1.0 \times 1.0 = 1.0$. Add this new amount of area to what we had prior to this (up to point A), getting $0.0 + 1.0 = 1.0$. Plot this value in Fig. 9-1B and label it b.

The next data point will be obtained by first calculating the area between points B and C. This increment will be $1.0 \times 1.0 = 1.0$. Add this to the total area obtained up to point B and get $1.0 + 1.0 = 2.0$; plot this as data point c. Continue in this manner to get as many additional data points as are necessary. The procedure is to take each new increment of area, add it to the total of all area computed previously, and plot the point.

Now that the curve is completed, notice that it is a ramp-type curve. If you remember introductory *calculus,* the foregoing description has followed the rules for *integration.* In other words, we *integrated* the curve in Fig. 9-1A to obtain the curve in Fig. 9-1B. In terms of calculus:

$$\int_0^t K \, dt = Kt$$

where Fig. 9-1A is a plot of $K = 1.0$ (for all time) and Fig. 9-1B is a plot of Kt (where $K = 1.0$, also).

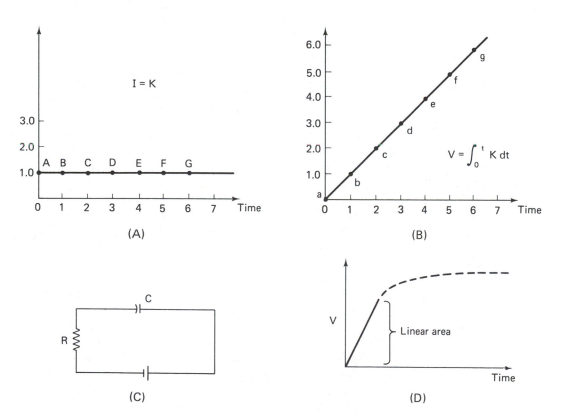

Figure 9-1 DC input *RC* circuit charging characteristics.

Now that we have discussed how to plot the integral of a curve, the next step will be to relate the curves to electronic components. Obviously the component of interest here is the capacitor.

Refer back to Fig. 9-1 and label the vertical axis of Fig. 9-1A "current" and the vertical axis of Fig. 9-1B "voltage." Now consider the schematic of Fig. 9-1C; the plot of Fig. 9-1A represents the current flowing into the capacitor while the curve of Fig. 9-1B represents the voltage that would be measured across the capacitor as it charged.

Note: It is very important to note that these relationships occur only as long as the capacitor is being used on the *linear* portion of its charging curve (Fig. 9-1D). In considering the use of capacitors in this type of application, they will *never* be used on the nonlinear portion of their charging curves. Therefore, we have now shown that a capacitor is an electronic component whose I/V characteristics are described by calculus. The use of a capacitor in any electronic circuit causes the circuit to respond in a manner that can be mathematically described in terms of calculus. This is why calculus has been stressed in your formal education.

As a further example, refer to Fig. 9-2. Integrate the curve in Fig. 9-2A to obtain the curve in Fig. 9-2B. In mathematical notation

$$\int_0^t Kt \, dt \, = \, \tfrac{1}{2}Kt^2$$

where curve Fig. 9-2A is a plot of $I = Kt$ (where $K = 1.0$) and curve Fig. 9-2B is a plot of $V = \tfrac{1}{2}Kt^2$ ($K = 1.0$ here also). Figure 9-2C is the schematic of a circuit where these curves could be generated and observed on two oscilloscope traces, one monitoring the current flowing in the circuit and the other monitoring the voltage across the capacitor.

As a final example, refer to Fig. 9-3. Here a pulse train is applied to a circuit containing a capacitor. Integrating the curve in Fig. 9-3A, the total area under the curve up to time-point 1 (point *B*) the total result is zero. Therefore, zeros are plotted on Fig. 9-3B at points *a* and *b*.

As we progress from point *B* to point *C*, an increment of area equal to $1.0 \times 1.0 = 1.0$ is included. Add this to the previous total area (up to point *B*, which was 0.0) and plot $0.0 + 1.0 = 1.0$ (point *c*). Progressing to point *D,* another increment of area is included $1.0 \times 1.0 = 1.0$); add this to the previous total area ($1.0 + 1.0 = 2.0$) and plot this as point *d*.

Thus far the entire process was similar to the past examples; however, now the input signal (illustrated in Fig. 9-3A) goes to zero (open circuit) again. What would you expect the capacitor to do? Hold its charge, of course! Now let us see if calculus and integration bear out what we intuitively know to be correct.

Taking the increment of area under the curve between points *D* and *E* (zero area) and adding it to the total area up to point *D* (2.0) we get 2.0. Therefore, the voltage on the capacitor has not changed and it is holding its charge. The process, and results, are identical up to point *G* (and *g*). Then there is an increment of area between points *G* and *H*. This increment (1.0) is added to the total area up to point *G* (2.0) and the result, 3.0, is plotted (point *h*). And the process continues.

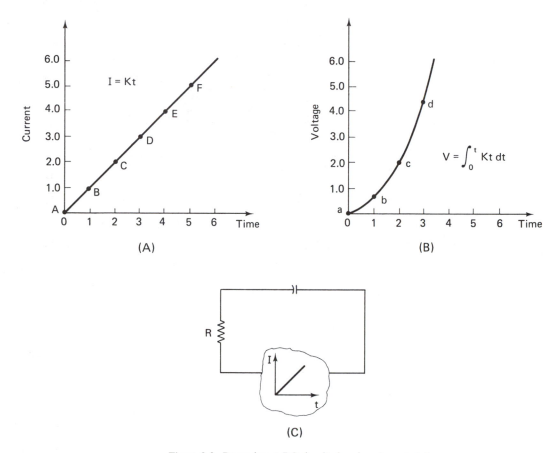

Figure 9-2 Ramp input *RC* circuit charging characteristics.

OP-AMP INTEGRATORS

In the foregoing discussion the relationship between the mathematical world of calculus and the practical world of electronics has been demonstrated. However, the capacitors used were "ideal" in that they did not lose (or leak) any of their charge. Even though a capacitor can be used to simulate the process of integration, there can be no practical use for such a device, unless it can be hooked up with another electronic circuitry.

Unfortunately, as soon as the integrating capacitor is connected so that it provides an input to another circuit, it must supply some power to that circuit. When this happens, the power drawn off by that circuit causes an inaccuracy in the charge on the capacitor. This inaccuracy is caused by the loading effects of the following circuitry upon the integrating capacitor. The problem here is very similar to the

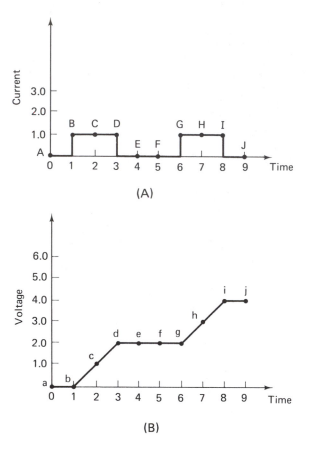

(A)

(B)

Figure 9-3 Pulse input *RC* circuit charging characteristics.

loading effects experienced when using inexpensive VOMs; the solution is similar, also.

The solution to meter-loading problems is to supply the meter circuit with a high impedance input circuit. The solution to the problem of using the integrating capacitor is similar; use of a high-input impedance circuit. Remember that one of the primary advantages of using op amps is for their high (nearly infinite) input impedance. Therefore, this may be an ideal device to use with the integrator.

Figure 9-4 illustrates the capacitor connected in the feedback branch of a closed-loop amplifier circuit. Let us investigate what happens in this circuit as the waveform of Fig. 9-1A is applied to the input. According to our previous discussion, the output had better be similar to Fig. 9-1B. This may best be shown with an example.

Initially, the input, SJ, and output voltages are zero; then (at time zero) $+1.0$ V is applied to the input. Instantaneously the voltage at the SJ starts to rise. The extremely high open-loop gain amplifies the SJ voltage and the output voltage begins to increase, in a negative direction. This causes a current to flow through C_{FB}

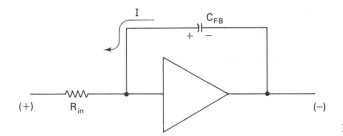

Figure 9-4 Op-amp integrator.

(as illustrated), causing it to charge with the polarities as noted. This same current *must* also flow through R_{in} (a very large value resistor) since the amplifier itself draws negligible current. This current causes a voltage drop across R_{in} which must be equal to 1.0 V under steady-state conditions; since the SJ approaches ground potential, the capacitor obviously must be charging toward 1.0 V also (the values of R_{in} and C_{FB} being properly chosen for a proper RC time constant).

If, at the end of time period 1 (points *B* and *b*), the input voltage were disconnected, all currents would cease to flow and the capacitor would attempt to discharge itself into the nearly infinite impedance of the amplifier. Therefore, it would remain with a charge equivalent to 1.0 V across it; and the output voltage of the op amp would remain at -1.0 V.

The input voltage is, in fact, not disconnected in our example problem; therefore, there must be a continuous amount of current flowing through R_{in} to cause a 1.0-V drop across that resistor. All of this current must flow through C_{FB} such that it constantly increases the charge on C_{FB}; and therefore the voltage across its plates. The voltage across the capacitor, and therefore the output voltage of the op amp, continually increases as per Fig. 9-1B only in the negative direction.

Therefore, the output of the op amp will be the negative of the voltage which would be developed across the capacitor in an ideal circuit. The important advantage, however, lies in the fact that any current required by following circuitry will be provided by the amplifier itself and not the integrating capacitor. Therefore, loading effects are minimized to the point of being negligible in most well-designed circuits.

The values of R_{in} and C_{FB} are chosen such that C_{FB} always charges only along the linear portion of its charging curve and such that the RC time constant will be adequate for the input signal frequencies involved.

Note that (in the example circuit) the output voltage will continue to rise until the op-amp output saturates. This characteristic is frequently employed in automatic control systems and is termed *integral feedback*. However, the input voltage must either go to zero or reverse polarity in order to keep the circuit from saturating. If the input voltage goes to zero, the voltage at the op-amp output will remain at its most current value at the instant the input voltage went to zero. Well-designed integrator circuits will be able to maintain this value to $\pm 0.1\%$ for periods in excess of 12 hr. This characteristic is employed in all electronic controllers.

Integrator Applications

Op-amp circuits with capacitors in the feedback circuit will be found in nearly all automatic control systems which use any type of electronic controller. The op-amp integrator circuits are used to perform many functions, a few of the most important of which are listed below:

    Sample-and-hold circuits

    Analog-to-digital converters

    Digital-to-analog converters

    Electronic controllers

    Filters and signal-conditioning circuits

    Function generators

CAPACITORS AND DIFFERENTIATION

Since capacitors are electronic components whose active I/V characteristics can be described by calculus, you might expect that they also can be used for the mathematical process of *differentiation*. This is possible; however, the use of op-amp circuits in which capacitors are included (in series) in the input circuit is not nearly as great as the use of capacitors in the feedback circuit. Therefore, we shall discuss their operating characteristics only briefly.

First, remember that a derivative refers to a *rate of change*. If there is no rate of change, the output (or answer) from a differentiator must be zero, indicating a zero rate of change. Remember also that, when referring to graphs, the rate of change is called the *slope* of the curve.

Referring to Fig. 9-5, the rate of change of the curve in Fig. 9-5A is zero; therefore, a plot of the derivative of this curve would be zero (Fig. 9-5B). Mathematically, this would be expressed as

$$\frac{d}{dt} (K) = 0$$

where Fig. 9-5A is a plot of $K = 1.0$, and Fig. 9-5B is a plot of

$$\frac{d}{dt} (1.0)$$

Now label the vertical axis of Fig. 9-5A capacitor "voltage" and the vertical axis of Fig. 9-5B "current." Referring to the circuit in Fig. 9-5C, the voltage vs. time plot in Fig. 9-5A is a plot of voltage across the capacitor *once it reaches steady-state conditions*. The plot of Fig. 9-5B is therefore a plot of current through the capacitor under the same conditions.

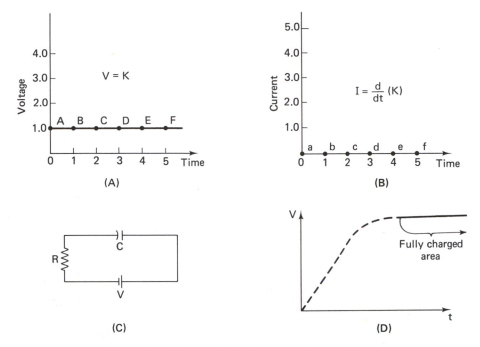

Figure 9-5 DC input *RC* circuit charging characteristics.

Note: Whereas the integrator circuits never permitted the capacitor to operate on any portion of its characteristic curve other than the linear charging portion (illustrated in Fig. 9-1D), the differentiator circuits make use of the fully charged portion of the charging curve (Fig. 9-5D). Practical use of the differentiator actually uses the entire curve. The valuable information (rate-of-change information) makes use of the linear portion also; however, the capacitor is normally fully charged, whereas the integrator circuit makes use of the capacitor while it is *never* fully charged.

Keeping the foregoing information in mind, refer to Fig. 9-6. Figure 9-6A is a plot of one cycle of a sawtooth waveform (a ramp). If such a voltage is applied to a series capacitor (as in Fig. 9-5C), the rate of current flowing into the capacitor will be a constant (Fig. 9-6B); in other words, the rate of charging (slope of Fig. 9-6A) will be constant. Note that at any instant in time, the capacitor is fully charged up to the value of input voltage at that particular instant.

Mathematically, the curve in Fig. 9-6A is a plot of the equation $V = Kt$, where $K = 1.0$. The curve in Fig. 9-6B is the plot of a constant vs. time, where the value of the constant *(K)* is 1.0. Taking the derivative of the equation plotted in Fig. 9-6A, the result will be a constant:

$$\frac{d}{dt}(Kt), \text{ where } K = 1.0, \text{ is equal to } 1.0$$

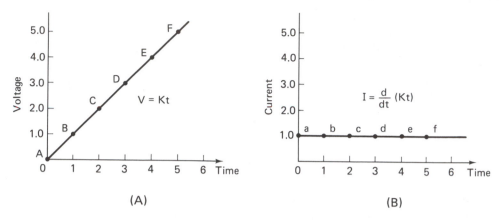

Figure 9-6 Ramp input *RC* circuit charging characteristics.

Therefore, the capacitor is again acting in a manner that can be mathematically described by calculus; by a derivative.

As a final example, refer to Fig. 9-7, where the same circuit is used but the input voltage has been changed to a square-wave pulse train. As with the square-wave example for integrators we shall not bother with the mathematical equation for the pulse train (because of its complexity) and will simply try to reason through the action of the circuit.

At instant *A* (Fig. 9-7A) there is no voltage (open circuit) applied to the capacitor; therefore, no current flows (point *a* in Fig. 9-7B). At instant *B*, 1.0 V is instantaneously applied to the circuit. Therefore, instantaneously the rate of voltage change, which appears across the capacitor, is nearly infinite. The capacitor starts to charge itself up (it is a relatively small value capacitor) at the maximum possible rate (limited only by *R* and the internal resistances of the capacitor and voltage source). Since it is a fairly small value capacitor, and furthermore since very heavy current is flowing, it charges rapidly and the charging current drops to zero when it is fully charged.

The plot of current versus time (Fig. 9-7B) is essentially a very high and narrow spike, the height and width being proportional to the values of the capacitor and resistor and their *RC* time constant. A fraction of a second after time *B* (or *b*), the current ceases to flow and no more current will flow until the voltage applied to the circuit changes value again. This occurs at time *D* and the voltage goes to zero. At this instant the capacitor must discharge through the voltage supply, so let us consider it to be a short circuit for this example. In a practical circuit there would be a short placed across the capacitor by another circuit in parallel with it, in order to accomplish this discharge function.

The capacitor attempts to discharge at its maximum possible rate, and the current therefore flows in the opposite direction for a fraction of a second (time again determined by the *RC time constant* of the circuit) and falls to zero when the

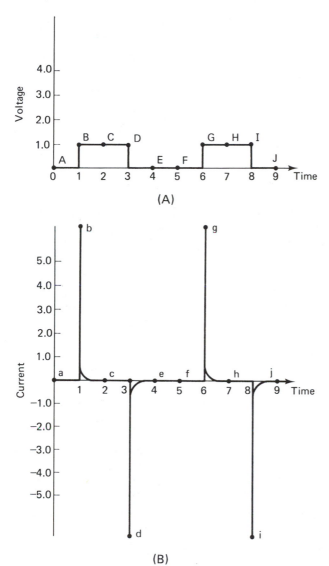

Figure 9-7 Pulse input *RC* circuit charging characteristics.

capacitor is fully discharged (or charged to zero volts). Again the plot of current flow versus time (Fig. 9-7B) is a spike, only in a negative direction. This same process repeats itself for the next pulse of input current.

It can be shown that the derivative of the equation which mathematically describes the curve in Fig. 9-7A will plot as illustrated in Fig. 9-7B. There is no need to go through the mathematical rigor of proving this if you will note and accept the following phenomenon. When changing the voltage across a capacitor, the value of current flowing through the capacitor is proportional to the rate of change (slope) of the applied voltage. This is mathematically described by the derivative operation in calculus. Again, one of the reasons for including calculus in a technical curriculum.

OP-AMP DIFFERENTIATORS

Again, the problem is to use the capacitor in a circuit without altering the charge on the capacitor. Figure 9-8 illustrates the use of a capacitor in an op-amp input circuit. Let us apply the voltage waveform in Fig. 9-7A to the input and determine what the output from this circuit would be.

With no input voltage, the SJ and output voltages are all zero, and no current flows through R_{FB}. At time B the input voltage goes to 1.0 V. Instantaneously the voltage at the SJ starts to rise toward 1.0 V also (the capacitor acts as a short circuit for this high-frequency change). The voltage on the SJ is amplified by the open-loop gain and appears as a large negative voltage at the op-amp output. Therefore, current will flow from the output through R_{FB} to charge the capacitor (since the op-amp input draws negligible current).

Large values of current will flow through R_{FB}, which is a large-value resistor; therefore, the output voltage will instantaneously cause the amplifier to saturate in the negative direction. As the capacitor charges toward 1.0 V, the current flowing through R_{FB} decreases and therefore the voltage at the op-amp output falls to zero. The time required to reach zero is determined by the RC time constant of C_{in} and R_{FB}.

Since there is no voltage change until time D, there will be no currents flowing (C_{in} is charged to 1.0 V) and therefore zero output from the amplifier. At point D in time, the voltage on the input is removed (falls to zero through a short circuit). Since C_{in} is charged to 1.0 V, the SJ now becomes instantaneously -1.0 V with respect to ground. This causes the op-amp output to saturate in the positive direction. The combination of -1.0 V at the SJ and positive saturation voltage at the op-amp output causes very high currents to flow through R_{FB} until the capacitor is discharged (again time determined by C_{in} and R_{FB}). When it is discharged, the voltage at the SJ goes to zero and the op-amp output returns to zero volts also. The circuit remains in this condition until a voltage change again appears at the input.

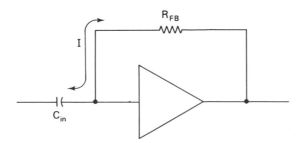

Figure 9-8 Op-amp differentiator.

Note that a plot of the charging (and discharging) currents flowing through C_{in} and R_{FB} would look exactly as the curve in Fig. 9-7B. Furthermore, note that the voltage appearing at the op-amp output would be exactly the same plot, but the negative of Fig. 9-7B. Since the plot of Fig. 9-7B has already been shown to be the derivative of the plot in Fig. 9-7A, obviously this op-amp circuit has an output that can be described as the mathematical derivative of the input. Any power necessary to drive the circuitry following this circuit will be supplied by the op amp and *not* by the capacitor.

Differentiator Applications

As previously noted, op-amp differentiator circuits are not used as frequently as the integrator circuits, the primary reason being the sensitivity of the output to electrical noise. Since the output of this circuit responds only to *changes* in input voltage, any electrical noise appearing at the input will cause an undesired output. Therefore, their use must be carefully considered and compensation filters are normally required to reduce the sensitivity to this noise.

Common applications would include:

Automatic control systems—derivative (velocity) control
Noise amplifiers
Filters
Anticipation circuits
Function generators
Unipolar-to-bipolar signal converters
Trigger circuits

ANALOG COMPUTERS

Basically, an analog computer is a hardware cabinet in which numerous operational amplifiers (and other equipment which will be discussed later) have been mounted. The amplifiers are completely independent of each other insofar as their input and output terminals are concerned. These terminals are connected directly to terminals which are available to the programmer at the front console of the computer. Therefore, the front console of an analog computer is basically a series of connectors (frequently common banana-type connectors) to which are connected the summing junctions and output terminals of the op amps located inside. There are normally schematics pictured on this console which illustrate the internal connections as indicated in Fig. 9-9, and there is normally more than one connector provided for each amplifier terminal.

Therefore, the computer programmer must add the necessary input and feedback components in order to have each op amp perform the desired function. Since

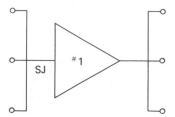

Figure 9-9 Basic analog computer op-amp schematic.

it would be very inconvenient to hang these components on leads dangling from the console of the computer, many commonly used components are also built into this (computer) cabinet: primarily fixed-value resistors, variable resistors (potentiometers), and capacitors. The terminals for these devices are also brought out to connectors on the programming console. Like the op amps themselves, there will be schematics printed on the console to illustrate the internal connections, Figure 9-10 illustrates how a typical analog computer console layout for a single op amp might appear.

Figure 9-11A illustrates the *programming* necessary to connect one op amp, using the internal components, into a circuit with an overall gain of 1. Note that the unused components simply are not connected into the circuit at all. Therefore, the programmer has an amplifier with a gain of 1 available at this point, with the amplifier input available at terminal *A* and the output available at terminal *B*.

Figure 9-11B illustrates another (identical) op amp connected into an integrator configuration. This amplifier circuit has its input at *C* and its output available at terminal *D*. There will be no interaction between these two op-amp circuits unless the programmer subsequently interconnects them from the console.

The amplifiers can be used as independent circuits, connected to equipment not included in the analog computer, or the various op-amp circuits within the analog computer can be jumpered together to perform more complex mathematical signal-processing functions. Figure 9-12A illustrates two amplifiers interconnected such that the first amplifier is an integrator and the second is a linear amplifier with

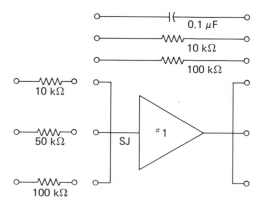

Figure 9-10 Complete analog computer op-amp schematic.

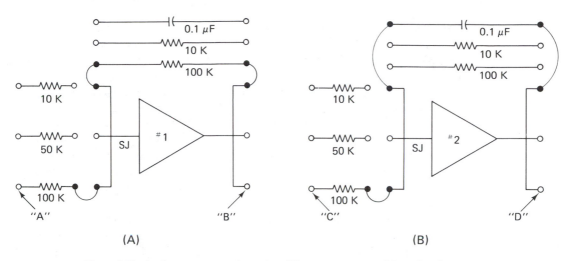

Figure 9-11 Analog computer schematics: (A) op amp connected for gain of 1; (B) op amp connected as integrator.

a gain of 10. The overall processing done on the signal appearing at the input terminal X will be the processing done by amplifier circuit 1 multiplied times the processing done by amplifier 2. Therefore, for this illustrated example, the output signal Y will be the integral of the signal X multiplied by a gain of 10; or $Y = 10 \int X \, dt$.

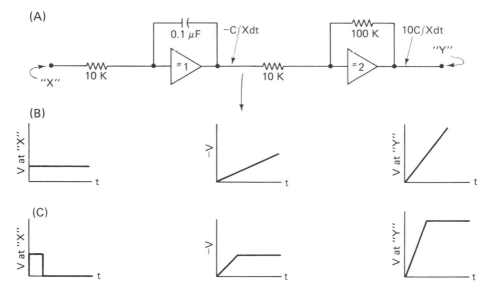

Figure 9-12 Interconnected op amps with output voltage waveforms due to "X" inputs.

Note that the unused terminals and components are not normally shown on the sketches. Figure 9-12B and C illustrates the approximate sketches of the voltages at each amplifier input and output terminal for the signals assumed to be input at terminal X.

Frequently, amplifier gains other than the values available using internal computer components are desired. For this situation, analog computers contain many variable-resistance devices (potentiometers). The input signal is connected to the input of one of these devices (as illustrated in Fig. 9-13) and the potentiometer is adjusted such that the appropriate *percentage* of the total applied voltage appears at the "wiper" output. Illustrated in Fig. 9-14 are several instances where the output voltage is some percentage (always less than 100%) of the (input) voltage applied to the "top" of the potentiometer.

The input would come from one op amp's output terminal and the output would "feed" the input to another op-amp circuit. The gains of each of the circuits would be multiplied together to get the final output voltage. Figure 9-15 illustrates a situation using a potentiometer in conjunction with an amplifier having a gain of 2, in order to achieve an overall gain equal to 67% of 2, or 1.33.

Therefore, analog computers are programmed by interconnecting the various circuit elements and amplifiers together to perform an overall mathematical signal processing. Only the most basic circuit elements commonly included within analog computers have been discussed here; normally other elements and circuits are also included, such as electronic multipliers, nonlinear function generators, diodes, and so on. These are all available to the programmer who must determine their use and interconnections to solve his problem. The diagram illustrating the interconnections is the *analog computer program* itself. Therefore, Figs. 9-12A and 9-15 are analog computer programs.

Frequently, the equations to be *simulated* (modeled electronically by the am-

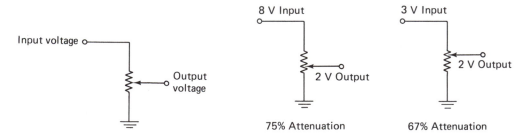

Figure 9-13 Schematic of variable attentuator (potentiometer).

Figure 9-14 Attentuator examples.

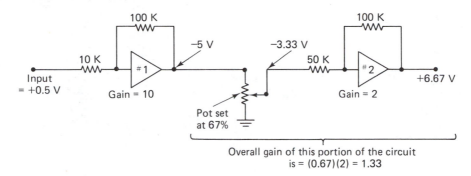

Figure 9-15 Analog computer program illustrating variable op-amp gain.

plifier circuits) are *differential equations* whose solutions would be extremely difficult to calculate manually. These equations are *programmed* onto analog computers and the solutions, in the form of voltages, can be observed or even used to operate or control other circuits or even systems.

Programs on analog computers are normally used to help design dynamic systems, such as automobile suspension systems, aircraft control systems, spaceship control systems, and, of primary interest here, industrial automatic process control systems. These systems are very complex and must respond to many, many external variables or conditions. Each component of each system must be chosen to *optimize* the system response to *all* the expected conditions. An analog computer simulation of the effects of these conditions as they vary can be observed instantaneously and the circuit (or controller) parameters easily altered so as to obtain the desired response.

Since *all* processes can be approximated (simulated) by differential equations, it is possible to simulate not only the processes themselves, but also the equations that approximate (simulate) the controller (as defined in the succeeding chapters). Therefore, the entire process may be simulated using analog computers and using the same computer the controller can be designed. Not infrequently, the analog computer itself is permanently hardwire-programmed to be the industrial automatic control system controller.

Figure 9-16 illustrates a general purpose hybrid computer of the type used for process simulation and control system design. Figure 9-17 illustrates the use of this type of computer system for simulation of an industrial plant for training purposes. Use of training simulators is an economical and effective technique used, for example, by the military in nuclear reactor system operating training and aircraft flight training.

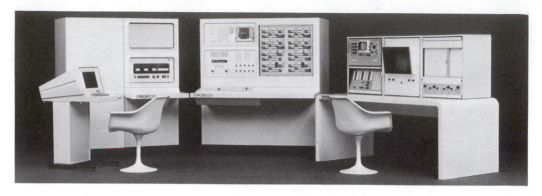

Figure 9-16 General-purpose hybrid (part analog, part digital) computer of the type typically used to simulate processes for the design and preinstallation checkout of automated process control systems. (Courtesy of Electronic Associates, Inc.)

(A)

(B)

Figure 9-17 (A) Electronic Associates designed and built a fossil-fueled power plant simulator for American Electric Power Company. Installed since 1973 at AEP's John E. Amos plant near Charleston, West Virginia, it is used to train operators for four 1300-MW coal-fired plants. Shown here is only part of the 240 ft of panels incorporated in the simulator; (B) instructor's console for the fossil-fueled power plant simulator pictured above. (Courtesy of Electronic Associates, Inc.)

QUESTIONS

1. Describe the effect of integration using any example that comes to your mind (other than the ones used in the text).

2. Explain the relationship between the current-voltage charging characteristics of a capacitor and calculus.

3. Why are op amps necessary when using a capacitor as an integrating circuit element in a practical circuit?

4. Repeat Question 1 for the effect of differentiation.

5. What is the major anticipated problem when using an op amp in a differentiating circuit?

6. Given that the curve in part (A) of the diagram is an input to an op-amp circuit, what type would the op-amp circuit have to be in order to obtain each of the other waveforms (curves) at its output: integrator, differentiator, gain < 1, or gain > 1?

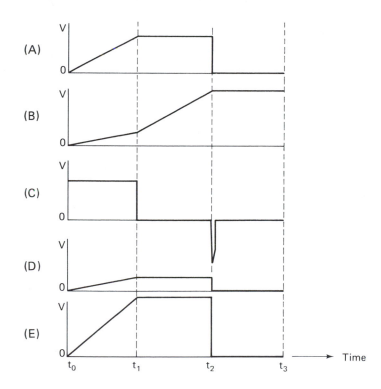

7. What is an analog computer?

8. What is an analog computer used for?

9. What is an analog computer program?

10. How is an analog computer programmed?

11. Referring to the analog computer program shown below, sketch the output waveform that you would expect to see on an oscilloscope at the output of op amp 3.

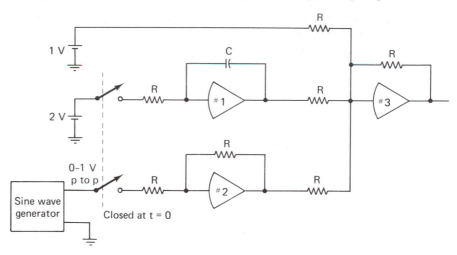

12. Referring to the analog computer program shown below, what is the algebraic equation for the output voltage in terms of the variables illustrated?

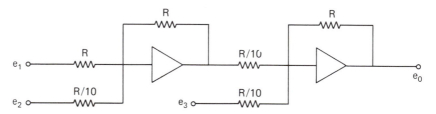

The next five questions refer to the waveforms shown on page 229.

13. Given the designated input waveform, which would be the correct output waveform from an integrator?

14. Given the designated input waveform, which would be the correct output waveform from a differentiator?

15. Given the designated input waveform, which would be the correct output waveform from an inverting amplifier?

16. Given the designated input waveform, which would be the correct output waveform from a noninverting amplifier?

17. Given the designated input waveform, which would be the correct output waveform from a unity-gain amplifier?

18. An op-amp differentiator:
    A. Produces an output voltage proportional to the input and time.
    B. Produces an output proportional to the difference in input voltages.
    C. Smooths out an applied waveform.
    D. Contains a high-pass filter.
    E. Contains a low-pass filter.

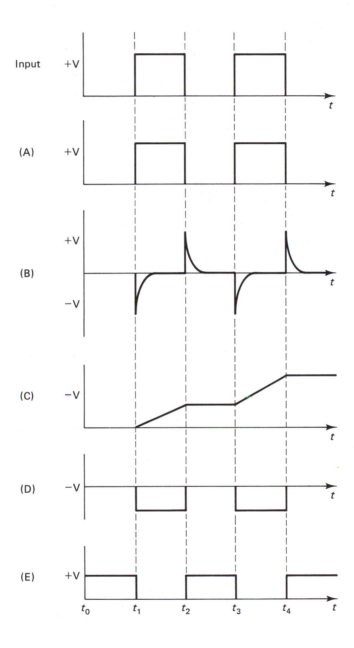

In the following five questions, match the mathematical equation with the op-amp circuit which illustrates the solution to that type of equation.

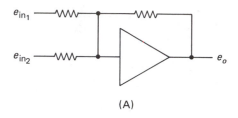

(A)

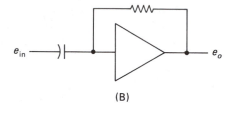

(B)

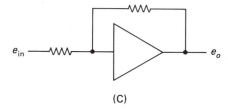

(C)

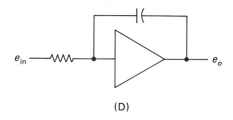

(D)

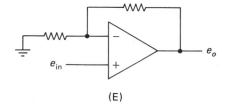

(E)

19. $e_o = -Ke_{in}$
20. $e_o = -K \int_0^t i \, dt$
21. $e_o = -K \frac{\delta e}{\delta t}$
22. $e_o = -(K_1 e_1 + K_2 e_2)$
23. $e_o = +Ke_{in}$
24. The circuit below is

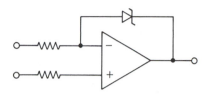

    A. An integrator.
    B. A differentiator.
    C. A differential amplifier.
    D. A sample-and-hold circuit.
    E. A comparator.

25. Which of the following statements is/are true?
    I. An integrator smooths input changes.
    II. A differentiator accentuates input changes.
    A. Both I and II are false.
    B. Only I is true.
    C. Only II is true.
    D. Both I and II are true.

26. Common-mode error is a problem only when using the noninverting input to an op amp. True or false?

GLOSSARY

Analog Computer Program   A schematic electrical circuit block diagram illustrating the interconnections between (and discrete components to be used with) the electronic elements normally included within an analog computer in order to solve a particular problem.

Calculus   The branch of higher mathematics which deals with variations in one parameter with respect to (as a function of) another parameter. Integration and differentiation are the two most basic techniques for describing these interrelated variations in calculus. In process control, we normally are interested in variations in parameters with respect to passage of time.

Differential Equations   Mathematical equations in calculus using derivatives of various orders to solve a particular problem.

Differentiate (Differentiation)   The inverse calculus operation from integration. Differentiation of a signal results in an output which is proportional to the rate-of-change of the input variable with respect to another variable (usually *time* in process control systems).

Digital Computer Program   A series of sequential instructions or commands which are to be interpreted eventually by a digital computer in order for it to perform a particular series of operations or solve a given problem.

Indicator   A device which displays (in a form which can be interpreted by humans) information which exists in a form which cannot be directly interpreted by humans.

Integrate (Integration)   The naturally occurring phenomenon by which products (or people) are a result of *everything* that has happened to them up to a particular point in time (or in a process). The mathematical process which simulates this universal, naturally occurring phenomenon.

Interface   As used herein, this term refers to the circuitry necessary to electrically adapt two (or more) pieces of equipment so that they can be used together.

Optimize   To determine and implement the single most effective or most favorable technique or system for the solution of a particular problem.

Programming (a computer)   The process of originating a detailed plan (series of steps, procedures, or interconnections) in order to solve a particular problem on a computer.

RC **Time Constant**   The time constant of a circuit or system having both resistance and capacitance. In the electrical energy system, it numerically equals the product of the two (resistance measured in ohms and capacitance measured in farads) with the resulting dimensions of seconds.

Steady State   The net (long term) output of a circuit or device sometime after having been disturbed by a transient input condition (after the output stops changing in response to that transient input).

Time Constant   The period of time required by the output of a reactive circuit or system to respond (exponentially an amount equal to 63.2% of the final change that will occur) in response to a change or disturbance at its input. A system is commonly assumed to have reached its final value or steady state output value in an amount of time numerically equal to five *time* constants.

10

THE OPERATOR INTERFACE

INTRODUCTION

Industrial process control systems are frequently very complex, spread out over wide areas, and consist of a multitude of individual control loops, each of which requires measurement and control of at least one process variable. *Control* is normally exercised by equipment and personnel located in centralized and frequently remote control centers. Into these control centers run the cables that convey all of the information concerning the current instantaneous state of the process variables and also of the various control elements which are controlling the process. From this control center issue the cables that carry the controller commands to the actuators being used to actually control the process. Furthermore, normally a fairly elaborate communications network exists to keep the control center operators in contact with the operators in the plant, both for normal operations and in the event that emergency operations become necessary.

The important concept here is that of remoteness and isolation from the physical process. This gives rise to the requirements for long-distance information transfer (via transmitters) and remote indication and alarms for critical variables and controls.

The operators (regardless of the type of control system used) must have all of the required information immediately and conveniently available in order to

1. Start up or shut down the process (these two operations normally are accomplished manually rather than automatically)

2. Determine if and when to have the automatic control system take over and exercise control

3. Assume manual control under unusual or emergency conditions

The equipment necessary for the (human) operators to be effective falls into several categories; analog signal transmitters, alarms, displays and status indicators, recorders, controllers, and possibly computers. Each of these categories is discussed in general in this text; however, do not lose sight of the basic reasons for including this equipment in process control systems. The quantity, mix, and complexity of the equipment will, of course, vary widely depending upon the specific process.

ALARM UNITS

There are numerous circumstances in process control systems where certain variables are so critical that regardless of whether they are connected to other instruments such as indicators, controllers, recorders, or even computers, they are connected to their own individual alarm circuits. The purpose for these alarm circuits is to provide continuous (and sometimes redundant) monitoring of the variable's values and to cause some type of automatic alarm to be activated whenever individually preset limits are exceeded.

Alarm units are basically analog-input, digital-output devices, which are operating in either the alarm state or the normal (monitoring) state. They accept any of several standard analog signal levels at their inputs and provide some type of electrical switching at their output. Internal circuitry (within each alarm unit) continually compares the present value of the process variable with manually preset upper and/or lower limits. Whenever the variable exceeds those limits a high-gain amplifier causes an output to switch positions. Normally, this output will cause some type of an alarm to be activated in order to alert the operator.

The output alarm can be in any of several forms, typically horns, bells, warbalarms, claxons, flashing lights, and/or backlighted displays. These alarms can be connected in any combination and the alarm unit can even cause a sequence to be performed (i.e., warning light, then bell, then a warbalarm). The alarm unit normally has an acknowledge switch such that the operator can cause any audible alarms to be silenced; however, normally lighted alarms still indicate that the alarm has occurred and that the alarm condition still exists.

The alarm unit outputs are normally energized (pulled in for relay-type contacts) and deactivated under alarm conditions. This provides a measure of "failsafe" operation in that if the unit itself fails, the digital output will become deenergized, thereby causing the alarm to be set off. Therefore, the alarm will be sounded either if the variable exceeds limits or in the event of alarm unit failure.

Manufacturers of this type of equipment also frequently provide dual alarm modules where the same process variable can be monitored against both upper and lower set points individually, causing different alarms under each condition.

A list of the different types of common, commercially available alarm units includes:

Current alarms, high- and low-level input types
Voltage alarms, high- and low-level input types
Thermocouple alarms
Millivolt alarms
RTD alarms
Strain gage alarms

What is inside these alarm modules? Figure 10-1 is a block diagram illustrating the internal (electronic) workings of an alarm module. Schematically, they normally include a signal-conditioning unit whose circuitry would vary depending upon the sensor it is designed to interface with. Standard interfaces are normally available to interface to several different ranges of RTDs, different types of thermocouples, and various dc voltages and currents. The units frequently have a bridge circuit of the type described for transmitters, including the necessary power supplies and *isolation* and compensation circuits.

The current value of a process variable (after suitable signal conditioning) is continually compared to the preset setpoint(s) by a comparator circuit. When the setpoint(s) are exceeded, internal circuitry initiates a logical alarm sequence, which in turn activates the appropriate alarm devices.

Some modern annunciator units have been designed to go beyond merely tripping at static setpoints. Some units monitor the trend or rate of change of a process variable and alarm based upon the magnitude of this trend as compared to preset limits. Other units provide for some level of user customization in that a selection of operational sequences are available; a wide variety of displays and alarms can be configured within standard modular chassis.

Typically, any of several alarm units are wired to cause the same alarm to be energized (visual or audible). Therefore, each of the individual alarm units normally

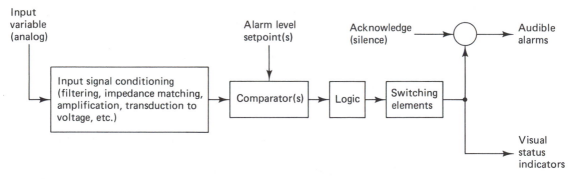

Figure 10-1 Block diagram of an annunciator.

has their own lighted and identified displays. These lighted displays can be as simple as a single incandescent lamp (possibly with a colored filter over it) and a label describing the reason for that particular alarm, all the way to elaborate multicolored system schematics. We will first discuss the simpler displays and progress to the more elaborate ones.

SINGLE-VARIABLE DISPLAYS

Figure 10-2A illustrates several basic styles of the backlighted digital displays frequently used. There can be several colored lamps behind the windows: a green one indicating that the circuit is energized and functioning normally, an amber lamp to indicate an abnormal operating condition (a warning), and a red lamp to indicate that the variable is in an alarm condition. As an example, assume a backlighted window engraved **Water Level # 1 Boiler.** A blank (unlighted) display indicates that the boiler is not in operation, a green light could be wired to indicate normal operation, a steady-amber lamp might indicate a low level (also sounding an audible alarm), a flashing-amber lamp to indicate that the audible alarm has been silenced (acknowledged) by the operator but that the alarm condition still exists, and a red lamp to indicate a power failure to the boiler's feedwater pump. Furthermore, the entire display could be mounted with a switch, so that depressing the switch could cause some other normal operation, such as starting the feedwater pump, or abnormal operation, such as silencing the audible alarm.

Several other common backlighted displays are illustrated in Fig. 10-2B, where two, three, or even more engraved windows are included within the same display, each window engraved with a code that identifies its significance and possible backlighted with colored lamps, as described previously.

Figure 10-3 illustrates a typical industrial alarm panel. It is a more complex (and integrated) system where each process variable input has its own individual isolated signal conditioning circuitry, its own alarm circuitry which controls the power to energize an audible alarm, as well as its own individual visual alarm(s). The alarm trip-point values may be engraved directly on the displays if desired. The audible alarm can be silenced by the process operator even though the alarm condition still exists by the silence switch. This will prohibit that module from causing a further audible alarm until after the abnormal condition has been corrected; however, any other alarm module can independently still cause an audible alarm.

There is an infinite variety of other visual alarm displays, consisting of LCD (liquid-crystal display) displays, LED (light-emitting diode) displays, and gas plasma displays, among others.

When displays and alarms are grouped together they are frequently collectively referred to as annunciator modules. This term is normally used throughout the industry.

Other features commonly incorporated into annunciator modules include *intrinsically safe* operation and ground-fault detection. The intrinsic safety feature is

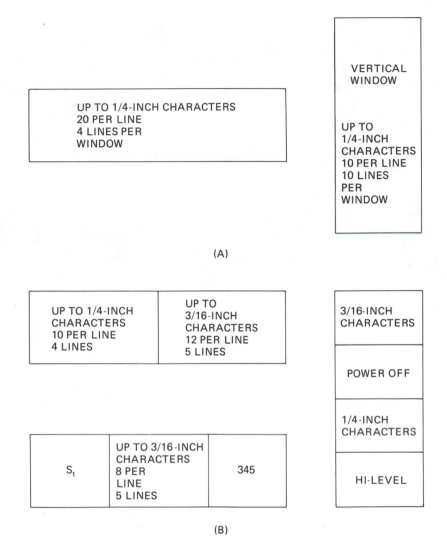

Figure 10-2 Common backlighted display types used for alarm and display modules.

accomplished by limiting the energy available to field contacts to values below the approved safe levels. The ground-fault detection systems are included to detect the presence of undesired grounds due to field wiring errors, insulation failures, and mechanical damage to the wiring.

The field wiring is the most prone to failure because the rest of the system (alarm system and protective devices) are normally operated from ungrounded dc power sources, and because this field wiring can be quite long and run in areas where

Figure 10-3 Typical annunciator/monitor unit. This unit accommodates multiple inputs with mixed monitoring capability and a flexible output selection. (Courtesy of AMETEK Inc., Panalarm Division.)

it is subject to mechanical wear and abuse. The details of the ground-fault detection systems vary between manufacturers and will not be further investigated here.

GRAPHIC DISPLAYS

The next level of complexity (and cost) beyond the annunciator panels is the custom, hardwired, graphic display. Figure 10-4 illustrates a typical application of a custom graphics display of the type being discussed here. Figure 11-14 also illustrates a graphic display.

This type of display is characterized by a graphic (normally multicolored) presentation of a portion (or all) of a specific process. It typically illustrates critical material and/or energy flow paths through major hardware and functional elements in the process. Also (typically) illustrated on these presentations are the major flow control elements, emergency equipment, and sensors. Frequently, the sensors will have a numerical display associated with them indicating current values, or at least multicolored lighted displays indicating conditions of operation, and flow control elements will have indicators depicting their current operational status.

Frequently, these displays will also have lighted indicators illustrating which loops are currently active, including direction of flow if applicable, and also the levels of material contained in any container may be indicated.

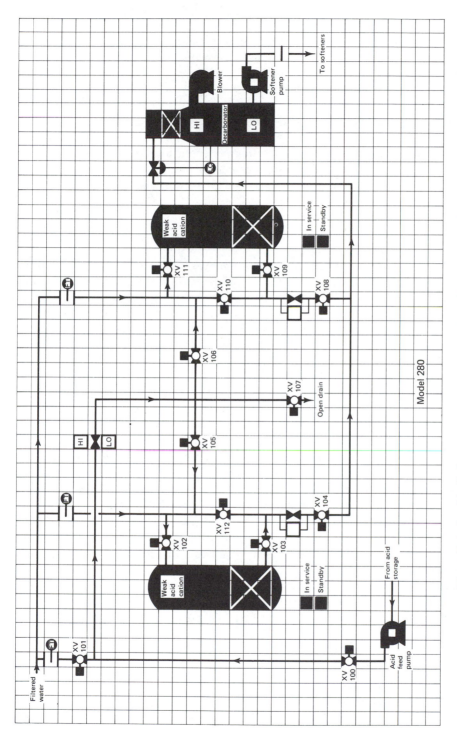

Figure 10-4 Graphic display system constructed of multiple individual mosaic tiles. The tiles are available with standard engineering symbols, some of which can be backlighted, and custom symbols can also be provided. Digital readout modules and pushbutton switches (that will fit into the same matrix) are available. This particular system can be updated and modified if necessary. (Courtesy of AMETEK Inc., Panalarm Division.)

With such a display the operator gets an up-to-date visual summary of the current status of all critical elements in the process. On other, normally adjacent, panels additional equipment is available for the operator to get more specific readings from the process sensors (i.e., chart recorders) and to manually effect control of critical or emergency control elements.

This type of display is costly and normally results in a fixed presentation. Modifications to the process (i.e., adding an additional loop) typically require costly modifications; however, it has found extensive application in industries such as electrical power generation and distribution, nuclear power control, gas pipeline flow control, and petrochemical processing plants, to name a few.

CRT DISPLAYS

As mini- and microcomputers have found their way into the field of process control, they naturally brought along the CRT displays with which they are programmed, debugged, developed, monitored, and controlled. Simultaneously, the personal computer market has provided the impetus and capital for development of very sophisticated graphics packages. These packages have been conveniently adapted to process control and can combine the functions of annunciator panels and fixed graphic displays with the ability to "zoom" in on any portion of the process to get a detailed picture of the current (or past) status of variables of interest.

Basically, the computer programmers put a number of pictures of the process into the computer's memory and then, as the process runs, the operator can select the particular view (from this stored repertoire of views) that is of current interest. Meanwhile, since that same computer system is simultaneously controlling the process, all variable values are available and can be programmed to be displayed where appropriate. In effect, the operator has several pages of displays available to him at all times, from overview pages of major portions of the process down to specific pages of the individual loops comprising that process. The flexibility arises because the same hardware can be multiplexed for all of those displays; also, modifications are relatively simple and inexpensive to make.

The CRT type of display has become almost universal in its acceptance in the industry. It is typically so cost-effective, so versatile, and so functionally effective that it is included to some degree in most automated processes. The other methods mentioned previously are still also included in processes where no digital computer is used and for manual and emergency backup purposes.

RECORDING AND INDICATING EQUIPMENT

Although not actually a part of any controller, there are normally any of several different types of indicating and recording devices connected between the transmitter and the controller. These devices are commonly designed to operate from the

ISA standard transmitter output signal levels and provide the operator(s) at all times with a visual indication of the status of the controller input variables. Frequently, a past history of the variations of some process variables is required, and in these applications recorders are used.

The recorders normally consist of a pen mechanism which is driven by a servomotor and whose instantaneous position is directly proportional to the value of the process variable being monitored. A chart is moved beneath the pen, based upon a time scale. The resulting trace on the chart is therefore a time history of the value of that variable.

These devices are necessary in order for the process control system operator to assume manual control of the process, if required, for him to check on the status of the process while it is under automatic control (i.e., verify that the controller is working), and for testing and system-adjustment purposes. Not infrequently the indicator and/or the recorder may be built into the controller itself; the resulting piece of equipment is commonly called a *recording controller.*

Figure 10-5 illustrates several commercially available recorders. Any particular system may have quite a large amount of this type of equipment installed.

TRANSMITTERS

The process variable signals may go directly to the controller if it is physically located close enough to the process, without any further signal handling. However, so many controllers are physically remotely located from the process that special pieces of interface equipment have been designed whose purpose is specifically to receive a process variable signal at the transducer (sensor) and give this signal enough power so that it can be transmitted over the necessary distance without the addition of excessive noise or reduction in accuracy, sensitivity, and other factors.

The Instrument Society of America (ISA) has addressed itself to the problems of long-distance signal transmission and has developed several standard signal levels. If the signal is to be transmitted in pneumatic form, the standard levels are 3 to 15 psi. The maximum pressure (15 psi) would represent the full-scale signal deviation (100% signal amplitude), and 3 psi would represent the minimum possible signal amplitude (0% of signal amplitude). As the process variable varies (in an analog fashion) from 0% to 100% or any place within this span, the pressure out of the transmitter will be proportionately between 3 and 15 psi.

A list of some of the more popular (unofficial) industry "standards" includes:

Current ranges	1–5 mA, 4–20 mA, 10–50 mA,
Low-level voltage ranges	0–50 mV, 0.25–1.25 V
High-level voltage ranges	0–1 V, 0–5 V, 0–10 V
Pressure ranges	3–15 psi

(A)

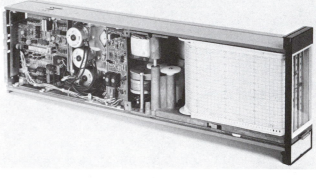

(B)

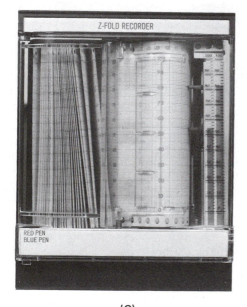

(C)

Figure 10-5 Commercial recording instruments: (A) circular chart pneumatic recorder; (B) narrow-roll electronic pen recorder; (C) multipen recorder; (D) multipoint electronic chart recorder capable of simultaneously recording up to 30 separate channels of information; (E) circular chart recorder integrated with pH measurement system; (F) portable multifunction test recorder. [(A) Courtesy of Foxboro Company; (B) courtesy of Taylor Instrument Process

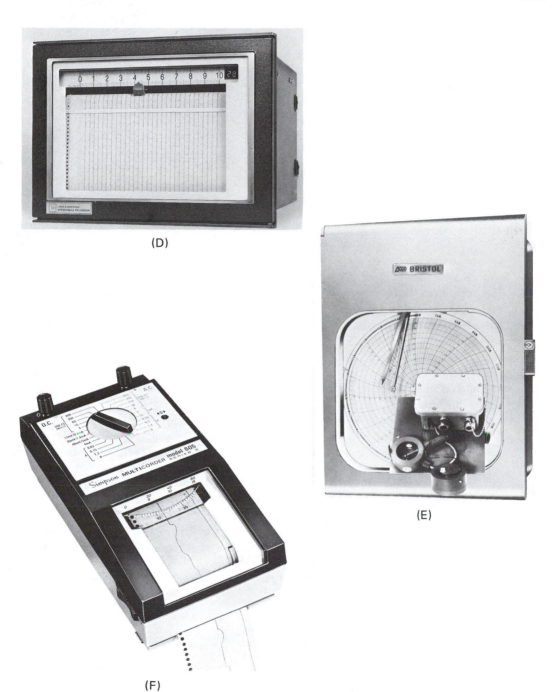

(D)

(E)

(F)

Control Division/SYBRON Corp.; (C) courtesy of Bailey Meter Co.; (D) courtesy of Leeds and Northrup; (E) courtesy of Bristol Division of ACCO; (F) courtesy of Simpson Electric Co.]

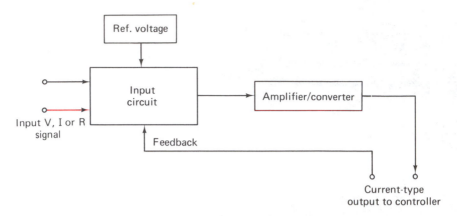

Figure 10-6 Block diagram of electronic transmitter.

Signal transmission may also (and more commonly) be in electrical form. The ISA standards for electrical signal transmission are primarily 4 to 20 mA of current and secondarily 1 to 5 mA of current. Here 4 mA (or 1 mA) represents the minimum value of the process variable, and 20 mA (or 5 mA) represents the maximum value of variable information (full scale or 100% deflection).

Transmitters (Fig. 10-6) also commonly include the hardware necessary to transduce variable information from one energy system to another for subsequent signal transmission. For example, a V-to-P (voltage-to-pressure) transmitter will accept a nonstandard (or standard) dc voltage signal at its input and transduce this voltage to the standard 3- to 15-psi pneumatic pressure signal for transmission. There are many other types of transmitters commonly available which accept voltage, current, resistance, or pneumatic signals in many different ranges at their inputs, transducing these signals to the ISA standard levels for transmission to the controller.

The transmitters themselves frequently contain the Wheatstone bridge equipment and/or amplification hardware referred to in Chapter 7. Figure 10-7A is a block diagram of an electronic transmitter, receiving a voltage input signal and transmitting an electrical current signal to the controller. Figure 10-7B illustrates a variable resistance-to-variable current transmitter and Fig. 10-7C a variable pressure-to-current transmitter.

Figures 10-8 and 10-9 illustrate typical commercial sensor/transmitter units.

Types of Transmitters

A list of the different types of commercially available transmitters includes:

1. Thermocouple transmitters
2. Millivolt transmitters
3. RTD transmitters

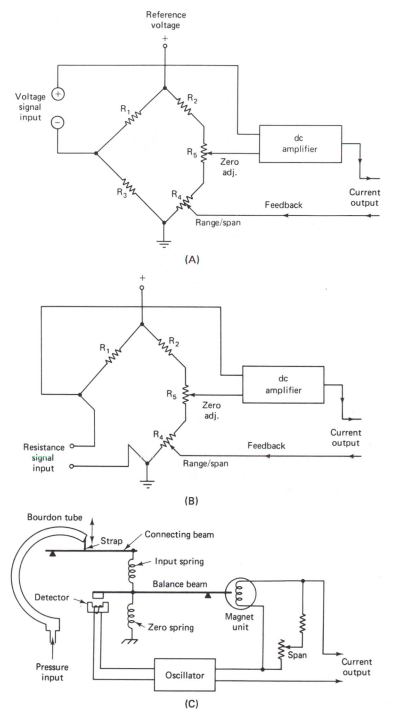

Figure 10-7 Simplified transmitter schematic diagrams: (A) voltage-to-current; (B) resistance-to-current; (C) pressure-to-current.

Figure 10-8 Commercial differential pressure sensor with a built-in solid-state current transmitter. (Courtesy of Bailey Meter Co.)

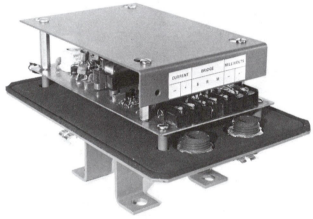

Figure 10-9 Internal view of the electronics for a temperature sensor (similar models for either RDT-type or thermocouple-type sensors)-to-current transmitter. (Courtesy of Leeds & Northrup Co.)

4. Potential-to-current transmitters
5. Strain gage transmitters
6. Isolated-type transmitters
7. Electrical-to-pneumatic transmitters
8. Ac current-to-dc current transmitters
9. Frequency-to-current transmitters
10. Square-root transmitters

Frequency-to-current transmitters are typically designed to accept ac or digital pulse signals at their inputs (i.e., from turbine flow meters), where the analog value of the process variable being measured is proportional to the instantaneous pulse rate or ac frequency of that input signal. Typical input frequency ranges are from 10 to 1000 Hz, possibly in several ranges. Other input pulse specifications typically include minimum and/or maximum values for pulse shape, pulse power, and pulse width (for digital signals).

Signal transmitters are also commonly called voltage-to-current converters. They exist on the marketplace to perform the obvious function of transducing from one electrical form to the other when interfacing requirements dictate. Sometimes they may be used strictly for the input/output electrical isolation that they commonly provide. For example, one might be used to convert a 4- to-20-mA input analog signal to a 4- to-20-mA analog output signal; however, there would be no electrical continuity (connection) between those input and output circuits to cause ground-loop-type problems. Ground loop problems are explained further elsewhere in this text, and Chapter 8 includes typical circuits of this type.

Isolation transmitters are similar to any of the transmitters described previously, with the added feature of providing complete electrical isolation between the input and output circuits. It is basically a combination of any of the standard transmitters (RTD, thermocouple, strain gage, etc.) with the added electrical isolation feature. Normally, adding the electrical isolation feature does not significantly affect the overall input/output specifications, only the cost.

TELEMETRY

What Is Telemetry?

Telemetry is a term which is loosely used when actually referring to radiotelemetry. In a process control context radiotelemetry is a technique used to link the information gathered by a remotely located sensing element or data acquisition system to the system which needs that information, using a radio transmitter and receiver instead of wires. The technique is just as useful in linking a control system with remotely located actuators. Figure 10-10 illustrates the type of application and includes a functional block diagram of the hardware.

Why/Where Is Telemetry Used?

There is no technical reason why this technique could not be used in most systems; however, its use is normally restricted to applications where it may be impractical or even impossible to run hard wire. Therefore, it may be considered to be a special type of transmitter. When it is used it introduces into the control system the additional problems associated with radio transmitters and receivers (such as reliability, weather dependence, and power dependence) as well as transmission problems (such as the introduction of additional noise sources).

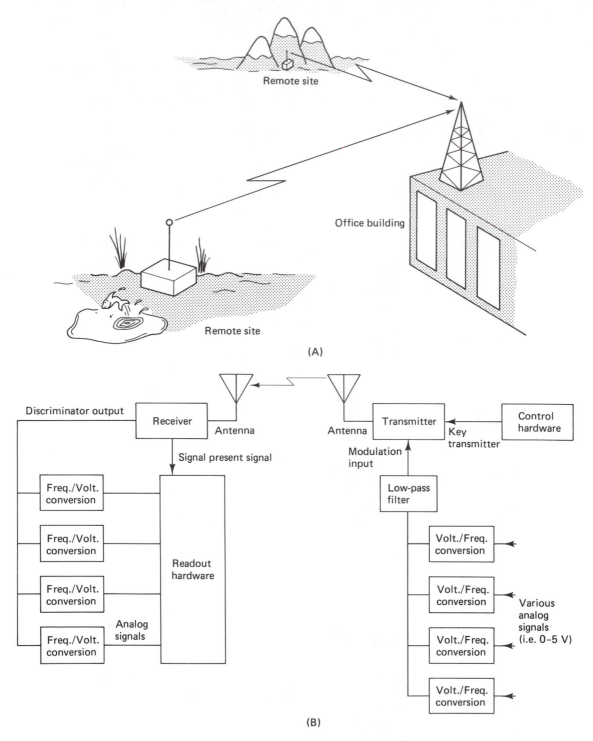

Figure 10-10 Radio telemetry system: (A) sketch of type of application; (B) block diagram of a complete telemetry system, including multiple-sensing provision. (Courtesy of Repco Incorporated.)

How Is Telemetry Implemented?

Modern radio transmitter/receiver modules suitable for radiotelemetry applications are available in small and rugged packages. Some are designed to be battery operated and for hostile-environment applications. They normally operate in the commercial FM broadcast band (88 to 108 mHz) using frequency-modulation (FM) techniques as they are less sensitive to the effects of RF noise, and over short distances at least, provide more reliable communications in a smaller and less expensive package than other techniques. Figure 10-11 shows a commercial radiotelemetry unit; physically, it measures only approximately 7 in. long.

Figure 10-11 Commercial telemetry units. The unit pictured is packaged for rugged usage, and is completely self-contained except for power source, antenna, and, of course, the measuring system. (Courtesy of Repco Incorporated.)

QUESTIONS

1. Describe in your own words the relationship of a control center to an automated process control system.

2. What functions must a typical control center perform in an automated process control system?

3. Describe the function that alarm units perform in automated process control systems.

Referring to Fig. 10-1, relate each of the functional blocks to

4. A furnace control system having thermocouple temperature sensors and a siren alarm.

5. A supermarket freezer control system using RTD sensors and an outside claxon.

6. Design an annunciator system control panel appropriate for each of the systems in Questions 4 and 5. Include pushbutton controls and current temperature readouts.

7. Design a graphic display system for an oil-fired hot water (base ray) home heating system. Illustrate the fuel tank (with level indication), burner, flame sensor, heat exchanger (hot air to water), firebox, expansion tank, at least two circulator pumps, two zone control valves, thermostats, and emergency switches. Illustrate the oil, water, and electrical systems in different colors.

8. What other figures in this book illustrate graphic displays? How about CRT displays?

9. What functions do indicating and recording equipment perform in automated process control systems? What is the difference between them?

10. What is a transmitter, and what is its function in an automated process control system?

11. What are the ISA signal transmission standards?

12. What function would a square-root transmitter perform? With what types of sensors?

13. List several applications for a frequency-to-current transmitter.

14. In your own words, what is telemetry?

15. List several commonly used commercial telemetering systems.

16. Which of the following functional circuits might you legitimately expect to find in a transmitter?
   A. A comparator circuit.
   B. A digital-to-analog converter.
   C. A sensor.
   D. A current-balanced Wheatstone bridge circuit.
   E. Any of the above are legitimate answers to this question.

17. The primary function of transmitters is
   A. Long-distance signal transmission of sensor output information.
   B. Recording past history of variations in some process variable signal.
   C. Indicating the instantaneous value of a process variable.
   D. Comparing the feedback (process variable) signal to the set-point signal to develop an error (deviation) signal.
   E. Operating on the error (deviation) signal to develop a command signal back to the process (servomechanisms).

18. Transmitters frequently include signal-conditioning equipment necessary to directly interface to sensors. True or false?

19. The primary difference between a recorder and an indicator is that
    A. The indicator does not give a permanent record.
    B. The recorder does not give a permanent record.
    C. There is no difference between them.

20. A recorder normally consists of a pen mechanism which is driven by a servomotor and whose instantaneous position is directly proportional to the value of the process variable being monitored. True or false?

GLOSSARY

Alarm Units   In process control systems these are the units which monitor process variables, notifying operators of abnormal values, regardless of the presence of other equipment monitoring those same variables.

Annunciator   A visual display having lighted and typically labeled indicators. The status of the light (i.e., color, intensity, or blinking state) indicates the status or condition of the variable or control element named on the label.

Control   Used in this chapter to refer to the ability of a human operator to modify the behavior of an element or a variable in any portion or portions of an industrial process control system.

Intrinsically Safe   Designed such that the circuit, device, or system referred to is incapable of releasing sufficient energy to cause a spark or otherwise ignite an explosive mixture, regardless of improper maintenance, maladjustment, or failure of any part of the internal circuitry.

ISA   The Instrument Society of America: An international technical professional society comprised of professionals with an interest in the field of automation and process control, including medical applications.

Isolation   Electrical separation between two systems. Used in process control systems to help control ground loop problems.

Telemetry   The process of sensing information at one physical location and transmitting (through the air via radio waves) that information to some remote location where it is needed.

Transmitter   Used herein in reference to process control systems, this term refers to a piece of equipment specifically designed to transmit (send) variable information from sensors or controllers in a system over long wires (normally), providing industry standard signals at their output.

11

PROCESS CONTROLLERS

This text up to this point has dealt with the problems of measuring and transducing process variables within acceptable standards, and properly conditioning the resulting signals (primarily electrical) in order that a "magic" device, called a "controller," could have this information available in a form compatible for its use. This chapter will explain in more detail just what this controller is, and how it works. We shall deal specifically with the theory of operation of controllers; the next chapters will deal with the "simpler" types of hardware controllers, including various types of digital controllers.

WHAT A CONTROLLER IS

The *controller* is the hardware piece of equipment (electronic, pneumatic, or mechanical) which is the practical result of applying automatic control system theory to industrial process control problems. It is the hardware that implements a comparison of the condition of the process material at the output with the desired condition or setpoint and computes and makes the correction, proportional to the difference between the two (the *actuating error*) to the input of the process.

The controller is, then, the "brain" behind the automation and control of any automatically controlled process. It may be as small and simple as a pneumatic relay or a spring-balanced mechanical level arm, or as complicated as a complete direct digital control computer. It may control one variable in a process, or it may control literally hundreds of process variables simultaneously. It may be analog, digital, or a combination, and, as previously noted, it may be electrical, mechanical, pneumatic, or any combination of these.

It is not a stand-alone device. It normally requires auxiliary "interface" equipment to be used with it. The more complex the control system, the more auxiliary equipment that will normally be required. Many of the typical types of auxiliary equipment were presented in previous chapters. This and subsequent chapters will deal with even more of this type of equipment.

USES OF CONTROLLERS

Controllers are used for any of the various reasons justifying process automation, as expressed in Chapter 1—essentially when there are physical constraints on the process itself, where the operator cannot be physically present, where the speed of response required of the operator is excessive for human operation, where the operations required are too complex for human operators to maintain adequate control, or where economics of the process itself require automation.

Reading between the lines of the preceding paragraph you can see that automation is an absolute requirement in many cases, not an embellishment. Without an appropriate controller, many processes would not exist. Frequently, the control exercised by a given controller will be so critical that one or more means of monitoring the performance of the controller itself are employed, possibly even with an automatic means of shifting to an alternative method of control in the event of controller failure. Most automatically controlled processes are designed with at least one backup means of control available in the event of a primary controller hardware failure.

So, why are controllers used? Because they are an absolute necessity if we are to have many of the products and services that we so routinely enjoy in our modern world.

Let us now take the conditioned signals that are representative of a process variable we are interested in and see what controllers do to these signals.

SIMPLE OPEN-LOOP CONTROL

As explained in preceding chapters, there are both open-loop control and closed-loop control. Although very limited control is possible using open-loop control systems, they are, in fact, used in many industrial situations, where there is no need for the more complicated closed-loop systems. The only open-loop types of control possible are "bang-bang" or on-off (digital) switching (e.g., light or motor switching); control of a rheostat or potentiometer movement (analog) based upon time or operator action (e.g., the volume control of a radio); or sequential switching control, where a sequence of events occurs once the controller has been activated, without any type of feedback which would indicate success or completion of any of those events (e.g., an automatic dishwashing machine).

SEQUENTIAL CLOSED-LOOP CONTROL

The sequential switching technique can be adapted to closed-loop (feedback) control where the advancing of the sequence of events that have been hardwired into the control sequence (programmed) is controlled based upon successful completion of certain key steps in the sequence. In other words, the sequence is held up at key steps until a sensor feeds back information to the controller, indicating successful completion of that key event. In either open- or closed-loop sequential control systems the outputs from the controller are either on or off (i.e., digital), not analog.

PROPORTIONAL-ONLY CONTROL

Beyond these most elementary types of control, the next more complicated type is analog *proportional feedback control.* The analog proportional feedback control system block diagram is essentially the one used in the definition of feedback control (reproduced in Fig. 11-1). The setpoint is compared to the analog feedback signal and the proportional controller responds only to the amplified difference between the two. The *error* is amplified and converted in form as necessary by the controller to drive the system actuator as required to alter the process output, bringing it into correspondence with the setpoint. Figures 11-2 and 11-3 are schematic diagrams of proportional-only controllers which are connected into systems. Let us examine the effect that the proportional-only controller will have on the process output.

A good example that can be used is the level control of a tank of liquid, where the flow out of the tank is regulated by a control valve responding to the level in the tank (Fig. 11-2). Here the *capacitance* of the system is the volume of liquid in the tank, and the gain of the system is determined by the relative lengths of the simple lever connecting the float to the valve stem. Note that both the comparator and the controller are purely mechanical.

Assume the system to be in a steady-state situation with flow into the tank at a steady value and the level at the setpoint, which is set by altering the vertical position of the fulcrum. As the flow input decreases to a lower steady-state value

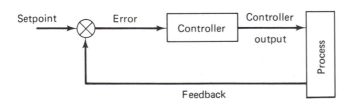

Figure 11-1 Analog proportional feedback control system block diagram.

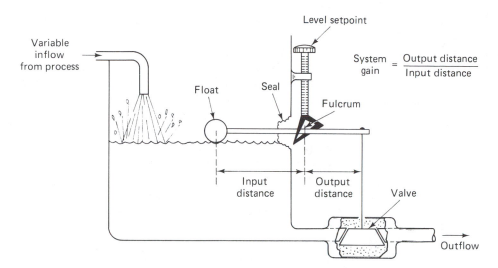

Figure 11-2 Hydraulic system proportional-only level control system.

due to some process disturbance, initially the amount of flow out of the tank is greater than the amount of flow into the tank. Therefore, the level in the tank will begin to lower. The rate of lowering will be proportional to the difference between input and output flow rates and the volume of the tank; the larger the volume the slower the decrease in level and the slower the response of the float-lever-valve system to the disturbance.

By the time the valve has responded by slowing down the outflow rate to match the inflow rate, the *level* of liquid in the tank (which is the variable being controlled) will be below its setpoint. The larger the tank, the longer it will take the controller to respond, and thereby the greater the final volume error. Note also that the lower the gain (as noted on Fig. 11-2), the greater the final volume error.

In order for the final position of the valve to differ from its initial position in response to the change in inflow rate, there must be a permanent change in the level of liquid in the tank or else the valve would revert to the same position it held before the permanent change in inflow. Then the outflow would not equal the inflow and the level would again change, but this time not because there was a change in the input.

If the capacitance (total volume of liquid actually in the tank) of the system is decreased, the level change would be felt sooner by the float, and therefore the system would react sooner to make the necessary correction to output flow and the smaller the resulting permanent volume change that would be necessary for the in- and out-flows to be equal. As the gain of the system is increased, smaller level changes would result in greater changes in output flow rate (changes in valve position). However, friction in the system, the forces necessary to cause the valve to assume a steady-state position, and the fact that the system would respond to roughness of the surface of the liquid due to splashing would all limit the maximum value of this gain.

One other, very common, example will be used to illustrate the limit of effectiveness of proportional (only) controllers: a temperature-control problem. Referring to Fig. 11-3, we have a reactor kettle that maintains a constant inflow and outflow (and therefore a constant level). The input variable in this system is the temperature of the liquid flowing into the kettle and the variable to be controlled is the temperature of the liquid leaving the kettle. Certainly, the degree of control possible will depend upon the type of liquid, mixing of the liquid, flow past the heater, size of the heater, and type and placement of the temperature sensor. However, regardless of these parameters, once they are determined, the degree of temperature control possible lies with the controller itself.

Begin with the temperature of the inflow liquid at a steady-state value and the heater (an analog heater) putting out a steady quantity of heat. As the temperature of the inflow liquid changes, there is a time lag due to mixing of the new fluid with the volume of liquid in the tank until the temperature of the liquid at the outflow changes and the temperature sensor's output reflects this change. The delay between the time the temperature of the inflow changes and the time the temperature sensor responds to this process disturbance depends directly on the volume of liquid in the kettle and the thermal capacity of that volume of liquid. The larger the mass of

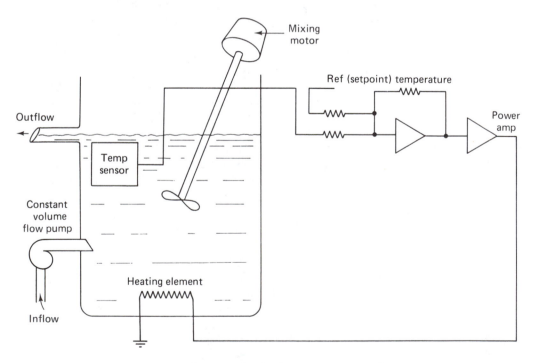

Figure 11-3 Proportional-only temperature control system.

liquid in the tank (capacitance), the longer it will take and therefore the larger the total heat-content change (not necessarily temperature change) of that mass before the controller can react.

Once the controller reacts, the magnitude of its reaction is directly controlled by the gain of the controller (i.e., small temperature variations from the setpoint cause large changes in heater output for high gain). The heater responds appropriately and starts to change the temperature back to the setpoint value.

If the temperature was brought back exactly to the setpoint, the error would be reduced to zero and the output from the heater would revert to exactly the same value as it was before the disturbance. However, the disturbance has not gone away; therefore, if that happened, the output temperature would respond exactly as when the disturbance was first encountered.

Actually, as the temperature of the outflow begins to change toward the setpoint, the error signal decreases and consequently the heater changes its output proportionately. Eventually a steady-state condition will be reached. However, the final steady-state temperature cannot possibly be exactly the same as it was before the disturbance. Increasing the gain of the system will decrease the difference between the new steady-state temperature and the setpoint, but if the gain is too high, the system will overcorrect or, even worse, respond to temperature variations due to imperfect mixing. Therefore, there is absolutely no way that this controller can reduce the steady-state error to zero.

Note the similarities in the control problem for both example systems. The capacitance in each limited the time response characteristics of final (steady-state) control that was achievable; the greater the amount of capacitance in the system the longer it takes to achieve the steady-state difference between setpoint and feedback (error). Increasing the gain of the controllers reduced the magnitude of the steady-state errors; however, the limit of gain was determined in each case by system stability considerations.

The following general conclusions can be drawn. Proportional-only controllers cannot reduce the steady-state error to zero. This type of control is therefore primarily limited to processes where the gain can be made large enough (with acceptable system stability) to reduce the steady-state offset (actuating error) to acceptable limits.

Figure 11-4 illustrates the time response of the output of all proportional-only controllers to a large change in input to the process (process load change). Figure 11-4A illustrates the output if the system gain is relatively low. The plot would look similar if the change had been due to a setpoint change, instead of a process disturbance as detected in the feedback signal by the controller.

The lower the system gain, the greater the steady-state error will be. The time difference between t_0 and t_1 is directly proportional to the capacitance in the system, larger values of capacitance resulting in larger time delays between the time the change is made and the time the system has reached steady-state in response to that setpoint (or feedback) change.

Figure 11-4B and C illustrates the effect of increasing controller gain. As the

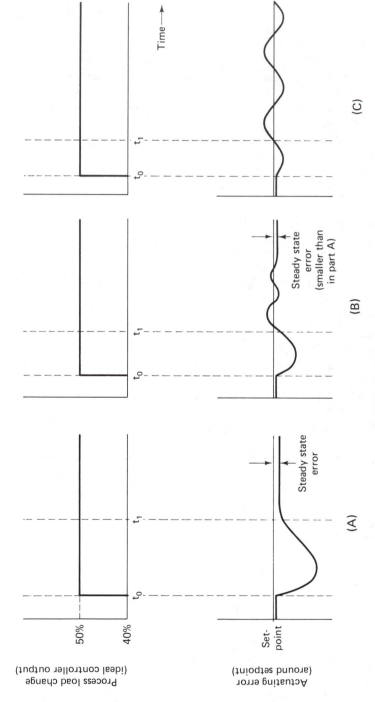

Figure 11-4 Proportional-only controlled response: (A) low gain; (B) critical (higher) gain; (C) gain too high (system unstable).

gain is increased, the overshoot of the output increases and it takes the system longer to settle down to a steady-state (final) value; the system is less stable (Fig. 11-4B); however, the steady-state error is decreased. Finally, gain is increased to the point of instability (Fig. 11-4C). Note also that the time delay (difference between t_0 and t_1) decreases as the gain is increased. Therefore, the proportional-only controller is designed to have a gain that optimizes the tradeoff among steady-state error, time delay for the system to respond to disturbances, and system *stability*.

PROPORTIONAL PLUS INTEGRAL CONTROL

In many systems the degree of control achievable using proportional-only control is not acceptable. Specifically, the steady-state error is not tolerable in numerous systems; therefore, a technique has been developed to respond primarily to the steady-state error and cause the system to respond to drive that error to zero.

An example schematic diagram of this system design is shown in Fig. 11-5. It illustrates the comparator, which is the same as in the proportional-only controller; however, there are additional operational amplifier circuits in this illustration. The controller now consists of a proportional-only circuit, which acts identically in all respects to the controller described previously, plus an additional operational amplifier circuit which was described in Chapter 9.

This second op-amp circuit is an integrator. If the steady-state error (by itself) is considered alone, it is essentially a dc signal, and the output from an integrator that has a dc input will be an ever-increasing output. Therefore, once the proportional portion of the controller has responded to a disturbance and driven the error to a small value (a value too small for the system to respond to), the integrator takes

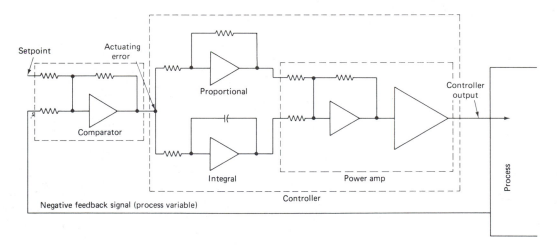

Figure 11-5 Schematic illustration of a proportional-plus-integral controller.

over, integrating the error signal and causing the controller output to increase until the system responds by driving the error to a smaller value.

The integrator continues to integrate the ever-decreasing error signal until it actually goes to zero, at which point the command output signal (due to the output of the integrator) remains at a constant value, thereby maintaining the process variable at its setpoint value. The integrator portion *resets* the value of the process variable to the setpoint value after a disturbance. It is therefore frequently referred as the reset mode of *control* rather than *integral control,* although the two terms are synonymous, reset being the older term. In reality, the controller does not require separate op-amp circuits for the proportional and integral portions of the controller. Figure 11-6 illustrates an op-amp circuit that combines both functions in a single circuit.

This type of combined control is called *proportional plus integral* (PI) *control strategy* and is used on systems where proportional-only control is not capable of reducing the steady-state error to acceptable levels. It accomplishes essentially the same effect as an extremely high gain proportional-only controller without the *initial* adverse effect on system stability.

There is, however, a longer-term undesirable side effect that does affect system stability when using PI control. It is easiest explained by using the curves illustrated in Fig. 11-7. The curve of Fig. 11-7A illustrates the command output of the controller, which initially is stable at some value. Then, at t_0 a disturbance to the system causes the generation of an actuating error signal with polarity and magnitude as illustrated in Fig. 11-7B.

Instantaneously, the proportional portion of the controller reacts to this actuating error signal, forcing the command output from the controller to some new value. Since there is an error signal, the integrator portion of the controller commences to increase its output at a rate determined by its *RC* time constant. For awhile, the command output continues to increase in magnitude (due to integrator) before the system begins to respond and the actuating error is reduced in magnitude

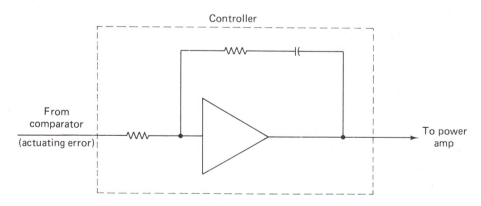

Figure 11-6   Proportional-plus-integral controller using a single op amp.

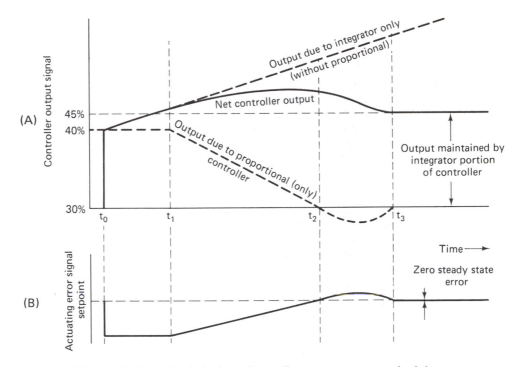

Figure 11-7 Proportional-plus-integral controller response to a process load change.

starting at t_1. As the error reduces, the integrator continues to increase its output, but at an ever-decreasing rate, because the error signal input is decreasing the whole time (and so, therefore, is the proportional signal).

Finally, at t_2, the error signal is at zero and the output from the proportional portion of the controller is therefore zero also; however, there is a significant output from the integrator.

For those with inquiring minds who are too practical to accept a circuit that maintains a constant output (which can be other than zero) while a condition of zero input exists, this certainly does sound like perpetual motion or creation of energy. However, if you will remember from Chapter 9 that if zero input is defined as zero current input, a capacitor will maintain the charge existing at the instant the charging current went to zero. This is exactly what happens in the integral portion of the controller. The error signal is used to generate a current which subsequently charges (or discharges) the capacitive "hold" element in the controller. Therefore, when the error goes to zero the controller will maintain that exact value it had at the instant that the error was forced to zero. This, of course, is when steady-state conditions exist in the process being controlled, which is normally at best an instantaneous (and transient) condition.

In practical systems the error never truly goes to zero and remains there for any significant length of time. Therefore, in any practical system, there will always be some component of the command output from a controller due to the fact that the error is not zero. Under normal conditions this component cannot be made large enough to maintain system stability at an acceptable level; therefore, the special "hold" circuit is included within each multimode controller; it would not be needed in a proportional-only controller.

The integrator output causes the error to cross zero and create an error in the opposite direction. If the RC time constant is short enough, the output from the integrator (at t_2) will be adequate to cause the error to build up enough (with appropriate direction) after the error crosses zero to cause the proportional portion to respond. This situation leads to system instability (an overshoot). The overshoot tendencies will increase as the proportional gain is increased and the integral time constant decreases. In practical systems this results in the gain of the proportional portion of the controller being less than for a proportional-only controller, and there are also definite limits placed on the integration time constant of the integrator portion of the controller.

With the gains properly set, in our example, the proportional portion of the controller would not respond significantly to the overshoot. The overshoot would be relatively small and, as the integrator then integrates the opposite polarity error signal, it eventually drives the error to zero (at t_3).

The example illustrated the effect of PI control when confronted with relatively large and rapid system disturbances. For control of this type of disturbance, proportional-only control has the major effect of getting the error close to zero. Actually, the integrator portion was no asset until the error became close to zero—in the deadband—which the proportional-only portion of the controller could not effectively cause the system to respond to. Actually, since the gain of the proportional portion of controller had to be reduced, the overall effectiveness was reduced for this type of disturbance as compared to a proportional-only controller.

Therefore, PI control would be used more on systems where load disturbances occur frequently, and setpoint changes are infrequent. It would also be used where load changes are slow and considerable time would pass before the error would be adequate to cause a system correction without the integrator included and, as previously noted, where proportional-only control results in unacceptable steady-state error signals.

DERIVATIVE CONTROL

Considering proportional control as *the* basic mode of control (which, in fact, it is), integral control can be added to decrease long-term or steady-state error. However, as previously shown, this is at the expense of timely correction to larger and quicker system disturbances. The faster the process error increases (rate of change of error), the greater the initial error (overshoot) will be, and neither the proportional portion nor the integral portion of the controller can respond to that. Furthermore, the greater the rate of change of error signal, the longer it will take the process to settle down to acceptable error levels.

Therefore, it is desirable that another mode of control be available if necessary; this mode is designed to specifically respond to rate of change of actuating error. Having read Chapter 9, it is obvious that the circuit required for this type of response must be designed to take the derivative of the error signal.

Therefore, this third mode of control is called *derivative* or *rate control* (since it only responds to rate of change). Since this mode of control is responding to the rate of change of error, it is in essence actually predicting future values of error if no change occurs in the error rate of change. It will, in effect, provide an output from the controller in anticipation of future values of output by the proportional portion of the controller; therefore, it is also correctly thought of as a predictive type of control. Figure 11-8 illustrates a practical differentiator circuit (which is insensitive enough to electrical noise to be useful) and Fig. 11-9 illustrates a control circuit using *proportional-plus-derivative* (PD) features.

Derivative control provides an extra component to the command signal from the controller. The addition of this component has the system effect of reducing the amount of time required for the controller to return the output of the process back to the setpoint; as compared to the amount of time that it would take the proportional-only controller to exercise similar control. It has the further effect of reducing the initial overshoot and has an overall effect of stabilizing the system by damping out any oscillations of the controlled variable. One additional side benefit of derivative control is that its tendency to reduce system oscillations permits the proportional gain to be set at higher values than could be permitted without derivative control, thereby further increasing the speed of response of the controller to system disturbances.

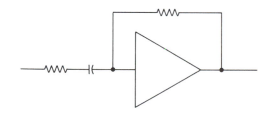

Figure 11-8 Practical differentiator.

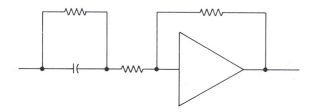

Figure 11-9 Proportional-plus-derivative controller.

Figure 11-10 is a set of curves illustrating the (output) response of the proportional and derivative portions of a two mode controller. The increased speed of response and improved stability (reduced overshoot and subsequent oscillations) should be obvious (note, however, that there is a steady-state error). It must be noted that, as with the proportional and integral modes of control, there are definite limits of gain for the derivative portion of the controller. The general rule is that *a little bit of derivative control goes a long way.*

Derivative-mode control is never used alone, as it is not capable of maintaining the error signal under steady and acceptable control. It is always used with the proportional mode and may be additionally combined with the integral mode of control in a *three-mode* (PID) controller. It is used primarily on processes that ex-

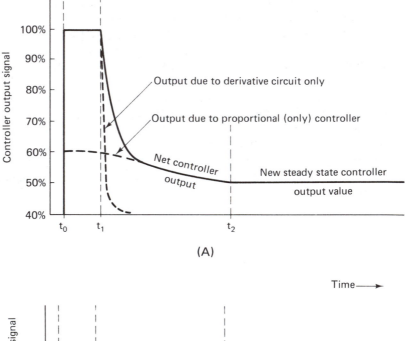

Figure 11-10   Typical output signals from proportional-plus-derivative controllers.

perience sudden and relatively large disturbances, where the proportional-only mode of control does not maintain acceptable finished product.

THREE-MODE CONTROL

As has been mentioned in preceding discussions, each of the three modes has a specific effect on the output of a controller. The proportional mode is the basic mode of control. It is complemented by the integral mode of control, which has the major effect on the long-term errors, which the proportional controller cannot respond to. The derivative mode of control, on the other hand, has its major effect during process transient disturbances which the proportional controller cannot respond to fast enough without excessive gain, and a resulting tendency toward system instability.

Three-mode (PID) control would be used on processes that exhibit rapid (taken care of by the derivative mode) and large (the resulting large steady-state errors are taken care of by the integral mode) disturbances where two-mode controllers are not capable of maintaining adequate system control. In using three-mode controllers there is considerable interaction between the system effects from the various modes in the same controller. Figure 11-11 illustrates a schematic for a three-mode controller in which the three control functions are separated in hardware. Note that each of the modes has its own adjustment. The interaction referred to is among the relative system gains according to each of the three modes of control. Figure 11-12 shows a more practical three-mode controller schematic, where all three functions are combined in a single amplifier circuit.

As noted previously, when integral mode control is added to a proportional controller, the proportional gain must be reduced. Also, if the integral portion of the controller has too large a reset rate (smaller time constant), it tends to become less stable. Adding in the derivative mode of control also affects the proportional gain, permitting it to be increased. Exactly how much of each mode is to be allocated is different for each system, and the final tuning of any specific system is still more of an art than a science.

Tuning a controller refers to the adjustment of its modes. This is done *on-line,* while the process is actually running. Many rules of thumb are used to determine initial settings that will at least allow the process to operate; then control systems engineers fine-tune the three modes to obtain the best control.

ANALOG METHODS OF CONTROL

The modes of control we have been discussing can be implemented in hardware using analog techniques, digital techniques, or a combination of analog and digital (*hybrid*) techniques. The illustrations and discussions in this chapter have used analog techniques exclusively. Also, these illustrated modes of control have used only

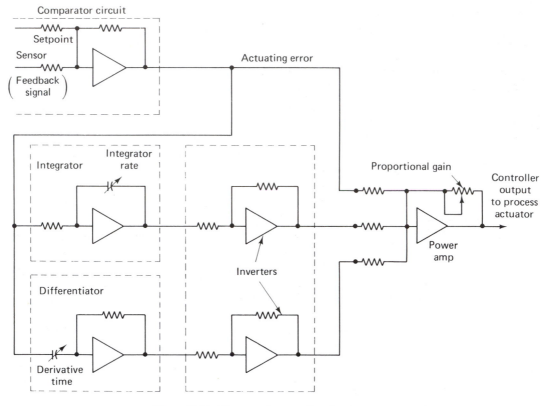

Figure 11-11 Three-mode controller showing separate sections.

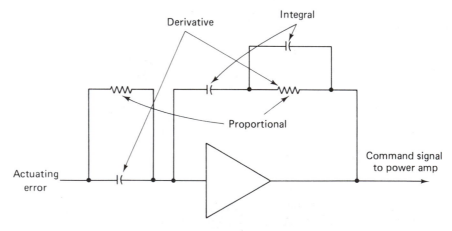

Figure 11-12 Three-mode controller using a single amplifier.

single controllers acting independently. The following discussion deals specifically with application of analog control techniques previously described, but in other forms and combinations.

Figure 11-13 illustrates a process wherein the process itself can be broken down into several identifiable subprocesses (or steps), each of which is under the control of independent controllers.

Controller 1 is illustrated as maintaining conventional single-loop feedback control over the first step. The output of this step is measured and provided as feedback for control over the output product from this portion of the system.

Controller 2 is associated (only) with control over the output of product from the second step; it is connected differently, though. Here the setpoint is compared to the *input* of the second step in the system rather than to the output. Very frequently in practical process control systems the control of one portion of the process depends upon conditions occurring *earlier* in the process. This is a single example of *predictive control*. The characteristics of the processing accomplished in the second step of our example system are such that the output can be adequately controlled based upon knowledge of the condition of the product entering this portion of the system.

Predictive control is not used as commonly as regular feedback control; however, there are instances where the variable that needs to be measured for normal closed-loop control is either not available or else no suitable sensor exists for its measurement, and therefore this type of control is necessary. In other instances it

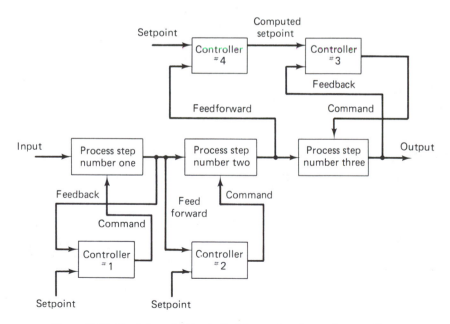

Figure 11-13 Single-loop feedback, feed-forward, and cascade controllers.

is simply the best way to control that portion of a system. The controller used would not differ in theory or construction from the controller used for step 1 (both would be single-, two-, or three-mode controllers, as necessary for the particular application).

Control over step 3 of this process uses *cascade control*. Here the setpoint for the conventional single-loop feedback controller 3 (connected as for step 1 of this process) is calculated by a separate controller (4). Both controllers would be similar to the ones described earlier in this chapter, each having the necessary modes of control. Control over the final end product from step 3 requires not only feedback from the output of step 3, but also some knowledge of the input to step 3 (in this example). The devices that actually control the processing in step 3 cannot adequately exercise control based only on feedback from the output of that step; therefore, it has been necessary to add the second controller (4), which calculates the continually changing proper setpoint for controller 3 to handle process disturbances. It is a combination of *feed-forward (predictive) control* with conventional feedback control. Figure 11-14 illustrates the analog control instrumentation for a typical industrial process, along with a schematic diagram of the process itself.

Figure 11-14 Control panel for saturated gas concentration unit. A typical process control panel consisting of an assortment of controllers, recorders, indicators, and controls necessary for the operator to control the process and monitor its condition. (Courtesy of Foxboro Company.)

DIGITAL METHODS OF CONTROL

When getting into the area of digital control, several problems immediately arise. The problem of resolution and conversion of analog signals will be discussed in Chapter 13. However, the problem of how frequently to sample a given signal must be discussed now. The series of digital samples taken by an *analog-to-digital converter* bear little resemblance to the original analog signal. How valid the digital representation is depends on how well it can be used to reconstruct the original signal.

Very sophisticated theory has been developed to address the problems of digital controller design using a finite number of samples of the various analog signals whose intelligence is required for the controller to make competent control decisions. This theory is well beyond the scope of this text; however, generally acceptable control can be achieved with samples of the analog signal taken at a frequency that is 5-10 times the rate of the highest frequency component of interest in the original analog signal.

Digital computers can be programmed to exercise proportional, integral, and derivative control; and when properly done, the process cannot determine whether it is under digital or analog control. In general, for simple control schemes, the digital computer does not offer any significant advantages from a control systems point of view. The advantage of digital control generally lies in its economics in larger systems, not small ones.

In implementing analog control, a separate analog controller is required for *each* control function; 100 control functions would require 100 separate controllers. With digital systems the computer *appears* to be connected to each control loop continuously, where, in fact, it is operating at a high speed and is sequentially dealing with each loop one at a time. Therefore, only one computer is necessary for the entire 100 control functions. The break-even point for total cost of equipment purchased is approximately 10 closed-loop functions; in general, above this number of loops, digital control is more economical; below, analog is more economical.

This sounds like a simple way to make a decision as to whether to implement a process with digital or analog control, however, the decision is far more complicated than simply a comparison of the initial hardware costs. Other factors that must be considered are how frequently changes must be made to the system (it is easier to alter a digital program than to rewire controllers), what plans there are for future expansion (it is generally easier and less expensive to expand the digital system), what other functions could be accomplished in the *digital computer* that would result in savings (i.e., production scheduling and control, inventory, payroll, etc.), whether a backup automatic method of control is necessary (implement primary control in the digital computer with backup automatic analog controllers), whether there are limitations in the process itself that would make either method more practical, whether there is a digital computer program already in existence that would help reduce programming costs in this application, and many other considerations.

Besides the basic PID modes of control, predictive and cascade control are easily implemented in digital computers. In fact, any combination of any modes of control discussed thus far is capable of being implemented in digital computers. However, many additional control schemes are available through the proper application of digital process controllers.

Any combination of digital combined with individual analog controllers is possible. In some instances, the process controllers are all analog controllers and the digital computer primarily monitors and reports on the performance of the independent analog controllers; this is called *data acquisition* or *data logging,* and the computer itself does no actual control. In other applications the digital computer monitors the performance of the analog controllers, computes the setpoints necessary for each controller, and subsequently provides setpoints to the various analog controllers; this is called *setpoint* or *supervisory control.*

In still other applications, there *may* be a full complement of independent analog controllers installed; however, the controllers simply follow the process variables, and their command outputs are not connected to the process. In this case the entire feedback control job is accomplished directly by the digital computer; this is called *direct digital control* (DDC). If analog controllers are installed, they are there for *automatic backup control* in the event of failure of the primary controller, the digital computer.

In many systems certain portions are under analog control while other por-

Figure 11-15 Control room for a modern industrial process control system. (Courtesy of Leeds & Northrup Co.)

tions are simultaneously under direct digital control. In these systems there normally are channels of communications between the analog and digital portions of the systems; these systems are referred to as *hybrid systems*.

Since digital computers have such fantastic computational capabilities, they are capable of being programmed such that they alter the system gains and other parameters based upon past history of the product manufactured; these systems are called *adaptive systems*. Furthermore, very sophisticated control system *optimization techniques* can be implemented in digital computer control schemes, whereas this is not possible with analog systems.

In the past few pages we have been discussing the theory and reasoning behind the various techniques used in automatic process control. Figure 11-15 illustrates an industrial control system which uses many of the techniques discussed in this chapter. The next chapter will be devoted to hardware that operates based upon these principles.

QUESTIONS

1. In your own words, what is a controller?
2. Why are controllers used?
3. What is the primary difference between open-loop and closed-loop control systems?
4. What is a proportional-only controller?
5. What does the proportional-only controller consist of (functionally)?
6. Describe the effect on a system of a proportional-only controller.
7. What limits the degree of control achievable when using a proportional-only controller?
8. Explain the general relationship between system stability and the controller gain.
9. Why would an integrator be added to a proportional controller?
10. What is the system effect of adding an integrator to a controller?
11. What is the primary purpose of adding the derivative control mode to a controller?
12. What effect does the derivative control mode have on a system?
13. Describe feed-forward control.
14. Describe cascade (or setpoint) control.
15. What system parameter normally limits the gain for a controller in any system?
16. What does *tuning a controller* mean?
17. What does the term *direct digital control* mean?
18. Which of the following parameters is dominant in a given system in determining the control system characteristics?
    A. System capacitance.
    B. System resistance.
    C. System voltage.
    D. System current.

19. The three basic modes of control are _____.

20. Which of the following *modes* of control is considered to be the basic mode (i.e., always used)?
  - A. Direct digital central.
  - B. Proportional.
  - C. Integral.
  - D. Derivative.
  - E. Pre-act.

21. "Tuning" a controller refers to
  - A. Selecting the proper controller for the process.
  - B. Adjusting the gains for each mode of control.
  - C. Programming a drum sequencer.
  - D. Selecting the sampling frequency for a direct-digital-control system.

22. Cascade control
  - A. Refers to one controller in parallel with another for added output.
  - B. Refers to one controller receiving its set point from the output of another controller.
  - C. Can be implemented only with analog controllers.
  - D. Can be implemented only with digital controllers.
  - E. Must use three-mode controllers.

23. Which controller type combines quick response with reduced overshoot and instability but suffers from noise problems and steady-state errors?

24. Which mode of control is specifically designed into controllers in order to improve transient response (response to sudden disturbances)?

25. Which mode of control is specifically designed into controllers in order to deal with the long-term or steady-state offset (error)?

26. In a given controller increasing the gain will generally result in improved performance. What limits the maximum practical gain settings?
  - A. Whether the controller is electronic or pneumatic.
  - B. Whether the controller is analog or digital.
  - C. The system transmitters.
  - D. System stability considerations.
  - E. There are no practical limits to the gain.

27. In a low-gain proportional-only controller, increasing the system gain:
  - A. Reduces the time to reach the desired value.
  - B. Reduces the steady-state error.
  - C. May cause overshoot.
  - D. May cause oscillation (instability).
  - E. All of the above.

28. Which of the modes of control would you expect to be included in a controller (not necessarily the only mode) to exercise acceptable control over a process subject to frequent, fast-acting disturbances?

29. Which of the modes of control would you expect to be included in a controller (not necessarily the only mode) to exercise acceptable control for a process subject to occasional, slowly acting disturbances?

30. Which controller type is also known as a rate controller?

31. Which controller type is also known as a reset controller?

32. The basic difference between feed-forward and feedback control, when using single-loop conventional analog controllers, is in reference to where the feedback (process variable) signal is measured relative to the servomechanism. True or false?

GLOSSARY

Actuating Error Signal   *See* Error.

Analog-to-Digital (A/D) Converter   A device designed to accept a continuous (analog) quantity at its input and to convert each analog value (within its resolution capability) to a unique digital code.

Capacitance   In any energy system, capacitance is the natural ability of the system to statically store energy. Appendix A goes into the phenomenon of capacitance in various energy systems in detail.

Cascade Control   A control strategy where the output of one conventional analog controller provides the setpoint to a second similar controller, which is the one actually exercising control.

Controller   The hardware piece of equipment (pneumatic, electronic, and/or mechanical) which is the practical result of applying automatic control systems theory to industrial process control problems. Its function is to maintain a process control variable at a predetermined value by comparing its existing value to the desired value and controlling an actuator in response to the difference between them in order to reduce that difference to the smallest value within its capability.

Derivative Control   A control mode which only provides an output that is related to the *time* rate-of-change of the difference between the instantaneous value of the process variable and its setpoint. Always used in combination with proportional control. Its primary effect on a process control system is to improve the *time* response characteristics to a process disturbance, thereby reducing the maximum error that would occur without this mode.

Digital Computer (Binary Digital Computer)   A computer which operates on data (information) represented by combinations of binary digits. Normally, it has a stored program of steps which direct the computer to perform a series of operations on that data. Chapter 14 describes digital computer operations in detail.

Direct Digital Control (DDC)   A control strategy wherein a process control digital computer is in complete control of the process directly. It receives inputs from sensors, executes control programs and directly controls the system actuators.

Error (Actuating Error Signal)   As used herein, it is the algebraic difference between the desired value for a process variable (its setpoint) and its actual instantaneous value as measured by a sensor and provided to the controller as negative feedback.

Hybrid Systems   As used herein, this refers to process control systems where some portions of the system are under the control of a digital computer, while other portions of the same system are simultaneously under the control of analog controllers (or an analog computer). There are channels of communication provided between these various controllers.

Integral Control   A control mode which only provides an output that is related to the time integral of the difference between the instantaneous value of the process variable and its setpoint. Always used in combination with proportional control. Its primary effect on a process control system is to act on long-term (or steady-state) errors in order to reduce them to zero.

Off-Line   The opposite of on-line.

On-Line   A term used to describe the state of operation of equipment or systems when they are actively connected into and working with a system.

Predictive Control   A situation where the process controller receives its "feedback" signal from a point in the process prior to the point where the system exercises control over that variable. Therefore, the controller actually receives a feed forward signal and must exercise control while never receiving any information concerning the quality of control actually realized.

Proportional Control   The most basic control mode in which the output is a linearly related function of the difference between the instantaneous value of the process variable and its setpoint.

Proportional Plus Derivative (PD) Controller   A controller which provides an output which is a linear combination of proportional control and derivative control, thereby combining the advantages of both control modes into a single system.

Proportional Plus Integral (PI) Controller   A controller which provides an output that is a linear combination of proportional control and integral control, thereby combining the advantages of both control modes into a single system.

Proportional Plus Integral Plus Derivative (PID) Controller   A controller which provides an output that is a linear combination of proportional control, integral control, and derivative control; thereby combining the advantages of all three into a single system. The PID control strategy approaches the theoretically ideal control system.

Rate Control   Common industrial name for derivative control.

Reset Control   Common industrial name for integral control.

Setpoint   The desired value at which a process variable is to be controlled.

Stability   As used herein to describe process control system performance, it refers to the capability of the system to attain steady-state control of a process variable once it has responded to a process load change or a setpoint change.

Supervisory Control   A control strategy where the process control computer performs system control calculations and provides its output to the setpoint inputs of (normally) conventional analog controllers. These analog controllers actually control the process actuators, not the main-control computer.

Three-Mode Control   Control strategy which uses all three basic modes of control simultaneously (PID control).

Two-Mode Control   Control strategy which uses a combination of proportional control plus either integral control (PI control) or derivative control (PD control).

CONTROLLER HARDWARE

ANALOG ELECTRIC CONTROLLERS

There is a wide variety of competitive electronic analog process controllers currently used in industry. The devices being referred to here all have the capability of exercising *automatic control* over a single process loop. They all come with the proportional mode of control basically built in; integral and derivative control modes are also available in most models. They are all basically designed to work with ISA standard signal levels for input and output and most have a transmitter built into the output portion. They are all designed to give the system operator the capability of either controlling the process manually through controls on the controller front panel or of switching the controller into an automatic mode, wherein it automatically controls the loop. Many controllers have an additional operating mode available, the computer mode, which permits the controller either to receive its setpoint directly from a digital control computer and/or puts the controller in a standby (backup) tracking mode while the computer itself exercises direct digital control.

All these controllers are designed to be used in any type of control loop. Therefore, the various meters and scales on their front panels are normally calibrated to read 0 to 100% of signal variation (other scales can be specially ordered). It is up to the system designer to place signs near each controller specifying exactly what type of variable is being controlled (i.e., temperatures, flow, level, pressure, etc.) and the range and span of the controlled variable. It is very important to realize that there is no necessity for the controllers themselves to differ in construction from one application to another, since the primary difference between applications will be the relative gains used in the three modes of control (if, in fact, all three are even used). Therefore, all the controllers have convenient calibrated gain-control adjustments for each of the three modes: proportional (P), integral (I), and derivative (D).

All the controllers have front panel controls for entering the loop's setpoint (in per cent of full scale). This is the setpoint at which the controller will strive to maintain the process variable when in automatic control. They all have an additional control on the front panel which gives the operator direct manual control over the command signal that leaves the controller and actually causes an adjustment in the process.

The *manual control* mode is normally used during process startup and during process shutdown. It is also used in the event of controller failure to permit the process to continue operation with that particular loop under manual control while the failure is repaired; thereby complete process shutdown may be avoided for an individual controller failure. This gives rise to the need for being able to transfer control from manual to automatic, and back again, without causing a process upset.

Since, in all controllers, the manual and automatic signals are generated by different circuits within the controller, it is necessary for the operator to make them equal (balance them) prior to switching from automatic to manual or vice versa. If the two signals were not equal at the instant of switching, it would cause a step change, an instantaneous change in command output, which would cause an undesirable change in the process. Therefore, all the controllers have provision for the operator to view the manual output as compared to the automatic value of command output and a means of balancing (matching) the two before switching modes of operation. Some controllers actually are wired internally to automatically have the outputs balanced without any specific operator action required.

Therefore, the need has been established for five basic indications or displays on the front panel of all controllers. There is a need to display the setpoint, the actual instantaneous value of the process variable, the controller *command output* signal, the controller's error signal (deviation), and the difference between the command output as generated by the controller and as generated by the manual control circuit (commonly labeled *balance*). In addition to these displays, there must be manual controls for insertion of the setpoint, gain control for each of the three modes of control, manual command output signal control, and a means of switching between computer (or cascade), automatic, and manual control.

Now that the general requirements for all controllers have been presented, let us look at a couple of representative techniques used for implementing these requirements in hardware. There are many different manufacturers of this type of hardware; only a couple of the more common techniques will be discussed here. Figure 12-1 illustrates the general controller whose functional requirements have been discussed thus far.

Since there is a simple relationship among setpoint, process variable, and deviation (which is the actuating error signal or difference between the two), normally only two of the three are individually displayed, the third being implied by comparison of the two that are indicated. Different manufacturers of controllers have

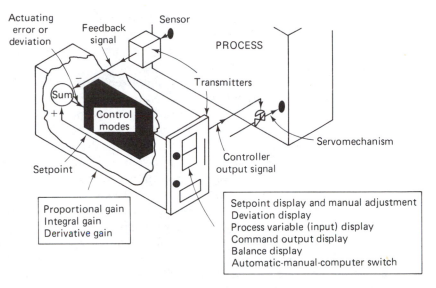

Figure 12-1 General controller.

chosen different techniques to display setpoint, process variable, and deviation, each of which has specific advantages and disadvantages.

One technique that seems to be widely accepted in modern controllers is illustrated in Fig. 12-2. It consists of a relatively long tape (approximately 6-10 in. in length) wound around spools such that only a small portion of its length is visible through a fixed window on the front of the controller. The tape is marked in 0 to

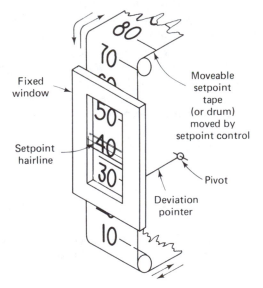

Figure 12-2 Centered expanded scale.

100% markings and its position is controlled manually by the operator. It is, in effect, the indicator for the setpoint control knob for the controller. The operator adjusts the setpoint knob until the desired reading on the movable tape lines up under the setpoint hairline on the fixed window.

Also under the fixed window, but completely independent of the tape, is the pointer for a meter that is wired to respond to the error signal in the controller; it is called the *deviation pointer*. When the deviation pointer aligns with the setpoint hairline, the process variable is at its setpoint. When there is a difference between the setpoint and the controlled process variable, the deviation meter pointer will be at a position other than under the setpoint hairline, indicating the magnitude and direction of the error (deviation). The deviation pointer is calibrated to the units on the movable tape such that its position over the movable tape indicates the actual value of the process variable (normally in per cent of full scale). Therefore, the three displays are combined into a single, very readable display.

A second very common display is illustrated in Fig. 12-3. It consists of a small (2- to 4-in.) fixed display, again with 0 to 100% markings. There are two movable pointers on this scale, one being manually controlled by the setpoint control knob and the other being the pointer of a meter that is calibrated to indicate the instantaneous value of the process variable. The difference between the two pointers is therefore the implied deviation. Again a very simple and readable display combines the three variable displays into a single display.

The requirements for displays of the controller command output and balance are normally combined into a single display. There are a great many variations in these displays from one manufacturer to another; therefore, only a single representative one will be discussed. Most controllers have a small (approximately 2 in.) meter which indicates controller output under all modes of control. The meter may have a single pointer which indicates either manual command output signal or controller automatic command output signal, depending upon the position of a selector switch, or it may have two independent pointers, each permanently connected to one specific command output signal source.

In the latter type of display (dual pointers), *balance* is achieved by manual manipulation of either the setpoint control (if the system is in manual and is going to be transferred to automatic) or manual manipulation of the manual control (if going from auto to manual) until the difference between the two pointers goes to zero. Once the balance is obtained, there will be no serious process disturbance created by switching from one mode to another.

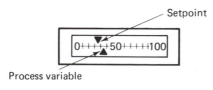

Figure 12-3 Typical dual moving pointer meter display.

In the single pointer display, there will be a manual switch associated with the display in order to alternately display the command outputs due to the automatic and manual circuits in the controller (Fig. 12-4). The switch is simply used to alternately display the two outputs until they are adjusted (as noted previously) to be the same and transfer from one mode to the other can be safely effected.

This brief discussion has presented the various features required of all controllers and how they typically appear in hardware designs. Figure 12-5 illustrates the front panel on several commercially available electronic analog controllers. Several of these models have the manual switches behind cover panels which swing out to permit the operator access to them. All the models have the gain control switches hidden behind the front panel so that they are not easily accessible, since they are normally adjusted when the system is installed and may never again require readjustment.

There is (at least) one additional type of analog controller that is used throughout industry and which shall be discussed here. This particular type of controller does not have three-mode capabilities as do the analog controllers previously discussed; however, it does provide a convenient means of obtaining nonlinear control of a single variable, and it is used to provide a nonlinear setpoint to the other types of analog controllers discussed.

Figure 12-6 illustrates a modern commercial *cam-program controller.* The shape of the cam itself is determined by the user, and the cam follower simply follows the shape of the cam as it is rotated by a timing mechanism. A position sensor connected to the cam follower transduces its position to an electrical signal that is the time-varying setpoint signal for an analog controller.

A typical application of this type of controller is temperature control of an industrial oven, where the required time-temperature relationship, as determined by the user, is cut into the cam. The output from the cam controller would provide the setpoint to a typical one-, two-, or three-mode controller of the types previously discussed.

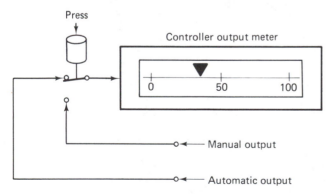

Figure 12-4 To balance automatic and manual outputs with this method, the operator repeatedly presses the button and adjusts either the manual output or the setpoint until the pointer stops moving.

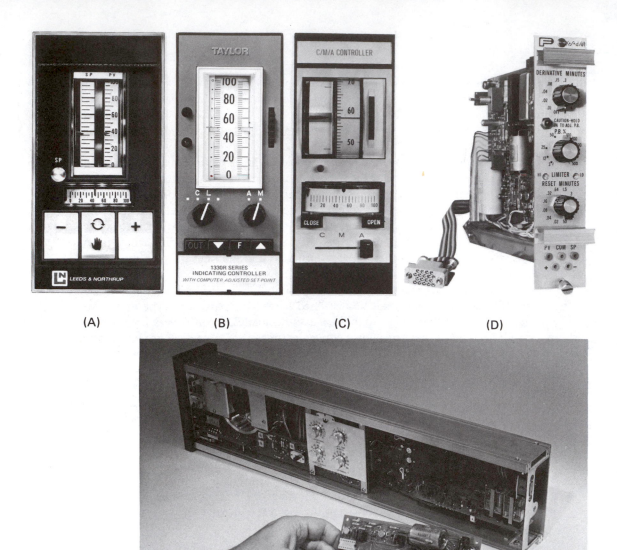

Figure 12-5 Commercial electronic analog (single-loop) controllers: (A) local automatic–manual indicating controller; (B) remote setpoint indicating controller; (C) direct digital control (backup) indicating controller; (D) three-mode gain control module from a controller; (E) rear view of typical controller illustrating general shape and the internal electronics. [(A) and (E) Courtesy of Leeds & Northrup Co.; (B) courtesy of Taylor Instrument Process Control Division/SYBRON Corp.; (C) courtesy of Bailey Meter Company; (D) courtesy of Fischer & Porter Co.]

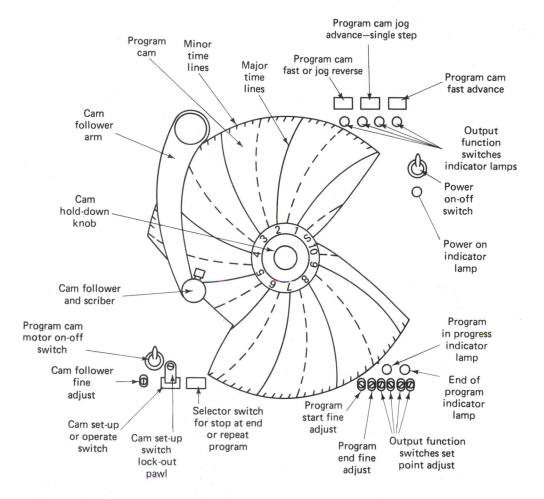

Program cam

Minor time lines

Major time lines

Program cam jog advance—single step

Program cam fast or jog reverse

Program cam fast advance

Cam follower arm

Output function switches indicator lamps

Power on-off switch

Cam hold-down knob

Power on indicator lamp

Cam follower and scriber

Program cam motor on-off switch

Cam follower fine adjust

Cam set-up or operate switch

Cam set-up switch lock-out pawl

Selector switch for stop at end or repeat program

Program start fine adjust

Program end fine adjust

Program in progress indicator lamp

End of program indicator lamp

Output function switches set point adjust

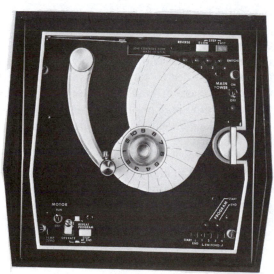

Figure 12-6 Solid-state cam program controller. Provides a nonlinear setpoint to a conventional analog controller or can be used as a single-mode controller. (Courtesy of Love Controls Corp.)

DIGITAL CONTROLLERS

Digital controllers range in complexity from simple relay circuits through process control computers. We shall discuss them here starting with the simplest and progressing through the various categories of digital controllers as they increase in complexity.

Originally, digital controllers consisted of banks of electromechanical relays, timing devices, and motor-operated cam switches. These were connected together primarily for control of large production machines where the only control necessary was capable of being achieved by switching controllers. The only control necessary was sequential and required no control system computations. The control strategy was hardwired in and very difficult to change, and the outputs were simply on-off.

For some applications, special devices called *meter relays* were used, either in conjunction with the other devices mentioned, or, in many cases, as stand-alone controllers themselves. The meter relay was an adaptation of the conventional D'Arsonval meter movement, where the pointer was part of a separate electrical circuit and (frequently) had contacts mounted on it. As the indicator moved across the face of the meter, it would come into physical contact with one or two other pointers (Fig. 12-7). The positions of these fixed pointers were manually preset by control knobs on the face of the meter, and a different electrical circuit would be completed when the movable pointer came into contact with each fixed pointer.

The position of the fixed pointer(s) would correspond to upper or lower (or both) limits of control for the variable being monitored by the meter movement. Completion of the appropriate electrical circuit, when the movable indicator came into contact with one of the fixed pointers, would then be used to simply cause alarms or even initiate corrective control actions to bring the variable back into acceptable limits. In modern meter relays the contacts are magnetically or optically operated, rather than mechanically.

As the relay logic systems became more complicated, devices called *drum sequencers* or *drum controllers* were developed. Essentially, they consisted of a cylindrical drumlike surface which rotated around its axis based upon time (Fig. 12-8). Fixed very close to (but not touching) the rotating surface of the drum were located a number of spring-loaded switch contacts. In order to cause the switches to be activated in any sequence, small protuberances were affixed to the surface of the drum such that when the protuberance was rotated into position under the appropriate (fixed) spring-loaded switch, it would make contact with the switch and cause it to be activated. The protuberances could be affixed next to each other such that the duration of activation of each switch could also be controlled.

Therefore, the drum controllers could be used in many applications to replace relay banks. The drum itself was programmed by affixing the protuberances such that each of the switches would be operated in the proper sequence, and for the proper duration, to achieve the desired control. The drum program could be added to or altered relatively easily simply by changing the patterns of plug-in protuberances.

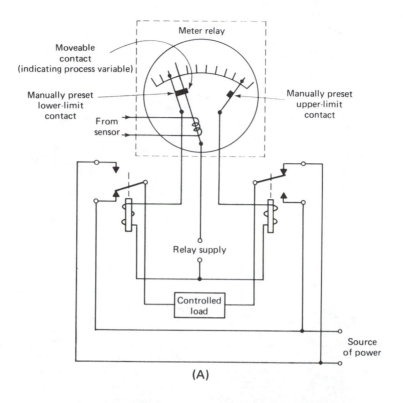

(A)

(B)

Figure 12-7 (A) Meter relay (employed as a high limit–low limit controller); (B) dual-setpoint meter relay (indicating controller). [(B) Courtesy of Jewell Electrical Instruments, Inc.]

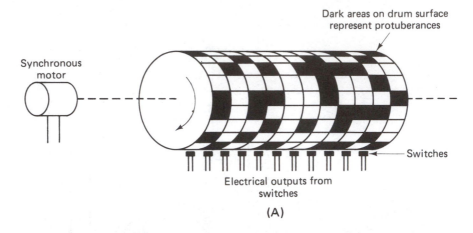

(B)

Figure 12-8 (A) Drum controller; (B) commercial sequence controller. [(B) Courtesy of Eagle Signal Division, Gulf & Western Manufacturing Co.]

In the late 1960s a newly designed solid-state controller was introduced into the process control market. It was aimed at replacing relay-logic control and also some of those applications where drum sequencers had been used. But it went much further than either of these two types of controllers were capable of going. This new controller, called a *programmable logic controller* (PLC) was a hybrid device with capabilities far in excess of either the drum-type controllers or relay logic controllers, but less than the digital computers, which will be discussed later in this chapter.

The *programmable logic controller* is functionally divided into four parts: the input, the output, the logic unit, and the memory unit (Fig. 12-9). The older PLC input units accepted only electrical on-off (switch)-type contact voltages. It electrically isolates and conditions them to be compatible with the digital logic voltage levels used in the logic unit. The output unit does just the opposite; it accepts digital logic voltage levels at its input and provides electronically isolated switch contact action at its ouput.

These two units interface with a process exactly like the sequencing devices and the relay-bank controllers (on-off switching) except that the interface for the programmable controller is normally solid-state; the elimination of mechanical contacts results in more rapid operation and longer life expectancy. Some versions do use mechanical reed switches in their input and output interfaces; however, most programmable controllers have eliminated even these mechanical contacts.

The primary difference between programmable controllers and their electromechanical predecessors lies in the internal workings of the logic unit and memory. The logic unit is implemented with standard digital logic devices and is wired to perform all the various functions that could be performed by sequencing controllers or relays. Figure 12-11 is a listing of the functions performed by a typical industrial programmable logic controller. On many controllers, each of the functions is described both in terms of relay logic symbolism and also in terms of digital logic symbols. This makes it very easy for engineers and technicians already familiar with relay logic and drum sequencers to program these controllers without having to learn computer programming.

In addition to the various functions that would be performed by relays and sequencing devices, since the logic unit is implemented with standard digital logic, any other functions that can be performed by any digital device can also be wired into programmable controllers. Therefore, programmable controllers may have capability far in excess of the earlier devices, and the limit of the wired-in capability is restricted by cost, not by available hardware.

Some of these additional capabilities are shift registers, up/down counters, digital timers, addition and subtraction, and large memory capability. In fact, mod-

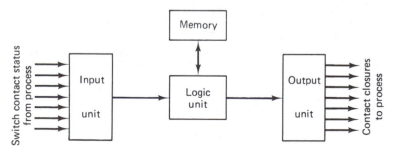

Figure 12-9 Programmable logic controller (PLC) block diagram.

ern PLCs include analog interface hardware in both the input and output units. Furthermore, the internal logic in some of these PLCs can implement 3-mode analog control. Therefore, these devices, in capability and complexity, approach the performance of small digital computers; in fact, they can be correctly described as special-purpose digital computers. Some new PLCs actually use microcomputers for the logic unit. Figure 12-10 is an illustration of such a device.

One of the primary reasons for the invention of these devices lies in the manner in which they are programmed. For sequencing devices and relay-bank controllers,

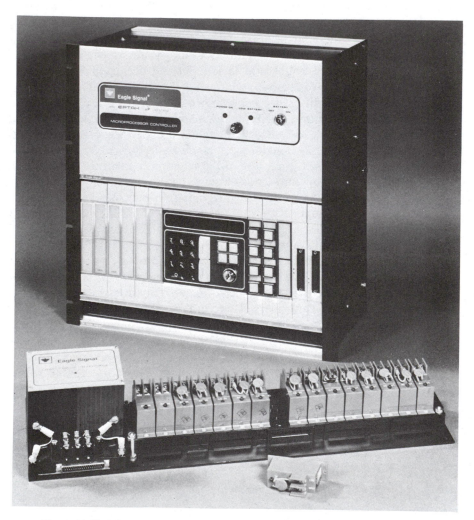

Figure 12-10 Microprocessor PLC controller. This model includes analog as well as digital capability. (Courtesy of Eagle Signal Division, Gulf & Western Manufacturing Company.)

the control strategy is hardwire-programmed in. Alterations and additions to the system capability require rewiring of the system. For programmable controllers, the control strategy is programmed in by a series of computer-type instructions. The sequence of instructions is determined by the system designers, entered through a keyboard (Fig. 12-11C) on the controller, and the entire sequence is transferred into a computer-type memory unit. The controller logic then retrieves these instructions (sequentially) from its memory and causes the coded operations (instructions) to be executed. The instructions themselves cause the logic to inspect the conditions of the switch contacts on the input, make logical decisions based upon the status of the input contacts, and cause output contacts to be operated based upon both the status of the input contacts and the logical decisions programmed in.

To modify or entirely change the sequence of instructions, the only operation necessary is to change the instructions in memory. Therefore, with the programmable logic controller, there are essentially no hardware modifications necessary, only programming changes. This is the reason they were developed. In the automotive industry the annual design changes required complete system and controller rewiring each year, using relay banks and sequence controllers. The programmable logic controllers were specifically designed to eliminate the rewiring expenses; here the only change necessary is a new computer program to be written and put into the controller's memory.

Since the programmable logic controllers are basically limited-capability special-purpose digital computers, they can be easily maintained and repaired using programmed troubleshooting routines called *diagnostic programs*. Also, since the logic unit is already implemented in digital logic, any interfacing required to larger digital computers is readily accomplished.

Digital Computers

The next step up the ladder of increasing capability is normally the top rung: *digital computers*. Digital computers as defined in this text include all general-purpose digital computational devices. They come in "micro," "mini," "medium," and "larger" sizes. However, they all operate basically the same.

What are the differences between the four sizes? In general, there are several basic differences; however, the dividing lines between the sizes are often very vague. The only sizes of interest in process control are the *microcomputers, minicomputers,* and *medium-size computers. Microcomputers* are the newest and least capable of the three sizes insofar as computing power, computational speed, and interfacing capability go. They are essentially designed such that the *central processing unit (CPU),* the unit that controls all the computer's operations, is physically located on a small number of integrated circuit chips (usually one). Figure 12-12 illustrates one such microcomputer.

The computer *word size,* which is the number of binary bits simultaneously operated upon by each computer instruction, is 8–16 bits (32 bits in the newer models). The cost for this central processing unit can be down to less than $5 when

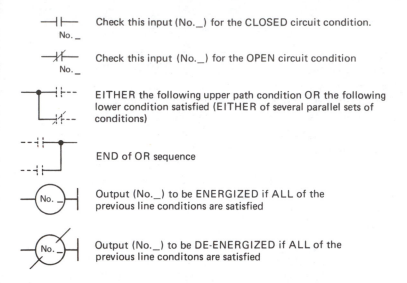

—| |—   Check this input (No._) for the CLOSED circuit condition.
No._

—|/|—   Check this input (No._) for the OPEN circuit condition
No._

EITHER the following upper path condition OR the following lower condition satisfied (EITHER of several parallel sets of conditions)

END of OR sequence

Output (No._) to be ENERGIZED if ALL of the previous line conditions are satisfied

Output (No._) to be DE-ENERGIZED if ALL of the previous line conditons are satisfied

(A)

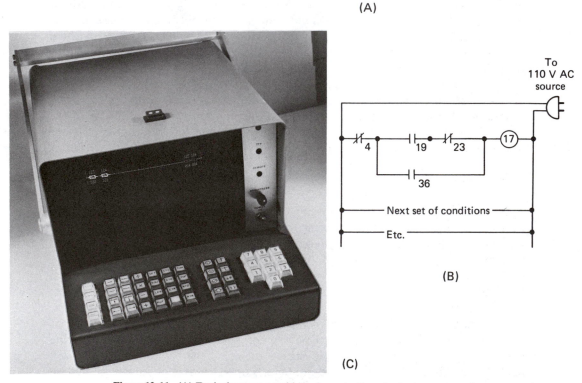

(B)

(C)

Figure 12-11 (A) Typical programmable logic controller relay logic programming commands; (B) example of PLC relay logic program: *energize* output 17 if input 4 is *open* and *either* input 19 is closed and input 23 is *open or* input 36 is *closed*; (C) the programming panel for a PLC, it connects to the PLC and serves to program the memory and to monitor the system operation. This model is capable of relay logic implementation, dealing with analog inputs and outputs, data manipulations, and a very flexible selection of timing and counting functions. (Courtesy of Allen-Bradley Co.)

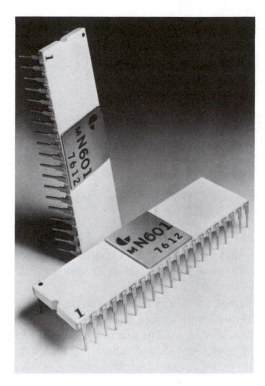

Figure 12-12 Central processing unit used in the Data General microNOVA computer family. It is a 16-bit microprocessor on a single NMOS chip, and is of the type used in process control systems. (Courtesy of Data General Corp.)

purchased in large quantities. Its primary application is as a direct replacement for hardwired binary digital logic.

The *minicomputer* is a general-purpose digital computer whose central processing unit is made up primarily of a large number of small- and medium-scale integrated circuits. Therefore, the central processing unit is physically much larger than the microcomputer's CPU but has much more capability than the microcomputer. The minicomputer normally has a 16-bit word length, although some are available in 32-bit word lengths. The cost for the minicomputer's CPU was originally (in the mid-1960s) less than $20,000; however, now the cost is closer to $1000. Figure 12-13 illustrates a commercial minicomputer.

A *medium-sized computer* is normally one with a very large input/output capability, extended computational capabilities, extended word lengths (18- to 64-bit words), and increased computational speed as compared to the minicomputers. The cost for these larger computer CPUs runs hundreds of times as much as for minicomputers; therefore, their use is economically justified only in very large process control systems.

The bulk of modern digital process control is being implemented with minicomputers. They are not exceedingly fast computers (in the area of $\frac{1}{10}$- to 1-μs basic instruction execution times). However, the speed and word lengths (16 bits) seem to be ideally suited to process control. These devices have a capability in excess of

Figure 12-13   Commercial minicomputers of the type typically used in process control systems. (Courtesy of Hewlett-Packard.)

64,000 words of primary (RAM) memory, which is normally more than adequate. They excel in their ability to interface to processes, with very large design flexibility built in.

The process control interfacing requirements have resulted in the inclusion of several functions which were originally process-control-oriented designs and have spread out to other computer applications. The primary examples of these functions are priority-interrupt capability and real-time clocks. They also have the ability to interface in a special way with certain types of backup or auxiliary memory devices. This ability is called *direct-memory access*.

Detailed discussion of their interfacing requirements and operation in processes is left as the subject for Chapters 14 and 15. This chapter will continue with other types of commonly used controllers, mostly nonelectronic types.

PNEUMATIC CONTROLLERS

Pneumatic controllers, controllers that operate on air pressure (only), are widely accepted and used in industry. They were exclusively used prior to the advent of solid-state electronic controllers. In many instances, they have now been replaced by more capable and maintenance-free modern electronic controllers; however, they are still used in *many* applications where electronic controllers, for any of several reasons, are not acceptable. Possibly the best advantage of penumatic controllers over electronic controllers is in acquisition cost, which is approximately a 2:1 advantage.

Pneumatic controllers are made in on-off (digital) models and in analog models. The analog models have one, two, and three modes of control available as described for the electronic controllers. The on-off pneumatic controller, using the *flapper nozzle,* is simpler to explain, so it will be presented first. Figure 12-14 il-

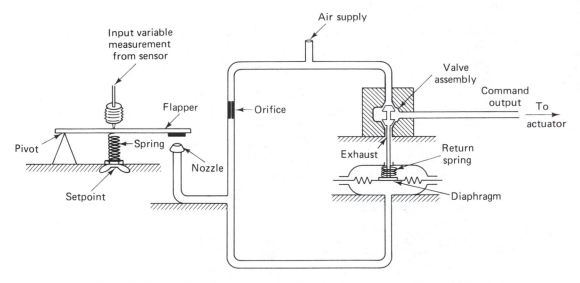

Figure 12-14 On–off pneumatic controller (showing flapper-nozzle and pneumatic relay).

lustrates a pneumatic on–off (flapper nozzle) controller using a pneumatic relay as the output device.

To understand the operation of pneumatic controllers, it is first necessary to understand the operation of the flapper-nozzle assembly (also called a *baffle-nozzle assembly*) and then to describe its effect on the system. The flapper itself can be thought of basically as a hinged simple lever arm. It pivots around one end while the other end moves to either close off or open the opening at the nozzle. Supply air is continually available at the nozzle, and it is permitted to leak out at a rate that is determined by the relative positions of the flapper and the nozzle.

The position of the flapper is partially controlled by the setpoint screw and spring assembly (in this schematic). The setpoint adjustment consists of controlling the spring force against one side of the lever (the lower side in the illustration). The setpoint spring force is balanced by the force exerted on the other side of the lever by the feedback bellows. The pressure that actuates this bellows comes directly from the sensor measuring the process variable that is being controlled. The spring and bellows characteristics are chosen such that when the process variable is at its setpoint the forces exerted by each will balance.

As an illustration of the operation of this device, consider a situation where the process variable is above the setpoint value. The force exerted by the bellows will overcome the setpoint spring force, causing the flapper to move closer to the nozzle, thereby reducing the volume of air leaking out. The reduced leakage causes a pressure increase in the tube supplying the air to the nozzle. If the process variable is below the setpoint, the setpoint spring force will overcome the bellows force and move the flapper away from the nozzle, thereby causing a decrease in the pressure in the tube supplying air to the nozzle.

Gain can be set in this system by moving either the setpoint spring or the bellows assembly in relation to the pivot point. Also, it should be obvious that the setpoint spring and feedback bellows could have been located on the opposite sides of the flapper lever in order to have the system work the opposite way.

Now let us turn our attention to the output portion of this pneumatic on-off controller. The output valve assembly is technically termed a *pneumatic relay*. It consists of a valve assembly in conjunction with a pneumatic diaphragm motor (which will be discussed again in Chapter 16). For the purposes of this discussion the pneumatic motor is a diaphragm that is mechanically coupled to the valve assembly. When the diaphragm experiences pressure on its lower side, the resulting force (force = applied pressure times area of the diaphragm) overcomes the spring force and causes the valve to move up. The upward motion of the valve simultaneously closes off the opening from the air supply *and* opens up an exhaust (or vent). Reduction of pressure on the lower side of the diaphragm will permit the spring to cause the valve assembly to move in the opposite direction.

Connecting the flapper-nozzle assembly to the pneumatic relay assembly, as illustrated, we have the on–off pneumatic controller. Notice the restriction (orifice) in the air supply line leading to the flapper-nozzle and diaphragm motor assembly. This restriction is to limit the maximum quantity of air that can flow to these assemblies. Note that there is no such restriction in the air supply line leading to the pneumatic relay valve.

Let us investigate the overall operation of this controller when the setpoint exceeds the process variable. Under this condition the flapper is moved away from the nozzle, thereby permitting increased leakage through the opening. The orifice in the supply tubing limits the available air flow such that when the flapper-nozzle assembly is in this condition, all the supply air leaks through the flapper nozzle, and there will be minimum pressure in the tubing leading to the *diaphragm motor*. Since there is minimum pressure acting on the diaphragm, the output valve is opened and the exhaust valve is closed. Therefore, there will be maximum pressure felt in the output line.

There would be some sort of pneumatic actuator connected to the output from this controller. The actuator would be installed in the system such as to cause the value of the variable being controlled to increase. The increasing value will be sensed and an increasing signal will be transmitted to the feedback bellows in the controller. Eventually the feedback signal will increase to the point where the force exerted by the feedback bellows will exceed the force exerted by the setpoint spring and the flapper will be moved toward the nozzle, decreasing the leakage rate through the nozzle.

As the leakage through the nozzle is decreased, the air pressure in the tube leading to the nozzle will proportionately increase. This increase in pressure will also be felt by the diaphragm motor, which operates the output valve. The increasing pressure will cause the diaphragm to overcome the spring pressure and move the valve upward (in the diagram). The upward motion will cause the valve assembly to close off the air supply and simultaneously open the exhaust valve. Not only will no more air be admitted to the output tube, but the air already in the tube will be permitted to leak out the now-open exhaust port. Therefore, the pressure in the output tube will drop to a minimum value. This reduced output pressure would be transmitted to the actuator, causing it to respond in such a manner as to effectively cause the controlled variable to decrease in value. Eventually, it will decrease below the setpoint and the process would repeat itself.

The effective gain in this controller is so high that it acts in an on–off fashion, with essentially no proportional control. To make a pneumatic proportional controller, basically a feedback loop is added to the on–off controller, as illustrated in Fig. 12-15. This controller works in most respects as the on–off controller previously described, with the single exception that the nozzle itself no longer remains in a fixed position; it is now free to move in response to the pressure at the output of the controller.

To understand the effect of this (negative) feedback to the nozzle, let us assume that the controller is in a steady state, and then the process is disturbed such as to cause an increase in the value of the controlled (feedback) variable. The resulting increase in pressure in the feedback bellows causes the flapper to move closer to the nozzle. This creates an increase in pressure in the tubing leading to the diaphragm motor, which causes the motor to move the pneumatic relay valve upward. The upward motion causes a reduction in pressure in the controller output exactly as in the on–off controller. However, in this controller, the reduction in output

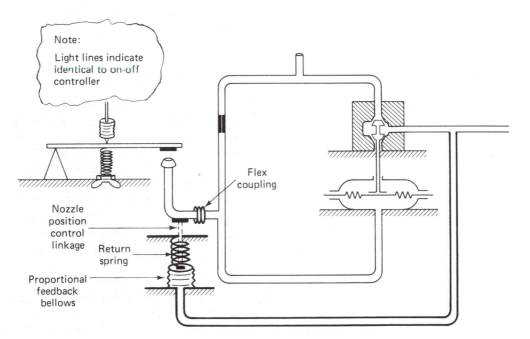

Figure 12-15 Pneumatic proportional (only) controller.

pressure is fed back to the nozzle positioning bellows. The reduction in pressure in this bellows reduces the force it exerts against the spring, and therefore the spring causes the nozzle to be moved downward, away from the flapper. This motion effectively causes a proportional reduction in pressure in the tube that supplies air pressure to the diaphragm motor.

The net result is that the position of the pneumatic relay valve can now be positively controlled at any point between the full-up and full-down positions. Therefore, the air pressure in the output line can also be controlled at any value between maximum and minimum and the controller exerts proportional-only control.

The flapper-nozzle (nozzle-baffle) technique is not the only one possible for the design of pneumatic controllers, but it is at present the most popular technique.

Frequently, pneumatic controllers (Fig. 12-16) look much like electronic controllers. There is the same need for manual setpoint entry; cascade control capability; cascade-auto-manual mode switching with balance capability; gain controls for each mode of control; and displays of setpoint, error (deviation), process variable value, and command output value.

Basically, the electronic and pneumatic controllers are very competitive as far as performance capabilities are concerned. The selection of one over the other is normally based on factors other than performance.

Pneumatic controllers can be designed to include integral (reset) and derivative (rate) control modes in addition to the basic proportional mode. Such two- and three-mode controllers are in common use, but detailed descriptions of their operation are beyond the scope of this text.

FLUIDIC COMPUTING ELEMENTS

The final category of controller to be discussed here is based on the use of *fluidic computing elements*. These elements can be combined to form *fluidic computers,* which are basically digital computers made up of binary logic elements that operate on air flows and pressures rather than electronic currents and voltages. They were developed in the early 1960s just after binary solid-state electronic digital logic was developed to an industrially acceptable level. Therefore, electronic digital logic was well on its way to commercial acceptance prior to the commercial availability of fluidic logic elements. Proponents of *fluidic logic* typically claim this to be one of the primary reasons for the acceptance of electronic logic as compared to fluidic logic. Other reasons include its susceptability to air contamination and difficulty of design as compared to electronic digital logic design.

Fluidic logic elements have one outstanding advantage over comparable electronic logic elements in many heavy industrial applications; they do not operate with electricity; therefore, they have no susceptibility to electrical noise interference. This means that fluidic computers can be physically located in close proximity to the process they are controlling, where the electromagnetic interference in the area would

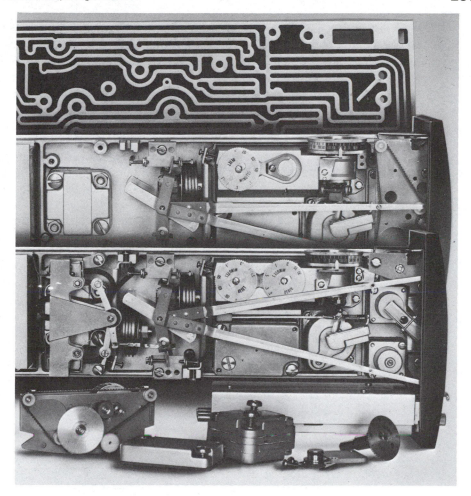

Figure 12-16 Cutaway of pneumatic three-mode controller. (Courtesy of Fischer & Porter Co.)

severely restrict the use of electronic computers. Therefore, fluidic computing elements have found natural application in the control of industrial machine tools.

These tools frequently require only very basic digital control, often requiring only a few logic elements. The user is faced with the choice of using electromechanical relay logic, binary electronic digital logic, or fluidic logic to implement the desired control. Certainly, cost is a major consideration governing which is selected. However, maintenance costs, repairability, reliability, and operational restrictions are frequently more important criteria governing selection. Fluidic logic is very competitive in many of these areas with electronic logic and certainly wins out where a high degree of electrical noise immunity is a requirement.

Some equipment (e.g., stamping machines) needs only very simple hardwired logic to control all its operations, and other equipment (e.g., multiple-axis multiple-function machine tools) requires a great deal of programmable logic. Fluidic logic elements are competitive for the smaller applications, but the more complex, stored-program controller requirements are more economically met using special high-noise-immunity electronic logic.

Fluidic logic elements operate on several principles, a few of which will be presented here. The first is the *turbulence amplifier,* shown schematically in Fig. 12-17, which consists of a nozzle on the left and a collector at the right. Fluid (air or liquid) coming in from the left will be squirted out the nozzle and most of it will enter the collecting tube on the right (Fig. 12-17A), thereby establishing a flow through to the output.

If flow is permitted to squirt out of one of the several (three are illustrated) nozzles located at right angles to this primary flow stream, it will cause such turbulence that relatively little of the main flow stream will pass into the output collector tube, illustrated in Fig. 12-17B. This is therefore a logic gate wherein, with flow entering, there will be *no* flow at the output if there is flow at X or Y or Z or any combination of X, Y, and Z. Therefore, this is a fluidic NOR digital logic element (for high-level logic) if the presence of flow is considered to be a 1.

The second fluidic logic element to be discussed is the *wall attachment device* (WAD), illustrated schematically in Fig. 12-18. As the fluid flow enters the nozzle, it squirts out into the output Y area. If you can picture in your mind the water squirting out of the nozzle on a garden hose, you can appreciate how it spreads out as it leaves the nozzle, as the liquid at the outer extremities of the stream is slowed

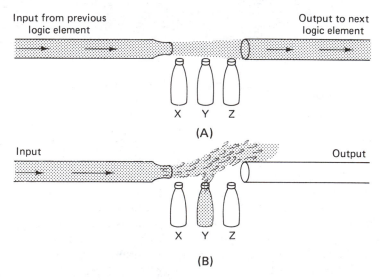

Figure 12-17 Turbulence amplifier.

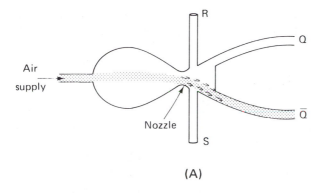

(A)

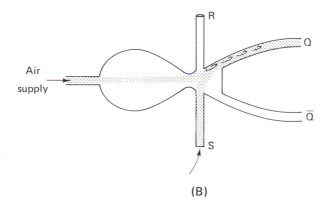

(B)

Figure 12-18 Wall attachment device (W.A.D.).

down as a result of friction with the air. A similar phenomenon occurs as the stream emerges from this nozzle, except that it occurs to an exaggerated degree on one side, the side nearer the top or the bottom walls, whichever the stream happens to be closer to at the time.

For the sake of this example, assume that it is nearer the bottom wall (Fig. 12-18A); therefore, there will be greater turbulence in the restricted area between the stream and the bottom wall than at the top of the stream. There is a basic law of physics which states roughly that where velocity is high, pressure is low, and vice versa. It is also a fact that turbulence in a fluid causes relatively higher velocity; the higher the turbulence, the higher the velocity. Putting these two (rather loosely explained) laws of nature together, it can be seen that there will be an area of higher velocity (and subsequently lower pressure) on the bottom of the stream than on the top. Therefore, the difference in pressure will cause the stream to literally attach itself to the lower wall, passing out the lower collector and not out the upper one. This logic element will remain in this state as long as there is no pressure introduced into the *S* input.

Assuming that pressure is momentarily introduced into the *S* input, it will cause an area of relatively high pressure between the stream and the lower wall which literally "blows" the stream toward the upper wall, where it will attach itself in a manner exactly as it did to the lower wall. The flow stream will now pass out the upper collector tube regardless of whether there is flow maintained through the S input. It will remain in this position until a "pulse" of fluid from the R terminal "blows" it back to the lower wall.

This fluidic logic element is therefore the equivalent of a binary *flip-flop* and it will "remember" the last state it was forced to assume by either the S or R inputs, even though that input subsequently is removed.

The wall-attachment principle is used in the design of hydraulic flow sensors and no-moving-part valves, as well as fluidic-logic elements. However, both of the techniques described have the common disadvantage that they require continuous airflow. Therefore, other techniques, which do not require the continuous flow, have been developed and find common application in the design of fluidic logic elements. These devices are compatible for use with WADs which operate on the same supply pressures.

Only one other of these techniques will be described here: it consists of a series of air passageways which either are blocked by diaphragms, or are not blocked by these same diaphragms, when in use. Figure 12-19 illustrates a cutaway view of a commercial OR element that operates using this switching diaphragm technique. This device will supply air pressure at its output port A if pressure is supplied to either input port X or input port Y (or both).

Referring to Fig. 12-19B, note that in the cutaway the element is split longitudinally into two halves. The left half depicts the position of the internal switching diaphragm when pressure is supplied to port X only, and the blocking portion of the diaphragm is pushed away up from the X port. This will permit the applied air to pass through openings in the outer edge of the diaphragm and to the output port A. The right half of the diagram depicts the position of the diaphragm when pressure is applied to the Y port only. For this condition, the blocking portion of the diaphragm is pushed down so that air can pass through the element and to the output port A.

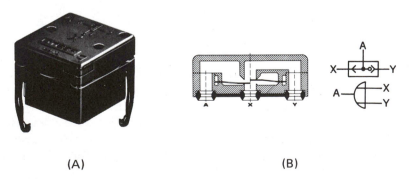

(A)   (B)

Figure 12-19 Commercial pneumatic OR logic element. (Courtesy of Festo Corp.)

If pressure is applied to both X and Y simultaneously, the higher pressure of the two will dominate and that pressure will be felt at the output. Therefore, this element performs the logical *inclusive* OR function for positive pressures.

Note that it does not amplify, it is a passive device. Other logic elements in the family have amplifiers included in them and do not attenuate the pressure from input to output. Other pneumatic logic elements available in this logic family include:

Inhibit elements	Clocks
Diodes	Amplifiers
AND elements	Timers
NOT elements	Card readers
Variable pneumatic capacitors	Paper tape readers
Pneumatic resistors	Switch input devices
Sequencers	Electrical-to-pneumatic devices
Registers	Pneumatic-to-electrical devices
Shift registers	Rotating air motors
Memories	Counters
Pulse generators	and other devices
Single shots	

Figure 12-20 illustrates a pneumatic controller that is a special purpose computer. Pneumatic controllers can be obtained in fixed-program versions, semiprogrammable versions (limited program size), or in fully programmable versions in which the program can be changed by inserting punched paper tapes or punched cards. They find common application throughout industry in controlling the operations of various types of machines.

For the more complicated machine tool control systems, more exotic programmable (not hardwired) controllers are necessary. They are implemented using electronic digital logic more because of the availability of complex functions than due to performance advantages. These machines are programmed to exercise control according to specifications fed into the system (normally) via punched paper tape and, so, are often called *numerically controlled machine tools*. Actually, in the more complicated numerically controlled machine tools the controller can be a mini- or microcomputer.

Ironically, possibly the greatest application of fluidic principles is to the field of valve design. Many valve designs on the market operate essentially on the WAD principle. The result is a fast-acting valve with no moving parts. There are numerous very successful industrial applications of such valves.

This nearly completes our discussion of controllers. One additional type of controller (the digital computer) has extremely complicated interfacing and applications considerations and is therefore discussed separately in Chapter 15.

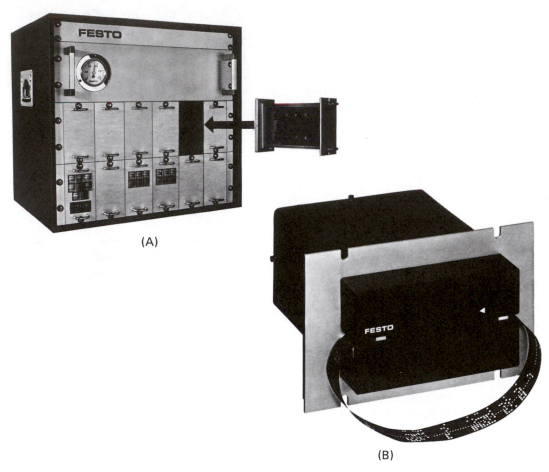

(A)

(B)

Figure 12-20 (A) Pneumatic controller; (B) paper tape input device for programmable pneumatic controller. (Courtesy of Festo Corp.)

QUESTIONS

1. What is the primary function of a controller?
2. In general, what capabilities do all electronic analog controllers have in common?
3. List the basic displays found on electronic analog controllers and the function performed by each.
4. Repeat Question 3 using the basic manual controls instead of displays.
5. How do electronic analog controllers differ in construction when used in different processes?
6. Draw a block diagram of a typical analog control system, illustrating the process, the

controller, the comparator, transmitters normally required, setpoint signal, deviation signal, feedback signal, a sensor, a servo, and the command signal to the process.

7. List the digital controllers referred to in the text from simplest to the most complex.

8. In your own words, what is a programmable controller?

9. How does a programmable controller differ from a drum controller (drum sequencer)?

10. Differentiate among micro-, mini-, and medium-size digital computers.

11. Which type(s) of digital computers would you normally expect to find in process control applications?

12. Explain the operation of a flapper-nozzle assembly as used in pneumatic controllers.

13. What factors would be considered in making the decision as to whether electronic or pneumatic controllers are to be installed on a particular process?

14. What is a fluidic computer?

15. How does a WAD work?

16. What does the term *numerically controlled machine tool* imply?

17. Controllers may be
    A. Analog devices.
    B. Mechanical devices.
    C. Pneumatic devices.
    D. Digital devices.
    E. Any combination of the above.

18. What is a meter relay?

19. The "setpoint" input to a controller can come from which of the sources listed?
    A. Supervisory controller (computer).
    B. Cascade controller.
    C. Cam controller.
    D. Manually inserted.
    E. All of the sources above are commonly used.

20. Since all commercially available "controllers" are designed to be applied to any process, what units are setpoint, etc., normally displayed in?

21. When we talk of a "controller" for a process, what portion of the system are we referring to?

22. Single-loop analog controllers used in process control systems differ when used in different processes in which way(s)?

23. Electronic analog controllers are all designed to
    A. Work with ISA standard signal levels.
    B. Give the system operator the capability of manual control of the loop.
    C. Be used in any type of process without modification.
    D. Exercise automatic control over a single loop.
    E. Electronic analog controllers meet all of these design objectives.

24. When analog controllers are used with digital controllers in the same system, the system is said to be a _____ system.
    A. Mostly analog.
    B. Mostly digital.
    C. Direct digital control.
    D. Hybrid.
    E. Confused.

25. How do electronic analog controllers differ in construction when used in different processes?
    A. They are basically the same in physical and electrical construction.
    B. They may differ in the gain settings for each mode of control (P, I, and D).
    C. They may differ in the modes of control necessary for different processes.
    D. They may have different electrical input ranges.
    E. All of the answers above are accurate statements of comparison.

The next eight questions refer to the following diagram.

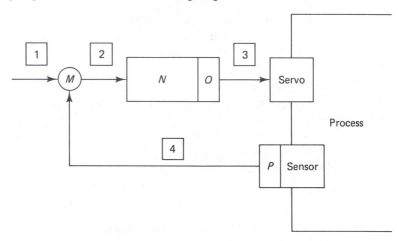

26. On the diagram, which boxed-in number indicates the (process variable) feedback signal?
27. On the diagram, which alphabetic letter(s) indicate(s) controller(s)?
28. On the diagram, which alphabetic letter(s) indicate(s) comparator(s)?
29. On the diagram, which boxed-in number indicates the setpoint signal?
30. On the diagram, which boxed-in number indicates the error (deviation)?
31. On the diagram, which boxed-in number(s) indicate(s) the longest signal cable run(s) (longest wires)?
32. On the diagram, which alphabetic letter(s) indicate(s) transmitter(s)?
33. On the diagram, which boxed-in number indicates the command signal?
34. What is the primary difference between feedback and feed-forward control strategies?
    A. Where the process variable is measured (before or after the controlled process).
    B. Whether the controller has two or three modes of control.
    C. Feedback control is only for analog controllers.
    D. Feed-forward control is only for digital controllers.
    E. All of the above are valid differences.
35. The decision as to whether to implement a particular process control system using analog controllers or a digital computer is based on
    A. How frequently changes must be made to the system.
    B. Plans for future system expansion.
    C. Desirability of a backup automatic method of control.
    D. Consideration of limitations of the process itself.
    E. All of the above (and more) are taken into consideration.

36. Which of the following functions would you normally expect to find on the front panel of a controller?
    A. A meter showing the value of the controller output signal.
    B. A meter showing the setpoint.
    C. An indication of the deviation between the setpoint and the actual process variable.
    D. A manual/automatic control switch.
    E. All of the functions above are normally provided on the front panel.

37. The primary function of the comparator is
    A. Long-distance signal transmission of sensor output information.
    B. Recording past history of variations in some process variable signal.
    C. Indicating the instantaneous value of a process variable.
    D. Comparing the feedback (process variable) signal to the setpoint signal to develop an error (deviation) signal.
    E. Operating on the error (deviation) signal to develop a command signal back to the process (servomechanisms).

38. Which of the following are characteristics of programmable logic controllers (PLCs)?
    A. They are programmed by relay-ladder-type program instructions.
    B. They were designed as replacements for hardwired relay logic.
    C. They are characterized by interfacing directly to 110 V ac on input and output.
    D. They are correctly described as special-purpose digital computers.
    E. All of the answers above are characteristics of PLCs.

39. Modern programmable logic controllers (PLC) are designed to interface directly with what types of electrical signals at their input and output terminals?
    A. Analog voltages.
    B. Digital logic level voltages.
    C. Raw 110-V, 60-cycle, ac voltages.
    D. High dc voltages from switches.
    E. Modern PLCs are designed to interface to all of the signals above.

40. The program for a programmable logic controller (PLC) is correctly called _____.

GLOSSARY

Actuator   A transducer whose output is mechanical motion. The output device in a servo-mechanism.

Automatic Control   As used herein, it refers to a mode of operation of conventional analog controllers wherein they actually are connected to the process actuators and are controlling the process variables.

Balance   As used herein, the procedure required for matching a controller's manual mode output signal with its automatic mode output signal when switching from one mode to the other for "bumpless" transfer.

Central Processing Unit (CPU)   That (functional) portion of a digital computer system which contains the main timing and control circuits, arithmetic unit, main memory, and provides device selection and data transmission paths for communication with peripherals.

Controller (Command) Output   The computed signal which goes directly to the process actuator from a controller.

Diaphragm Motor   An actuator whose input is pressure (normally pneumatic) which acts over a flexible diaphragm in order to produce linear motion of an output shaft which is fixed to the diaphragm and moves with it.

Drum Controller   A "programmable" industrial cam-operated multipoint switching device.

Drum Sequencer   *See* Drum Controller.

Flapper Nozzle (Baffle Nozzle)   A pneumatic-mechanical device used to control air pressure in response to pneumatic and/or mechanical inputs.

Fluidic Computer   A digital computing system whose logic elements operate in response to air pressures and flows rather than electronic voltages and currents.

Fluidic Logic   *See* Pneumatic Logic.

Manual Control   A mode of control, normally provided in systems as backup in the event of controller failure and for unusual operating situations, wherein the process actuators are manually controlled (remotely).

Meter Relay   A specially designed meter in which circuits (normally optical or magnetic) "watch" the position of its pointer, causing other circuits to be switched at either one or two manually adjusted setpoints when the meter's pointer indicates these setpoints.

Microcomputer   A general-purpose digital computer having all the major functions of the central processing unit (CPU) plus basic memory and interfacing circuits, on a single, printed circuit board. A microcomputer has a microprocessor as its CPU.

Microprocessor   A single large-scale integrated circuit containing all the binary logic circuits necessary to perform all the functions of a central processing unit (CPU) for a micro-computer.

Minicomputer   Originally, a general-purpose digital computer whose basic central processing unit (CPU) sold for less than $20,000. More modern interpretation includes all digital computers whose CPU is physically "small" (larger than a microcomputer, however), has word lengths of 16 to 32 bits, and operates in the sub-micro-second range for instruction execution. It is an economical general purpose computer which has found natural application in numerous process control applications.

Pneumatic Logic   Digital (on–off) logic elements which operate on air pressure and flows. The logic elements from which fluidic computers are constructed.

Programmable Controller   *See* Programmable Logic Controller *and* Drum Controller.

Programmable Logic Controller   A special-purpose programmable, digital, solid-state computing device, primarily intended to replace banks of electro-mechanical relays and perform their hardwired logic with an easily modified software program. Modern PLCs incorporate many functions not practically realizable with relays.

Wall Attachment Device (WAD)   A bistable, fluidic logic device. A pneumatic or hydraulic device which operates on the wall attachment principle, and has also been successfully applied to no-moving-part-valves.

Word Size   Referencing parallel digital computers, this term refers to the number of binary bits that the computer handles simultaneously (in parallel or as a unit) in its routine data processing operations.

13

ANALOG/DIGITAL/ANALOG CONVERSION TECHNIQUES

INTRODUCTION

When control systems engineers made their first attempts at applying the computational power, speed, and data-handling capability of digital computers to automatic process control systems, they were faced with the tremendous task of making the digital computer work with analog devices.

As noted previously, most naturally occurring phenomena are analog in nature, and this generally applies to manufacturing processes also. Even when these processes are active for only short periods of time, they normally act in an analog fashion while the process is actively working. Furthermore, nearly all available sensors and servomechanisms were (and still are) analog-to-analog transducers. Therefore, there was a pressing requirement for equipment to provide the conversion between the digital and analog "worlds."

In the process of designing *analog-to-digital* and *digital-to-analog converters,* many important trade-off decisions had to be made. Probably the most basic of all these decisions concerned resolution requirements.* The original analog signal had essentially infinite resolution, since it was analog or continuous in nature. The digital representation of this analog signal would of course decrease this resolution. An example may help to illustrate this.

Resolution is defined as the smallest detectable incremental change in the input signal that will cause a change in the output signal. A good-quality analog trans-

---

*The reader is referred to Appendix B for a brief presentation of the binary number system. Information covered in Appendix B will be assumed in the following chapters.

ducer with a 0- to 10-V input will be capable of detecting, say, a 1-m V change in the input signal. This means that its resolution is 10 V/0.001 V $= 10 \times 10^3$ or 1 part in 10,000. Referring to a table of decimal values for powers of 2, it would be noted that 2^{13} is 8192; this means that the resolution offered by a 13-*bit* digital signal would be one part (one digit) out of 8192 parts, which is significantly less than the resolution of the analog transducer.

From a practical standpoint, analog-to-digital converters with a 14-bit capability are normally prohibitively expensive; here "expensive" refers not only to dollar cost but also to other system parameters, such as speed and maintenance. Therefore, a decision was necessary as to the minimum acceptable resolution required of an *analog-to-digital* (A-to-D or A/D) *converter;* and also of a *digital-to-analog* (D-to-A or D/A) *converter.*

At that time the question was not one that could be positively answered and backed up with conclusive evidence. So for many years the problem was discussed (often rather heatedly) at some of the technical society symposiums on industrial control which were set up to resolve the problem. Eventually, time and use seemed to indicate that a resolution of 1 part in 1000 was an acceptable value. This requires a digital representation of at least 10 binary bits (10 bits supplies a resolution of 1 part in 1024).

A resolution of 10 bits means that a 10-V signal can be resolved down to 10 V/(1024 bits), or 0.01 V (per bit). As the analog signal starts increasing from zero volts, the A/D converter will not have an output value until the analog signal equals or exceeds 0.01 V. After the A/D converter recognizes this minimum signal and produces the correct output, that output will not change again until the analog signal reaches 0.02 V. Therefore, the A/D converter cannot resolve, or recognize, changes in input voltage less than 0.01 V in magnitude. It is therefore said to quantize the input, or chop it up into chunks (quanta) 0.01 V in size.

Note that at any given time the analog input signal could be as much as 0.01 V different from the digital output indication; this error is appropriately referred to as a quantizing error. *Quantizing error* (sometimes referred to as *quantizing noise*) is defined as the error in output code due to limited resolution of the conversion system. Each 0.01-V chunk can be considered to be a *dead-band.*

Two other important concepts were covered in the past few paragraphs which must be completely understood; they are the concepts of scaling and of digital representation of analog quantities. The fact that each change in output from the A/D converter represents a change in voltage of 0.01 V gives rise to a useful concept called the *scale factor.* If the system keeps track of how many changes in output have occurred from the time the system was zeroed, the actual value of analog voltage at any time is approximately the product of this scale factor (0.01 V/change or 0.01 V/bit) times the number of changes that have occurred (keeping account of whether they were increases or decreases in voltage).

This product, as represented within a digital computer, can be used for control system calculations. However, the computer must have knowledge of how many changes have occurred; and by inference, it must have the capability of recognizing and dealing with the maximum possible number of changes.

As an example, suppose that the computer word (binarily coded piece of data), which represents the scaled value of the input voltage to the A/D converter, and the voltage itself are simultaneously zeroed. The technique to be used by the computer is to simply count the number of changes from the A/D converter, each of which represents a voltage change of 0.01 V, while maintaining an account of whether each change is an increase or decrease in voltage and appropriately *incrementing* (increasing by one) or *decrementing* (decreasing by one) the scaled computer word. This is basically how the *incremental (serial) system* of A/D conversion operates.

The process is started and some time later the computer program requires the value of this particular voltage. It inspects the computer word, which is the simple *binary* sum of the changes from the A/D converter. The value of the voltage is equal to the fixed scale factor (0.01 V/bit) multiplied times this count. Therefore, the computer representation of that voltage is a binary word which has meaning only when used with its appropriate scale factor. In some instances, the computer representation of the analog voltage is as simple as a binary count; in others it will be a series of several binary words with the value hidden in some unique coding scheme.

The design of A/D and D/A conversion equipment is based upon a multitude of other system specifications. The reasons for and definition of most of these other specifications are either fairly obvious or will be discussed in the appropriate portions of the text which follows. A partial listing of some of the more important of these specifications would include conversion time, conversion rate, quantum level, aperture, sampling rate, offset, drift, signal-to-noise ratio, input and output voltages and currents, and others. Many of these specifications have the same definitions as presented for op amps.

DIGITAL-TO-ANALOG CONVERSION TECHNIQUES

It would seem logical to commence the technical discussion of A/D and D/A conversion techniques with a presentation of A/D converters. In actuality, however, some popular A/D conversion techniques require an integral-functioning D/A converter. The reasons for this will be presented later; however, the presentation will therefore commence with D/A conversion techniques.

The purpose for a D/A converter is to convert the information contained in some binary digital (input) word to a unique dc (output) voltage corresponding to the bit pattern in that word. Do not forget the significance of each bit in a binary word. For the purpose of discussion in this text we shall assume that the leftmost bit is the *most significant bit* (MSB) and that it has a weight (maximum value) which is twice as large as the next most significant bit (next bit between the MSB and the binary point). Each binary bit has a weight (maximum value) which is twice the weight of the next less significant bit.

Let us apply this binary weighting scheme to a dc voltage. Regardless of how many bits there are in the binary word to be converted, the most significant bit (whether it is a 1 or a zero) has to have twice the weight, maximum value, importance, or voltage capability as the next less significant bit. This means that if the MSB can cause the output voltage to assume a value equal to one half of its maximum possible magnitude, then the next less significant bit must be able to cause the output voltage to change by one half of that value (the MSB's value); therefore, it must be able to change the output by one fourth of the maximum possible magnitude.

For example, if the D/A converter is to convert from binary to a 0- to 10-V maximum output, the MSB must be able to cause the output to change by 5 V. The next most significant bit must have a 2.5-V capability (or weight), the one next to it 1.25 V, and so on. Notice that the position of the MSB in relation to the binary point has not been a consideration at all; the binary point has no significance except when working with binary arithmetic.

The more bits the digital word has, the greater the resolution the output voltage will have. For example, if the digital word has only one bit, then the only values that the output voltage could have would be 5 V or 0 V. This is all the resolution (5 V resolution) that the converter would be capable of. The maximum possible error could be almost 5 V; for example, the actual digital representation of the output value could be 0 V, whereas more D/A converter accuracy would show that it should be 4.9 V.

To increase the conversion accuracy, which would mean that the resolution must also be increased, add another bit to the digital representation of the analog voltage. This second bit would have a weight of 2.5 V. Therefore, the ouput from the D/A converter could assume any one of four possible values: 0, 2.5, 5, or 7.5 V. The resolution is now 2.5 V, the weight of the *least significant bit* (LSB); and the analog ouput could be as much as 2.5 V in error. Adding a third bit would increase the resolution to 1.25 V and a fourth bit to 0.625 V. Each additional bit increases the resolution by a factor of 2 and simultaneously decreases the maximum error by the same factor (assuming that all the error is due to conversion resolution).

Based upon the foregoing theory, all that remains is to come up with an electronic circuit that will have the capability of monitoring the digital word and adding binarily weighted values of voltage based upon the bit pattern of that digital word. Previous text material has covered the capabilities of the operational amplifier and indicated that it is an ideal device to add together several voltages, at the same time according a gain to each of these voltages which is independent of all other inputs.

Binarily Weighted Resistance Ladder D/A Conversion

Figure 13-1A is a schematic of a typical binarily weighted resistance ladder type D/A converter. Note the application of the op amp, specifically noting that the

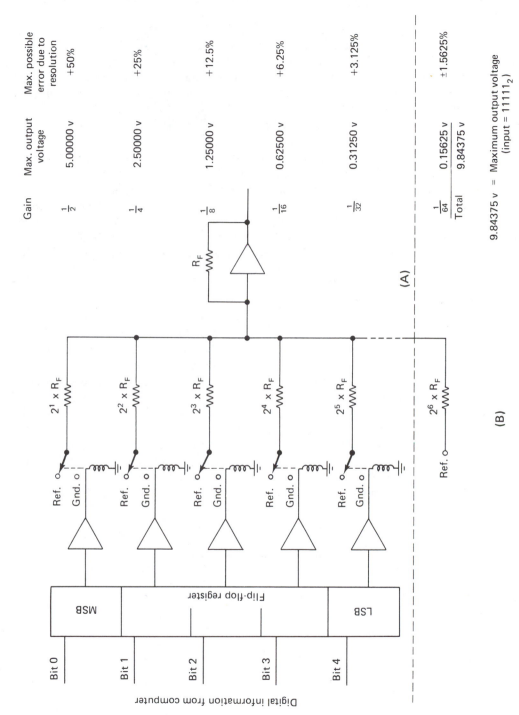

Figure 13-1 Block diagram of a binarily weighted resistance ladder-type digital-to-analog converter.

values of the input resistors are related to each other by powers of 2. This binary-weighted resistor network provides overall amplifier circuit gains which are also binarily related to each other and provide the proper output voltages, as previously discussed. Note also that the maximum output voltage from this system (a 5-bit D/A converter) is 9.6875 V. This means that if full output of 10 V is required, the converter will introduce a 0.3125-V, or 3.125%, error in the system. The magnitude of this maximum possible error can be halved by introducing one more binarily weighted *constant* input to the op amp.

Figure 13-1B shows this additional op-amp input. It does not depend upon the value of the digital word to be converted; it is a constant input. Therefore, it cannot reduce the maximum possible error, but it does change the error, from an absolute zero to a 3.125% error to a ±1.5625% error. For example, when the output should be zero volts, it will be 0.15625 V, a +1.5625% error. When the output should be 0.3124 V, it will still be 0.15625 V (rather than the zero volts it would have been without this final op-amp input). In this case the error is approximately −1.5625% rather than −3.125%.

R-2R Ladder D/A Conversion

Figure 13-2 is a simplified schematic for a 4-bit D/A converter that uses the R-2R ladder technique. This technique is very simple in operation and in construction, which is its primary advantage. Only two different values of resistance are required for all of the resistances needed, as compared to the many different values which are required to implement the binarily weighted ladder network design. It is considerably easier and less expensive to manufacture this simpler network, especially for the higher-resolution systems.

Basically, due to the way in which the resistors are interconnected, each of the *n* inputs supplies (or sinks) a binarily weighted value of current with respect to the next-higher and next-lower bit inputs. The op amp is in its summing configuration; therefore, it sums those binarily weighted currents in accordance with the energized

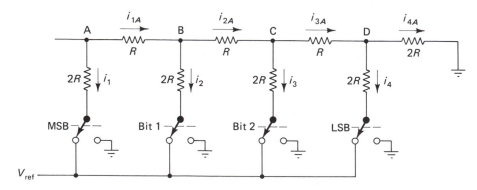

Figure 13-2 Four-bit D/A conversion circuit using the R-2R ladder technique.

sequence of input switches (again, these switches would be solid-state switches in practice). Therefore, other than being made differently, this system operates very similarly to the binarily weighted resistor ladder network discussed previously.

The following section is a more detailed explanation of the operation of the R-2R technique, and some readers may prefer to pass over these details. Be assured that the detailed explanation is not absolutely required for either a basic understanding of D/A conversion or the subsequent material on A/D conversion.

For nonelectrical readers, probably the simplest explanation is one that requires only a basic knowledge of Kirchhoff's current law and reduction of resistors connected in parallel to equivalent series resistances. One further concept is necessary—that the internal resistance of a voltage source is negligible. This means that whether the input switches illustrated in Fig. 13-2 are connected to earth common (ground) or to V_{Ref}, the resistance to ground (or common) is the same in either case. For each bit in that schematic a binary "1" results in the (electronic) switch connecting the appropriate input to V_{Ref}, and a binary "0" connects the input to common.

Begin by referring to Fig. 13-3A, which is simply the lowest portion of the resistor divider network and the LSB input; note that the input as illustrated is connected to electrical common since there will be no additional value of effective dc resistance added into the circuit regardless of whether the connection is made to ground or V_{Ref}.

At node D i_4 represents the current contributed by the LSB of the binary word to be converted to digital by this circuit. We are going to try to relate the value of this LSB's current to the values of current contributed by the other digital inputs in order to prove the binarily weighted ratio between those values. First let us express i_{3A} in terms of i_4; then we'll advance to node C, where i_{3A} originates, and relate the second input current i_3 to i_{3A} and subsequently to i_4; and so on for nodes B and A.

At node D of Fig. 13-3A,

$$i_{3A} = i_4 + i_{4A}$$

by Kirchhoff's current law; but since the resistances to common are equal, $i_4 = i_{4A}$. Therefore, $i_{3A} = 2i_4$ and we have the relationship necessary to advance to node C. Before we advance, though, we need the electrical equivalent of all resistances connected at node D and connected to common; this value will be necessary for the node C calculations. Note that at node D there are two resistors (each of value $2R$) connected to common and in parallel; the equivalent value is therefore simply R ohms.

At node C in Fig. 13-3B the equivalent resistance from node D to common is illustrated as resistor R_D with the value R ohms. This has been done only to simplify the correlation between this text and the schematics. Note that the resistance at node C (to common) again consists of two parallel paths, each with a total resistance of $2R$ ohms; therefore, the equivalent resistance from node C to common is again the

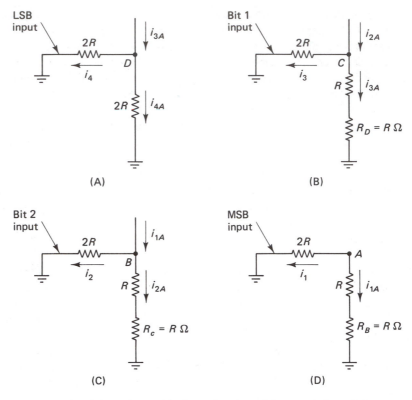

Figure 13-3 Schematics used in the explanation of the R-2R D/A technique.

value of R ohms. This is illustrated as resistor R_C in Fig. 13-3C. Kirchhoff's current law at node C in Fig. 13-3B states that

$$i_{2A} = i_3 + i_{3A}$$

but $i_3 = i_{3A}$ (due to equal resistance values in parallel) and, from the preceding paragraph, $i_{3A} = 2i_4$. Therefore, $i_3 = 2i_4$, showing a 2:1 current relationship between these two inputs. Note also that i_{2A} will equal $4i_4$ as we advance to node B.

     Advancing to node B as illustrated in Fig. 13-3C, we can, by reasoning as in the previous paragraphs, arrive at the fact that the equivalent resistance from node B to common is equal to R ohms. This equivalent resistance is referred to as R_B in future schematics. We can similarly derive the following current relationships. At node B

$$i_{1A} = i_2 + i_{2A}$$

where $i_2 = i_{2A}$ and $i_{2A} = 4i_4$ from the node C calculations. Therefore, $i_2 = 4i_4$, continuing the binary relationship between input currents that we started out to prove. Note also that $i_{1A} = 8i_4$ and advance to the final node.

Finally, for our 4-bit converter, anyhow, we have advanced to the MSB, illustrated as node A in Fig. 13-3D. Here it should be relatively easy to extend the previous discussion to derive the relationship $i_1 = 1_{1A}$, which is equal to $8i_4$ from the preceding paragraph. Therefore, we have an 8:4:2:1 relationship between the input currents i_1 through i_4, and this circuit will give us a true binary D/A conversion.

This discussion has covered the op amp and its external resistance networks necessary for D/A conversion. However, the connection of voltages to this op-amp input resistance network must also be understood. The computer is a very busy machine, it operates at such high speeds that it can service literally thousands of D/A converters so frequently that each one acts as though the computer were continuously servicing it. Since the computer does not continuously service each D/A converter, each one must have the capability of remembering what the computer told it to convert between the times that the computer actually does service it.

Normally, the simplest and least expensive form of memory unit that can be provided for each D/A converter is a simple digital *flip-flop register*. This register is a series of independent electronic circuits, each of which normally has a single input and, upon a given command, has the capability of storing the digital information present on the input line at the instant the command is received. The output from each of these circuits will be the same binary information that was on the input at the time the memory command was given. The output will remain constant regardless of what changes occur at the input until the memory command is again given; therefore, these circuits have a digital memory capability.

The output from each flip-flop is amplified and activates a small relay or switching transistor. This relay connects either the reference voltage power supply or electronic ground to the op-amp input resistors, depending upon the value of the digital word currently in the flip-flop register. The reference voltage is an exceptionally accurate and stable voltage reference source, used to decrease the errors in the op-amp output voltage due to errors present in the input voltage. This precision power supply is used only as a voltage reference for D/A converters and is normally not used by any other system circuits.

Therefore, connecting all these subsystems together (the flip-flop register, relays and relay drivers, *reference power supply,* binarily related resistor network, and op-amp), the complete system performs the function of converting the information contained in a binary digital word to a unique value of analog voltage. Frequently the op amp does not have the power to operate the device it is intended to control, so it normally supplies the input to an appropriate power amplifier.

The foregoing presentation has discussed the errors (and overall accuracy) based upon quantization and resolution. Other errors are introduced by an imperfect reference voltage supply, accuracy of the input and feedback resistors, and the imperfections present in the op-amp circuitry. Specifications for a D/A converter will include the effects of all these sources of error. Naturally, any errors attributable to the op amp will also affect the accuracy of the D/A converter.

As one final note on the subject of D/A converters, the memory function in

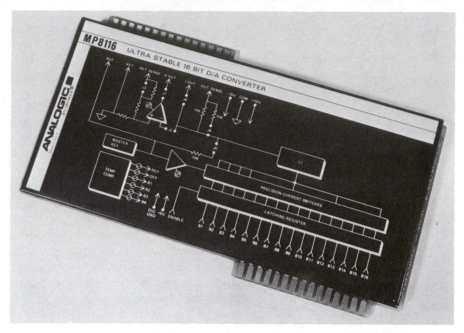

Figure 13-4 Commercial digital-to-analog conversion module. (Courtesy of Analogic Corp.)

the example was accomplished by a digital memory device, the flip-flop register. Memory can also be provided by an analog device, a capacitor. A circuit following the op amp can store the analog value on a good-quality capacitor and then provide the isolation and amplification as necessary. This is a possible, although not very popular (insofar as use is concerned), D/A converter design. Its main advantage would be so that the same D/A converter could be used for many analog outputs, each one being sequentially connected (multiplexed) to the D/A converter when required. In actuality, however, the circuits required to "remember" the analog voltage are expensive, and therefore this scheme presents little economic advantage. Therefore, D/A converters are usually dedicated to a particular output circuit, and each analog output is permanently connected to its own D/A converter. Figure 13-4 illustrates a commercial module which includes all necessary component parts to comprise a complete D/A conversion system.

ANALOG-TO-DIGITAL CONVERSION TECHNIQUES

Digital-to-analog conversion is a relatively straightforward technique, and the technique as described previously is the most popular in practice. The situation differs, however, when viewing the field of analog-to-digital conversion equipment. There are several quite different techniques which have found widespread acceptance in automatic process control systems. Only six are going to be discussed in this presentation.

The basic purpose of an A/D converter is to accept a *standard* range of input voltages and convert this analog signal to a unique digital representation for each possible analog value. Note that there may be some preinput signal conditioning (filtering, level shifting, impedance matching, etc.) required prior to input to the A/D converter, since it will deal only with a single range of input voltages.

Voltage-to-Frequency Conversion Technique

A block diagram of one of the simplest A/D conversion techniques to understand is illustrated in Fig. 13-5. It consists primarily of a voltage-to-frequency converter and a binary digital counter. The analog voltage is presented to the input of the voltage-to-frequency conversion circuitry. The circuit produces a square-wave output, the frequency of which is proportional to the magnitude of the input analog voltage. In actuality, it is an A/D converter in itself; however, it is normally associated with a binary counter in practical usage, as illustrated.

The binary digital counter receives a "start" signal from the computer, clears to zero, and begins counting the pulses arriving from the voltage-to-frequency converter. It counts these pulses, normally, for a predetermined interval of time, at the end of which it stops counting. Therefore, it has converted from a serial-frequency-modulated pulse train to a parallel digital word, the magnitude of which is directly proportional to the frequency of the input pulse train, and therefore to the original analog voltage.

This technique is not presented because of its popularity but because it is easy to understand. It does make the A/D conversion; however, it has several rather serious system disadvantages, primarily conversion speed and accuracy. Note that this particular technique is open-loop and does not require a D/A converter. Its primary application is in digital meters.

Linear Ramp Encoder Technique

The second basic method to be discussed employs the *linear ramp encoder technique,* which is in widespread use and capable of very accurate A/D conversions

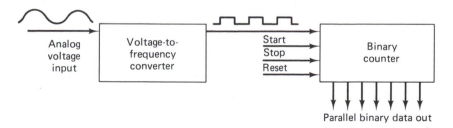

Figure 13-5 Voltage-to-frequency A/D converter technique block diagram.

and is relatively inexpensive to implement. The heart of the linear ramp encoder A/D converter is a circuit that generates a very linear increasing voltage (ramp) output, similar to one cycle of the waveform from a sawtooth or triangular-wave-signal generator. The output from this *ramp generator* is so critical that any non-linearities present in its output are limiting factors in the A/D converter's linearity. Any inaccuracies due to the op amp itself (offset, drift, frequency response, etc.) also affect the accuracy of this type of A/D converter. Fortunately, circuits can be designed such that the output voltage has excellent accuracy specifications, and therefore this technique is frequently used where conversion speed is not the primary consideration.

The output voltage from this linear ramp generator (see Fig. 13-6) is fed to one input of an operational amplifier voltage *comparator*. The other input to this comparator is the analog signal, the value of which is to be converted to digital form. The comparator's output is essentially a digital signal in that when the two (input) voltages are exactly equal (within the comparison capability of the comparator), the comparator's output rapidly switches rather than providing an analog output signal. This rapid switching is a signal to the rest of the A/D converter circuitry that the comparison is equal.

When the A/D converter receives a conversion command, the unknown analog signal must be present at the comparator input. Two functional circuits within the A/D converter are simultaneously commanded to "do their tricks." The ramp generator is started and the output from the digital clock is simultaneously gated to the binary counter. As the ramp generator builds up its output voltage, the binary counter counts the pulses from the digital clock. The digital clock is an exceptionally

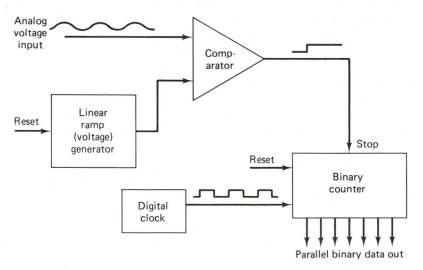

Figure 13-6 Linear ramp generator A/D conversion technique block diagram.

accurate and stable frequency source. Both the ramp generator and binary counter continue to operate until the instant that the comparator's output switches polarity—signaling that the ramp generator output is exactly the same value as the unknown analog voltage. This switching causes the binary counter to stop counting the clock pulses (and also signals the ramp generator that its output is no longer required).

At this point the binary counter has kept very accurate track of how long it took the ramp generator to generate voltage equal to the unknown analog input signal. The ramp generator's output is calibrated to increase at a specific rate (millivolts per second, for example). Therefore, knowing how long it took to build up to the same value as the analog signal, and applying the mV/s scale factor, the exact value of the analog signal can be calculated.

The counter will hold the value it had when the comparator signaled it to stop counting until the digital computer "looks" at its output and signals the A/D converter to reset itself. The reset command from the computer means that it has accepted the digital representation of the analog signal and that the ramp generator and binary counter are to reset and be prepared to receive the next "convert" command. Note that this A/D technique does not employ a D/A converter or feedback.

The resolution of the linear ramp type A/D converter is dependent upon the frequency of the digital clock and the number of bits that the binary counter has. For example, a "slow" clock frequency may cause the counter to accumulate a maximum of only four pulses of information in the amount of time that it takes the ramp generator to reach its full value output. A faster clock frequency will permit the counter to accumulate more than 16 (4 bits) pulses and, therefore, make a more accurate representation of the analog voltage.

A specific example should help to clarify the relationship among resolution, clock frequency, and number of bits. Assume a ramp generator whose output will go from zero to 10 V in 1 ms. Assume also a digital clock with a 10-kHz output pulse rate. This means that there will be 0.1 ms between each output pulse, and in 1 ms (the time the ramp generator takes to reach full scale) the counter will count only 10 pulses (10 pulses times 0.1 ms per pulse will equal 1 ms). This means in effect that this A/D converter can only resolve the 0- to 10-V input into 10 parts (less than 4 bits' worth of information); or the resolution is 1.0 V.

Change the clock frequency to a 1-MHz rate, and now the interval between pulses from the clock is 1 μs. During the time that it takes the ramp generator to generate its full-scale output, the clock will send out 1000 pulses. This means that the 10 V has been resolved to 10 divided by 1000, or 0.01-V increments. The binary counter must, therefore, have a 10-bit capacity (in order to count up to a maximum of 1000 pulses).

To look at this resolution from another point of view, during the interval between each clock pulse (1 μs in the last example) the voltage output from the ramp generator will change by 0.01 V. At some point between the generation of clock pulses the ramp generator output will exactly equal the analog input voltage, and the comparator will stop the counter before the next pulse from the clock is counted. Therefore, the scaled binary count could be as much as 0.01 V in error from the actual value, or the resolution is 0.01 V.

Digital Ramp Technique

A common variation of the linear ramp technique is the *digital ramp technique*. The primary difference is that the linear ramp generator (Fig. 13-6) is replaced by a D/A converter. The D/A converter receives its input from the counter, and the analog output from the D/A converter supplies the reference input to the comparator. The two systems operate in a nearly identical manner, except for the way the reference voltage is generated. Both techniques are relatively inexpensive to implement and are, therefore, commonly used in economy D/A converters.

Dual-Slope Integrator Technique

Another very popular (open-loop) analog-to-digital conversion technique is the *dual-slope integration technique*. In this technique, both the reference voltage and the unknown analog signal are sequentially fed to the integrator. The reference voltage and the unknown voltage must be of opposite polarity. It uses a comparator, binary counter, and binary clock in the same way as the linear ramp encoder A/D converter does; the operation differs somewhat, however. Figure 13-7 is a block diagram illustrating the functional pieces of hardware used to implement this technique.

When a convert command is received, the counter is automatically reset to all zeros and the switch driver causes the unknown voltage to be applied to the input of the integrator. The output from the comparator is designed such that at this time it will permit the counter to count up for a fixed period of time until it counts up to *all 1s*. During this entire period (since the unknown signal has been connected to the integrator), the output from the integrator will be steadily increasing in value. On the next count after the counter has counted all the way up to all 1s (i.e., the next count will cause it to go to all zeros and start over again), the switch will change positions, disconnecting the unknown voltage from the input to the integrator and simultaneously connecting the reference voltage to it (remember that the reference is of opposite polarity). Therefore, the integrator now integrates the opposite polarity voltage, which causes its output to decrease toward zero volts. Meanwhile, the counter is counting up from zero again.

When the output of the integrator goes to zero, it causes the comparator to switch its output, thereby stopping the binary counter. The binary number in the counter at this time is proportional to the amount of time that it took the integrator to integrate down from its starting point (which was determined by integrating the unknown voltage for a specific amount of time) to zero. Therefore, the binary count is proportional to the unknown voltage. Figure 13-8 illustrates the output from the

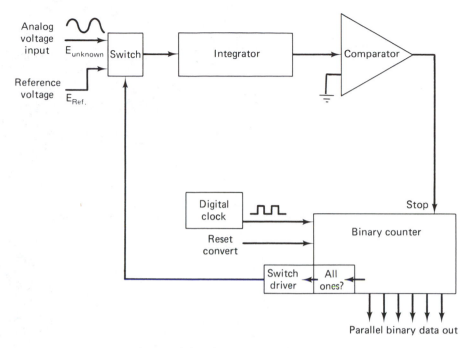

Figure 13-7 Dual-slope integrator technique block diagram.

integrator during a conversion cycle. The difference in the slopes of the curve is due to the difference in voltage levels between $E_{unknown}$ and E_{ref}.

For example, assume that the system was a 0- to 10-V A/D converter and was presented with an (unknown) voltage at its input of 10 V. Upon receiving a conversion command, the unknown voltage (10-V in this example) is connected to the integrator. Simultaneously, the counter begins to count (from zero) and the integrator commences integrating the unknown voltage. The integrator continues to integrate the unknown voltage until the counter has all 1s in it; that is precisely (by design) the amount of time that it takes the integrator to integrate up until its output is at its maximum value (since the unknown voltage was the maximum value).

Then, on the next count, the unknown voltage is disconnected from the integrator and the reference voltage (opposite in polarity to the unknown voltage signal) is connected to the integrator. Also on this next count, the binary counter overflows (goes from all 1s to all zeros) and commences counting up from zero again. The integrator will, therefore, integrate from the point that the unknown voltage was disconnected (maximum integrator output voltage in our example) down to zero volts at a predetermined rate.

When the integrator output finally goes to zero, the comparator output is designed to switch, thereby stopping the counter from changing any further. The

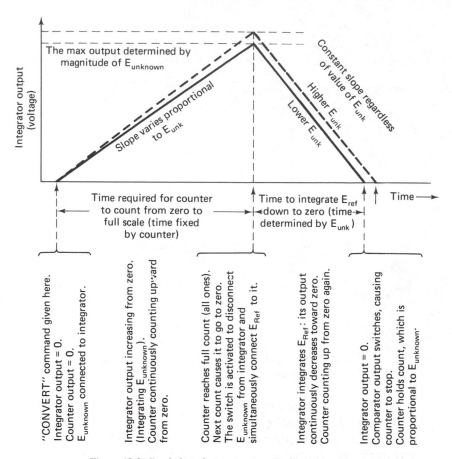

Figure 13-8 Dual-slope integrator waveform explanation.

output from the counter at that point will be proportional to the unknown voltage in a manner similar to the linear ramp encoder technique.

The advantages of this type of converter over the linear ramp type lie in the fact that the conversion accuracy is not dependent upon the tolerances (and linearity) of the integrator components, since they affect the output of the integrator both when it integrates the unknown voltage (in one direction) and the reference voltage (in the other direction). Thereby, these inaccuracies cancel their effects. Furthermore, long-term variations in clock frequency also will not affect the overall conversion accuracy, since they also affect both slopes and cancel out. Finally, the system will not be as sensitive to *electrical noise* present at the input terminal while the converter is integrating the unknown voltage, since it is being directly integrated and the process of integration itself acts as a filter. With all these advantages, it should be obvious that it is a very popular A/D converter technique.

Successive-Approximation Technique

The next A/D technique to be discussed is the *successive-approximation technique.* Some version of the ramp generator technique is probably the most popular technique, owing to its application in digital meters, but where conversion speed is a requirement, the successive approximation technique is used.

Prior to discussing detailed operation of a complete successive approximation A/D converter, let us begin with a general-feedback-type A/D converter block diagram. Figure 13-9 illustrates a very general block diagram of a feedback-type A/D converter. Generally, its operation commences with a ''guesstimate'' of what the analog voltage might be. This digital value is converted to analog, compared with the actual analog voltage, and the result, as generated by the comparator, is used to refine the original guess. This process of serially refining the existing value (from the original guess) continues until it is within acceptable tolerances.

As with the linear ramp technique, the successive-approximation technique uses a voltage comparator which essentially has a digital output rather than an analog output. Its operation requires that the output be (for example) the equivalent of a digital 1 if the analog input voltage is greater than the voltage from the D/A converter. The comparator output must, therefore, be the equivalent of a digital 0 if the analog input voltage is less than the D/A converter's output.

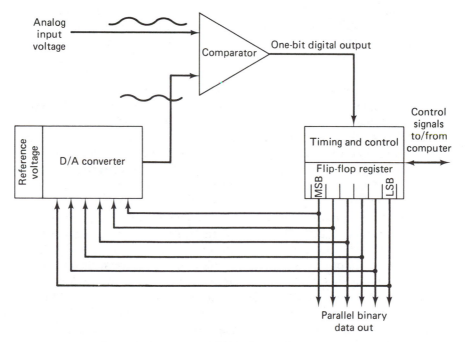

Figure 13-9 Block diagram of a feedback-type (specifically a successive-approximation) A/D converter.

Keeping the operation of this comparator in mind, probably the best way to understand the operation of a successive approximation A/D converter is by an example. For this example, assume that the actual value of the analog input voltage (to be converted to binary) is 8.0000 V; the 6-bit A/D converter having a 0- to 10-V unipolar input. There is also a series of digital flip-flops in which to store the binary value of the analog voltage as it is derived, bit by bit. The successive-approximation technique essentially makes a series of decisions, starting with the most significant bit and working down to the least significant bit, one at a time; therefore, this is a serial conversion technique as compared to the linear ramp generator technique, which performs the entire conversion all at once, or in parallel.

Referring to Fig. 13-9, initially the timing and control logic will make a "guesstimate" that the analog voltage is greater than or equal to 5.000 V. It does this by forcing the MSB to be a 1. This binary word (100000) is fed to the D/A converter, which feeds 5.0000 V to the comparator. The comparator senses that the analog voltage is greater than the voltage from the D/A converter, and its output goes to the digital 1 state. The timing and control logic interprets this comparator output as confirmation of its initial guess; therefore, it leaves the MSB's flip-flop set to a logical 1 state and turns its attention to the next MSB. The converted digital representation at this point is 100000 (with 6 bits of accuracy and resolution).

The timing and control logic makes a second guess, this time setting the second bit to a 1. The D/A converter output is now 5.0000 + 2.50000 V, which the comparator recognizes as less than the analog input voltage so it forces its ouput to a 1 state again. The timing and control logic again interprets this comparator output as meaning that its guess was correct. Therefore, it leaves the second flip-flop (next to most significant bit) set to a logical 1 and sequences itself to guess that the third bit is a 1. The converted binary value at this point is 110000, but the input to the D/A converter is 111000, since the timing and control logic has made its third guess.

The D/A converter's output is now 5.0000 + 2.5000 + 1.2500, or 8.7500 V. The comparator senses that the D/A converter output is too much and therefore outputs a logical 0. The timing and control logic interprets this 0 output as an indication that it made a bad guess about the third bit, so it resets the third bit to a 0 and goes on to guess that the fourth bit is a 1. The converted value is now 110000 and the trial input to the D/A converter is 110100.

This results in a D/A converter output of 5.0000 + 2.5000 + 0.625, or 8.1250 V. Again, the comparator signals the timing and control logic that this is too much, so it resets the fourth flip-flop to 0 and tries the fifth flip-flop in the 1 state.

The D/A converter output is now 5.0000 + 2.5000 + 0.3125, or 7.8125 V. The comparator now signifies that this value is less than the analog input voltage, so the timing and control logic tries the LSB as a 1 (the sixth and final bit in this sample A/D converter). The D/A converter now converts 110011 to 7.9688 V. The comparator recognizes that this is still less than 8.0000 V, and the timing and control leaves the LSB flip-flop set and signals the computer that the conversion is complete.

The flip-flop register holds the converted value until the computer gets around to looking at it, after which the computer gives the A/D converter a reset command.

The A/D converter resets its circuitry and awaits the next conversion command from the computer. The overall resolution of the successive-approximation A/D converter is strictly dependent upon the number of bits converted, by exactly the same reasoning that is behind the resolution specifications for the D/A converters previously described.

One of the primary advantages of this technique is its conversion speed. It is the fastest technique discussed so far and has an accuracy which is equivalent to that attained by the dual slope devices. However, it requires somewhat more complex timing and control circuitry, plus a D/A converter; therefore, it is more expensive than the other techniques but is one of the most popular of all A/D conversion techniques presently used in process control and the high-quality digital meters. Figure 13-10 illustrates a commercial successive-approximation A/D converter.

Flash or Parallel A/D Technique

For extremely high speed applications another technique is currently available, the "flash" or parallel technique. The engineering concept for this technique is not revolutionary or new, but its implementation was impractical until modern advances in large-scale semiconductor fabrication were realized.

Basically, the flash A/D system is constructed of a (very) large number of op-amp-type comparator circuits, each one having a fixed reference voltage on one of its (two) inputs and the voltage whose value is to be converted to digital form connected to the other input terminal. Each comparator compares the "unknown"

Figure 13-10 Successive-approximation analog-to-digital converter, completely contained in a single module. (Courtesy of Burr-Brown Research Corp.)

voltage to its own specific reference voltage and "switches" its output level in accordance with the result of that comparison. For example, the output may switch if the unknown voltage is greater than the specific reference and will not switch if it is less than its reference. Therefore, each comparator makes its own independent decision; it does not have to wait for any other circuits to make their decisions first.

Now refer to Fig. 13-11, which illustrates $2^n - 1$ of these comparator circuits connected into a flash A/D conversion system. The reference voltage to each comparator is one LSB different from either of the adjacent comparator's references. Therefore, if we want a 4-bit conversion system, 15 comparators will be required, along with a resistor divider network capable of dividing the entire voltage conversion range into $2^n - 1$ increments of voltage, each one spaced exactly one LSB worth of voltage apart from the next. This figure also shows that a digital logic encoding network will be required; this will be necessary in order to encode the $2^n - 1$ converter outputs into an n-bit parallel binary format. Therefore, this conversion process consists of two sequential steps; the comparators all (simultaneously) make their decisions and then the encoder logic performs its trick. These two time-delay periods basically determine how fast a conversion can be completed. Currently available flash A/D converters can make their conversion in a fraction of a microsecond. Flash A/D converters are capable of making complete conversions in less than 100 nano-seconds.

Figure 13-11 illustrates an example of a flash A/D converter having a 10-V full-scale input voltage range and 3 bits of conversion resolution. This requires seven comparators and eight resistors, plus the encoder to convert the seven binary outputs from the comparators into a 3-bit binary parallel code. Note how the first resistor spaces the other voltages into the center of their voltage weight range in a manner similar to the additional input shown for the D/A converter illustrated in Fig. 13-1B, for the same reason as explained in the text describing that figure.

As an illustrative example, if the unknown voltage happened to be 5.1 V at a specific instant, then all of the comparators from the one having a 4.375 V reference input voltage (and below) would be "switched," and all of those comparators with reference voltages above that value would remain "unswitched." The 7-bit code input to the encoder would convert the lower four "switched" inputs to a parallel binary word, probably a straight binary count of the number of comparators that have "switched," 4 (decimal) in this case, with a comparative binary output equal to 100. This binary value represents all input voltages between 4.375 and 5.624 V (for this example, conversion system), and our assumed value of 5.1 V certainly falls into that range.

For a practical flash A/D conversion system a limit of 4 bits of conversion resolution is typical; this requires $2^n - 1 = 15$ comparators and 16 resistors. For an 8-bit flash A/D conversion system (which would require 255 comparators plus 256 matched resistors) two 4-bit flash converters can be interconnected as illustrated in Fig. 13-12. This drastically reduces the amount of hardware required to perform the same function, but slows the overall conversion speed to less than one-half of the conversion speed of each of the 4-bit converters illustrated. In this system the

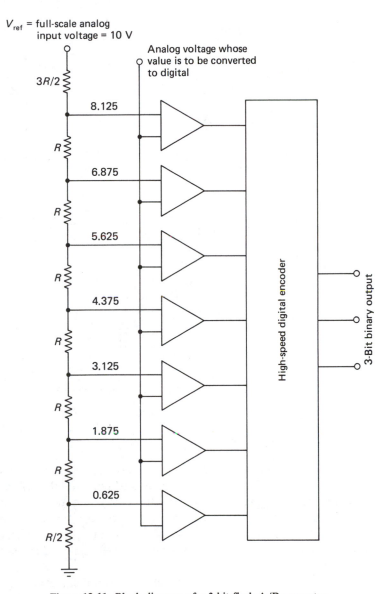

Figure 13-11 Block diagram of a 3-bit flash A/D converter.

first 4-bit flash A/D converter performs a conversion; its value is converted back to analog by the D/A converter and that value is subtracted from the unknown input voltage. Finally (and sequentially), the remainder of the voltage is converted by the second flash converter. The encoder/register formats the completed conversion value into a parallel 8-bit word.

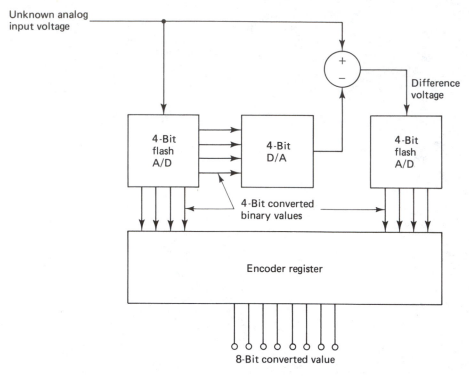

Figure 13-12 Eight-bit flash A/D converter using two 4-bit flash A/D converters.

SAMPLE-AND-HOLD MODULES

Modern analog-to-digital converters convert the value of analog voltages present on their input terminals very accurately and consistently—*if* the voltage does not vary too much during the time that the A/D converter is actively making a conversion. Significant electrical noise present on the signal, or the use of an A/D converter that converts too slowly as compared to the time variations of the signal being converted, can easily result in totally meaningless digital information from a properly working A/D converter.

To compensate for the electrical noise problem, a filtering function may be required prior to attempting a conversion. Depending upon the characteristics of the signal being converted and the characteristics of the specific A/D converter being used, a device called a *sample-and-hold unit* may be required. Referring to Fig. 13-13A, a relatively slowly (time) varying signal is illustrated with no electrical noise present on it. A conversion ordered at time t_1 should result in 8 V being converted. If the signal varies slowly enough as compared to the time it takes the A/D converter to complete its conversion, a successful conversion will result.

Now refer to Fig. 13-13B, where the same signal is again illustrated, only this

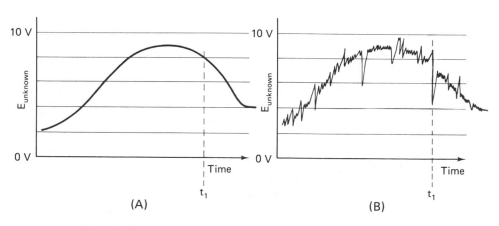

(A)   (B)

Figure 13-13 "Slowly" varying signals with and without electrical noise.

time with some electrical noise added. The noise could be due to *electromagnetic interference* created by nearby transmitting antennas, aircraft, or automobile ignition systems; it could come from power supply noise; or it could be due to pickup from switching circuits in nearby electronic equipment. There are innumerable possible sources of electrical noise, most of which are practically impossible to (economically) eliminate. Therefore, in all systems the designer must plan on running across electrical noise, and he must design the system to live with it.

Referring again to Fig. 13-13B, if the conversion is ordered on this noisy signal at t_1, and the converter converts fast enough, it may convert a voltage anywhere between approximately 5 and 9 V, whereas the correct value was 8 V. Therefore, in this system, a sample-and-hold unit is required, owing to the electrical noise present on the unknown signal to be converted. Without this sample-and-hold filter, the output from the A/D converter cannot be trusted. It must be noted that when using digital computers, the digital values from the A/D converter can themselves be subjected to filtering by the digital computer. However, it may be advantageous to have the sample-and-hold unit even in this case. Other systems considerations will dictate where the filtering is best accomplished.

The other condition under which a sample-and-hold unit may be required (whether digital filtering can be accomplished or not) is when the unknown signal can vary significantly during the period of time that the A/D converter is actually making its conversion. All A/D converters have a period of time during which they are "looking at" the unknown voltage to determine its value. This period of time can range from microseconds (for successive-approximation-type A/D converters) up to milliseconds (for slower types of linear ramp-type A/D converters). If the voltage being converted varies during this period of time (called *aperture time*), erroneous outputs may occur.

Figure 13-14A illustrates a signal that has the capability of changing significantly during the aperture time of the A/D converter being used (30% of full scale in this example). Therefore, the converted value may be in question by this amount.

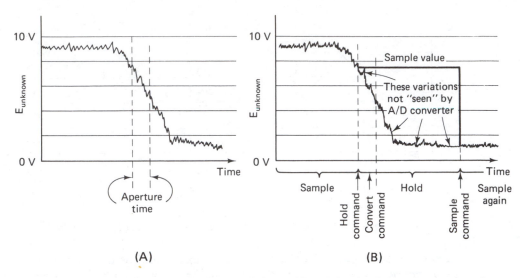

Figure 13-14 (A) Before and (B) after passing rapidly varying signal through a sample-and-hold unit.

In this example, two possible solutions are available: purchase an A/D converter with a faster aperture time or use a sample-and-hold unit. Assuming that the former solution is not necessary (because it is normally the more expensive solution), assume that a sample-and-hold unit will be installed.

Figure 13-14B illustrates the same signal as processed through a sample-and-hold unit. Note that when a conversion is required, first the sample-and-hold unit is commanded to hold; then the A/D converter is commanded to convert. Therefore, the A/D converter converts a dc value of the signal that existed at the time the S/H command was ordered, and further variations in the signal will not affect the A/D converter's output.

Now that we have seen why a sample-and-hold unit may be desirable in a system using an A/D converter, and what its effect on the overall A/D conversion system's output will be, what does a sample-and-hold unit look like? Figure 13-15A schematically illustrates several possible configurations for practical sample-and-hold units. They may be as simple as the flying capacitor or more sophisticated, to include op amps. The choice of type to be used must be based upon the overall system peformance requirements.

APPLICATION OF A/D CONVERTERS

Because of their expense, and to reduce the quantity of hardware included in a system, A/D converter systems are normally *multiplexed* to many input analog voltages. This means that the input to the A/D converter comes from banks of small

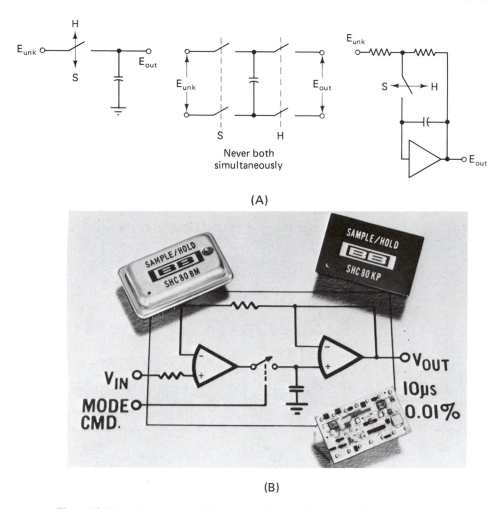

(A)

(B)

Figure 13-15 (A) Sample-and-hold module circuits; (B) commercial sample-and-hold module. [(B) Courtesy of Burr-Brown Research Corp.]

relays, only one of which connects a signal to the converter's input at a time. This relay network may have hundreds of analog signals connected to it, and only one A/D converter. Therefore, the control for these relays (or multiplexing network) becomes fairly complicated itself; however, the economics of the alternatives always justifies the use of an analog multiplexer for an A/D converter, where there are many analog signals to be connected. Figure 13-16 illustrated both relay and solid-state analog signal multiplexers used with A/D conversion systems. The decision as to which type to use is based upon the individual system requirements; both are in common use.

This creates an additional problem, though. For of the many signals that are

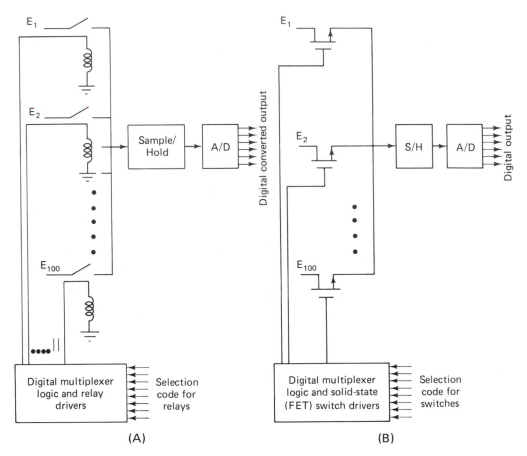

Figure 13-16 Multiplexed analog-to-digital conversion systems: (A) relay multiplexer; (B) solid-state multiplexer.

to be converted, there may be several different ranges of voltage to be dealt with. Therefore, the output from the multiplexer network normally goes first to a *buffer amplifier (programmable gain amplifier)*. The gain of this amplifier is controlled by the computer (through another switching network). The purpose of this buffer amplifier is to shift the level of the analog voltages to the standard input voltage level required by the A/D converter.

To obtain a feel for the complexity of this system, consider the amount of information required by the A/D conversion system, with its program-controlled amplifier and multiplexer network. The computer must tell the multiplexer which one of the 500 or so inputs to convert (possibly nine or more bits of digital information). It must also tell the amplifier which gain is to be provided for this particular analog signal (probably two more bits of information). Finally, after giving the relays time to activate, the computer can issue the conversion command to the A/D converter. This presentation has not considered the additional complexities required when dealing with the thermocouple-type inputs and their necessary compensation networks, even though this is a very popular type of input. Figure 13-17 schematically illustrates a complete A/D conversion system, including its information transfer requirements.

Therefore, the A/D conversion system represents a sizable quantity of electronic equipment. The overall system is very complex and an extremely sensitive portion of the overall computer control system. Nearly all the computer's process control functions rely on information provided by the A/D conversion system. Complete A/D conversion systems can be purchased which are physically located in a single module; Fig. 13-18 illustrates one of these modular systems.

Many industrial processes have *data acquisition systems* installed to monitor the performance of the system without directly exercising any control. These data acquisition systems are primarily based around A/D conversion systems, many of

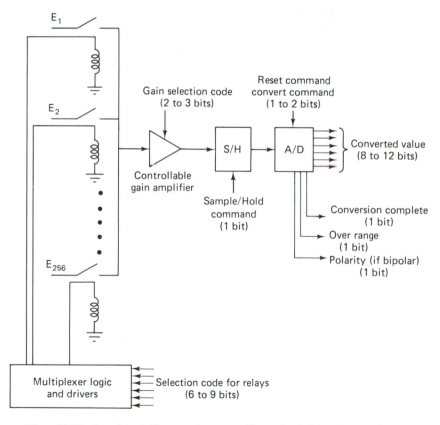

Figure 13-17 Complete A/D conversion system illustrating information transfer requirements between A/D system and the "controller."

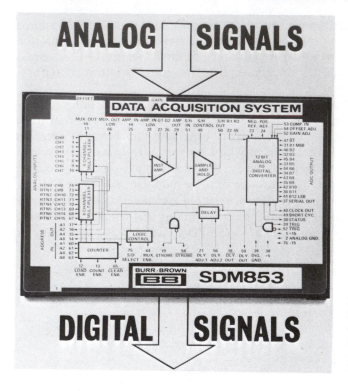

Figure 13-18 Complete analog data acquisition system (in a single module), capable of sequentially converting 16 analog signals to digital form with 12 bits of conversion accuracy. (Courtesy of Burr-Brown Research Corp.)

which are so complex that they may even require minicomputers to control them. Figure 13-19 is such a system.

QUESTIONS

1. Where did the requirement for analog-to-digital and digital-to-analog converters first arise?

2. What is the most basic A/D (or D/A) converter specification?

3. Explain resolution as applied to A/D (D/A) converters. Compare it to quantizing error and sensitivity.

4. Explain the concept of the scale factor as applied to A/D converters.

5. Briefly explain the operation of the D/A converter of Fig. 13-1. Use an example if you choose.

6. Briefly explain the operation of the linear ramp encoder A/D conversion technique. Use examples and sketches if you wish.

7. Briefly explain the operation of the dual-slope integrating A/D conversion technique. Use examples and sketches if you wish.

Figure 13-19 Re-zeroing a voltmeter in preparation for a performance reading of the condenser section of York's central residential air-conditioning unit is Stephen Burg, engineering associate in York's Instrumentation Group. Shown in the foreground is the Hewlett-Packard 9600 data acquisition minicomputer; the condenser section of the air conditioner, called the Champion IV, appears within the glass-enclosed test room. (Courtesy of Hewlett-Packard.)

8. Briefly compare the advantages and disadvantages of the linear ramp encoder A/D and the dual-slope integrating A/D techniques.

9. Briefly explain the operation of the successive-approximations A/D conversion technique. Use examples and sketches if you wish.

10. Briefly compare the linear ramp encoder technique and the successive-approximations technique, listing the advantages of each as compared to the other.

11. What is the systems function of a sample-and-hold module?

12. In your own words, explain the concept of multiplexing.

13. What is the major advantage of using a multiplexer in a system as compared to not using one?

14. What is a programmable-gain amplifier, and what is its function in an A/D conversion system?

15. Sketch the block diagram of a complete A/D conversion system, illustrating the analog signal multiplexer, programmable gain amplifier, sample-and-hold module, and the A/D converter.

16. Where digital memory is required in A/D and D/A converters, what functional piece of (digital) hardware is universally used?

17. How is the overall conversion accuracy of an A/D or a D/A converter implied?

18. A digital computer presents a D/A converter with the following digital word (11111). The D/A converter has a 0- to 10-V output range; what will be its output voltage?

19. Given a 10-V, 8-bit A/D converter, what is the minimum change in input voltage necessary to cause a change in the digital output (i.e., what is its sensitivity)?

20. An A/D converter makes a conversion after which its binary output is (10101). If its scale factor is 0.01 V/bit, what is the value of voltage that has been converted?

21. The following is a diagram of a successive-approximation type of analog-to-digital converter. What functions do the numbered blocks perform?

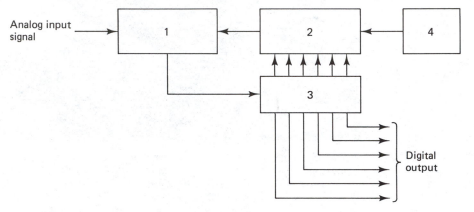

22. The basic principle of operation of the successive-approximations technique is
_____.

23. The microprocessor in the following simplified block diagram is connected to a D/A converter and is performing an analog-to-digital conversion. Which A/D conversion technique could be used?

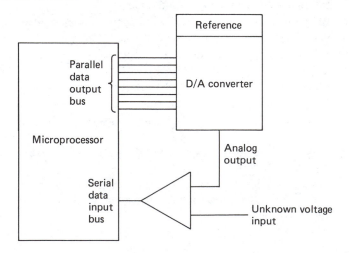

24. The following sketch is of a linear-ramp-type analog-to-digital converter. Name the numbered devices on the sketch.

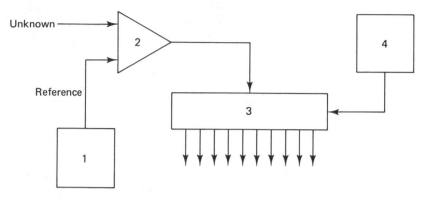

25. What is the primary consideration when selecting between a linear-ramp-type and a successive-approximation-type A/D converter for a given application?
A. Conversion accuracy.
B. Conversion speed.
C. Conversion resolution.
D. The need for a sample-and-hold unit.
E. Electrical noise on the signal.

26. The following block diagram is of a dual-slope-integrator-technique A/D converter. What functional pieces of hardware would you expect in each numbered block?

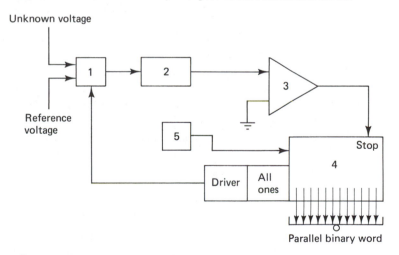

27. The overall conversion accuracy of a A/D or D/A converter
A. Is implied by the number of bits.
B. Depends upon its application.
C. Depends upon its maximum voltage.
D. Is determined by the weight of its MSB.
E. All of the above determine its overall conversion accuracy.

28. Where would a sample-and-hold unit be used in a process control system?

29. Data (information) transfer between A/D (and D/A) converters and the computer is accomplished by use of functional devices called _____.

30. The dual-slope-integrator A/D conversion technique is more complex than the linear ramp generator technique; what is its primary advantage over the linear ramp technique?

31. What is an analog-to-digital converter?

32. Almost all A/D conversion techniques have certain functional pieces of equipment in common; among the common elements is/are _____.

33. *Quantizing error* is a term closely related in concept to
    A. Resolution.
    B. Sensitivity.
    C. "Dead band."
    D. Minimum signal necessary to cause output to change.
    E. All of the above are similar in concept to the quantizing effect.

34. "The heart of a linear ramp encoder A/D converter is a circuit which generates a very linear increasing voltage (ramp) output" in conjunction with _____.

35. Digital memory is normally provided for a D/A converter. The memory device is most commonly
    A. A core memory.
    B. A common flip-flop register.
    C. A shift register.
    D. A digital counter.
    E. All of the named memory devices are commonly used.

36. Given a 10-V, 10-bit A/D converter, what is the minimum change in input voltage necessary to cause a change in the digital output? (What is its sensitivity?)

37. If an A/D converter makes a conversion and its binary output is 1010, if the scale factor is 0.01 V/bit, what is the value of voltage that has been converted?

38. What digital input signal will result in exactly 10.0000 V out of a 10-V D/A converter? (There is no offset voltage—i.e., all 0's in = 0 V dc out.)

39. You have a 0- to 10-V dc signal which is to be converted to binary. The system requires that the binary representation have better than (less than) a 5% conversion error. How many bits (minimum number) does the A/D converter require?

40. Frequently, it is necessary to stop an analog signal from changing value while an analog-to-digital converter goes through the process of converting it. The device used is called a sample-and-hold unit. Which of the following circuits could be used to perform this function?
    A.

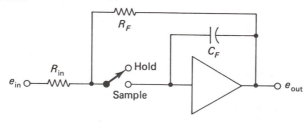

B.

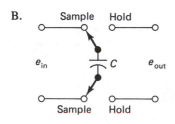

C.

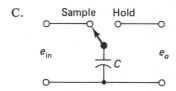

D.

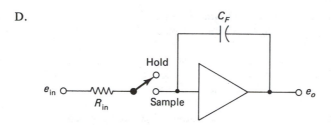

E. All of the circuits above can be used to perform this function.

41. In a common digital control system, you would expect to find very few _____ and probably (relatively) quite a few _____.
    A. A/D converters; D/A converters.
    B. D/A converters; A/D converters.
    C. You would normally expect to find only one of each (A/D converter and D/A converter).
    D. You would typically not expect to find either A/D or D/A converters.

GLOSSARY

Adaptive Control   An automatic control scheme in which the controller is programmed to evaluate its own effectiveness and modify its own control parameters to respond to dynamic conditions occurring in or to the process which affect the controlled variables.

Analog-to-Digital (A/D) Converter   A device which accepts an analog quantity at its input (normally electrical voltages), and produces a unique binary code at its output for each value of analog input (within its resolution).

Aperture Time   As used with A/D converters this term refers to the discrete period of time during which the A/D converter is actually in the process of converting the voltage at its input. During this interval of time, there are restrictions as to the permissible variations in the input signal which will still result in a valid conversion.

Binary   A number system using base 2. The system's digits can have only the symbols 0 and 1. The term is used to describe the operation of logic elements and computers which operate using two-state devices and which simulate operations performed in the binary number system.

Binary Number System   A number system based on the existence of (only) two digits, zero and one. It has been applied to electronic switching circuits which are forced to operate basically in two states, full on and full off: collectively termed binary digital computers.

Bit   A commercially accepted abbreviation of *Bi*nary Digi*t*. The smallest unit of (binary) digital information (1 or 0; true or false; go or no-go; etc.).

Boolean Algebra   A mathematical algebraic system (named for its discoverer, George Boole) which allows for the existence of only two possible values for each of the algebraic variables. Boolean algebra (logic) operators are AND, OR, NOT, EXCLUSIVE OR, and combinations of these operations.

Buffer Amplifier   An electronic amplifier which electrically conditions an electronic signal for use by following circuitry; primarily included to provide electronic isolation between the two circuits.

Byte   A group of binary bits which are to be interpreted together in order to decode the information they represent. Or a group of binary bits which are processed as a unit. Or both.

Cathode Ray Display Terminal (CRT)   As applied to digital control systems, this term refers to the class of peripheral equipment which has a cathode ray video display (a television-type display) as an output device, plus a keyboard (typewriter-type) input device.

Comparator (Voltage)   As used herein, this term refers to an electronic circuit which has two voltage input terminals, and the output of which is driven either to saturation or cutoff (in a "digital" fashion) based upon the relative magnitudes of the two input voltage signals.

Decrement   Decrease in magnitude by adding a negative 1.

Digital Computer   An electronic computing and processing device designed to operate on numbers and quantities which have been represented by discrete codes. These numbers and quantities are binarily coded, and the processing performed on them can be represented either by rules of binary arithmetic or Boolean logic equations. They are normally stored program computers, which means that the sequence of processing is controlled by a program written and appropriately represented within the computer's memory banks.

Digital-to-Analog (D/A) Converter   A device which accepts discrete binary codes at its input, and produces a unique analog signal (normally voltages) at its output for each discrete input value.

Electrical Noise   Any unwanted component of an electrical signal. Degrades the performance of an electrical system.

Electromagnetic Interference   Also called radio-frequency interference, one form of noise, and other terms. It refers to the property of wires in an electrical system to act as antennas, and in an electromagnetic field (caused by transmitting antennas, motors, car ignition systems, etc.) these wires "pick up" noise signals and couple them into the system.

Flip-Flop Register   *See* Register.

Increment   Increase in magnitude by adding a positive 1.

Least Significant Bit (LSB)   The binary bit which is closest (next to) the binary point in the integer portion of a binary value. Also the binary bit which is furthest from the binary point in the fractional portion of a binary value.

Most Significant Bit (MSB)   The binary bit which is furthest from the binary point in the integer portion of a binary value.

Multiplex   This term refers to the sharing of specific devices, circuits, or wires for many mutually independent purposes. Normally in digital systems the devices, circuits, and wires are *time*-multiplexed, or *time*-shared.

Peripheral   A general term applied to hardware which is connected to a digital computer and which communicates with the computer. Teletype units, cathode ray displays, A/D converters, etc., are common computer peripherals.

Programmable Gain Amplifier   An electronic amplifier which has several electronically selectable gains.

Programs (Digital Computer)   A series of appropriately coded instructions (or commands) which sequentially control the operation and data manipulations executed by a digital computer.

Quantization Error   The error associated with the limited resolution of digital codes when representing (essentially infinite resolution) analog quantities.

Reference Power Supply   An electronic power supply whose purpose is to provide reference values for other circuits, but NOT to provide the basic power requirements for those circuits. Reference power supplies are abnormally well regulated, extra stable, etc.

Register   A group of (binary) flip-flops so interconnected that their operation is coordinated (synchronized). They are used as temporary storage devices for bytes and words, and also provide buffering between asynchronous devices in digital systems. The binary outputs from a register are normally (but not necessarily) meant to be interpreted as a group.

Sample-and-Hold   An electronic circuit which performs the functions of (1) tracking or sampling—in which mode its output is an exact time replica of its input signal; and (2) hold—in which mode its input is electrically (temporarily) disconnected from the signal it was tracking and its output remains at the precise value it was at the instant the input was disconnected.

Scale Factor   A numeric quantity which describes the relationship between (the magnitudes of) two variables. In most cases, the dimensions associated with this numeric quantity will be those of the two variables represented (one set of dimensions in the numerator and the other set of dimensions in the denominator).

14

INTRODUCTION
TO PROCESS CONTROL
DIGITAL COMPUTERS

INTRODUCTION

Now that we have developed the theory behind the use of controllers in automatic process control systems and the types of hardware necessary to convert process variables from analog to digital, we can finally discuss the most complex, most flexible, and most capable of all process controllers—the digital computer.

Notwithstanding this glorious introduction, it must be noted that digital computers may not necessarily be the best choice as controller for any given process; they definitely have their limitations and disadvantages. The choice of whether to go to a digital computer as the controller for a process, or to use analog computers or any of the other types of controllers heretofore discussed, is a very complex question. The decision as to which way to go is based primarily on the economics of the particular situation; if a digital computer process controller is going to realize some economic advantage as compared to other types of controllers, it should be considered further; otherwise, it may not be the best choice.

Digital computers have found natural application in more complex systems, where their tremendous computational capabilities can be fully utilized. Because they are so complex, require so much signal conditioning of variables prior to their being acceptable for use in the computer, and are basically fairly expensive from a systems point of view, they have typically not been used in smaller systems; they are just too "expensive" to compete with the other types of controllers. It must be noted, however, that potential uses of microprocessors promise to cause a reeval-

340

uation of the actual cost of digital computer control even for single-loop control. Not uncommonly, they are used in combination with other types of controllers. For example, many modern electronic analog (single-loop) controllers are being manufactured with a computer-mode option. This option permits the (single-loop) controller to receive its setpoint from a supervisory digital computer. There is essentially no limit to combinations of types of controllers that may be used in any given process to achieve the most economical acceptable control.

In order to commence our discussion of the application of digital computers to process control, we had first better discuss the workings of digital computers in general, then discuss process control application of digital computers specifically. Appendix B has been included in this text as a review of the binary numbering system and common binary coding schemes for those readers either unfamiliar with this information or desiring a brief reveiw. The information contained therein will be assumed to be a prerequisite to the following chapters on digital computers.

WHAT A DIGITAL COMPUTER IS

A *digital computer* is a hardware electronic device which operates based upon principles and equations as expressed in the binary number system and Boolean algebra; it operates only on binarily (two-state) coded information. It has the internal capability of accepting binarily coded information on its input terminals from external devices and providing binarily coded data at its output terminals for use by other devices. It has the internal capability of performing binary arithmetic and (Boolean) logical operations on the information either presented to it at its input terminals or on information generated internally. Finally, all basic digital computers have the capability of indefinitely storing large volumes of information internally. This stored information can be of two distinctly different types. The first type is *data*, which the computer is going to use or to modify in its calculations. The second type of stored information consists of uniquely coded binary *words*, which are (only) meant to be interpreted by the computer's own internal logic circuits in order to cause the computer to do some useful operation. This latter type of stored information is the computer's program instructions.

The computer is not a stand-alone device. It requires a considerable quantity of auxiliary (peripheral) equipment in order to perform in a given system. As noted in Chapter 13 simply to convert analog information into digital form for use within a digital computer requires a considerable quantity of equipment. In most practical systems, additional equipment is also required in order for the system operators to evaluate the performance of the computer and to control its operation. It is not uncommon for this auxiliary equipment (everything except the basic computer itself) to be considerably more expensive and to occupy considerably more volume than the computer.

HOW A DIGITAL COMPUTER DIFFERS FROM AN ANALOG COMPUTER

Both analog and digital computers are used to simulate time-variant processes and mathematical equations. Since automatic control system theory generates the control requirements in the form of mathematical expressions, both types of computers are therefore adaptable to simulate the control equations and thereby act as process controllers. The basic difference lies in how they accomplish the solutions to those equations. Appendix B offers a comprehensive discussion of the basic differences between the terms "analog" and "digital" which will not be repeated here; of course, this is the most basic difference between the two types of computers.

Analog computers act directly on analog quantities, accomplish all the data (or information) processing in analog form, and provide the solutions at their output directly in analog form. To accomplish this, certain portions of the analog computers must be hardwire dedicated to specific control functions, and these portions of the computer cannot be used for any other purpose without rewiring. Therefore, as the system gets more complicated, the quantity of hardware required in the computer increases in direct proportions to the complexity. The analog computer has no true memory capabilities as far as data storage is concerned; therefore, its capability is limited to the processing of current information only.

The digital computer acts directly on digital information, does all its processing on information in digital format, and provides information at its output in digital format. This can be a disadvantage in that all analog quantities must be converted to digital before they can be operated on by the computer, and vice versa for the outputs from the computer. However, this can also be a distinct advantage, since once the information has been converted to digital, it can be stored indefinitely in that form without loss of accuracy.

The digital computer has certain basic built-in hardware, and then programs cause data to be manipulated by these pieces of hardware. To modify the solutions or add control functions, normally neither additional hardware nor rewiring of the system will be required. Modifications normally can be made directly to the program (only), which will result in the desired changes in the solutions to the control equations. Therefore, modifications to the system calculations are more easily (and economically) accomplished.

Techniques have been developed which permit the virtual unlimited storage of binarily coded information, both as far as time is concerned and as far as volume of stored information is concerned. Therefore, digital computers have the distinct advantage that the programs can use this stored information in equations in ways not possible with analog computers.

In analog computers, only equations for which hardware circuits and devices can be constructed can be solved. In digital computers, practically any equation, regardless of its complexity, can conceivably be solved by programming, without the need for any specially designed hardware. Therefore, the digital computer can solve more complex problems than is possible with analog computers.

Analog computers, since they consist of electronic amplifiers and other electronic components, each of which has a stated accuracy and each of which changes it characteristics with temperature and time, introduce inaccuracies due to drift of these components. The more of these components and amplifiers used in the solution to any problem and the longer since the computer was recalibrated (rezeroed), the worse these total errors become. Digital computers, since they operate on only two levels, do not suffer from drifting components in the same way as analog computers. Therefore, none of the errors due to drifting components affect the inherent accuracy of the digital computer's solutions; eventually they may drift far enough for the computer to fail, but until that point, none of the accuracy will suffer.

HOW A DIGITAL CONTROLLER DIFFERS FROM OTHER CONTROLLERS

Probably the most basic difference is that one single digital computer can accomplish the same overall control as literally hundreds of individual conventional controllers. This is basically why digital computers, although they are considerably more expensive than individual controllers, can be economically justified for a control system.

When using individual (single-loop) controllers, at least one controller will be required per control loop (in cascade control systems, more than one will be required). Each controller will be hardwired into its loop, and future changes will require rewiring. When using a digital computer, the sensors are run to the computer's input, and servomechanisms are connected to the computer's output; that is all the wiring there is. In order to change control functions (i.e., P, I, and D) and configurations (i.e., from cascade control to feed-forward control, etc.), the changes will be made to the computer's program and not necessarily to any hardware. The digital computer is so flexible in this regard that the control strategies can be changed by the computer program itself, based upon the past history of the measured output of the process. This is called *adaptive control strategy*.

Of course, there is a greater variety of display and other input and output equipment available for use with digital computers. Perhaps the most notable piece of display equipment currently finding increasingly popular application in process control systems is the *cathode ray display* (CRT).

Finally, owing to its high-speed computational power and the possibility for vast storage of information, process control computers can simultaneously be used to constantly compute inventories and even schedule production runs. Not infrequently, the same computer can be used for customer billing and payroll during its idle periods. Therefore, when a digital computer is used in a control system, there are considerable system differences.

HOW DIGITAL COMPUTERS EXERCISE CONTROL

In most process control systems, digital computers are not capable of exercising any better control. They are no better or no worse than other types of controllers; they are simply more economical for the specific applications where they are used. In other process control systems, conventional controllers, for one reason or another, simply could not be used to obtain satisfactory system performance. Examples would include some types of processes that react at very high speeds, where weight, power consumption, and volume of electronics are a major consideration; where the process control system is so complex that individual controllers are not feasible; processes in which the product produced varies considerably from time to time, and the necessary frequent resetting of individual controllers would be uneconomical; processes that are spread out over considerable distances (even hundreds of miles), where digital communications networks will be required; and processes where the system accuracy specifications cannot be economically met using analog control—to mention a few.

To present a fair picture, however, it must be noted that there are also process control systems where digital computers normally cannot be used. Examples might include processes that occur in potentially explosive atmospheres and no electronic equipment can be used, and processes that occur in environments where electrical noise potentially could render the digital computer useless.

As mentioned previously, digital computers have their place in the process control field, but so do the other types of controllers. There are many applications where any of several types of controllers could be successfully used. In these applications, factors other than ability to exercise adequate control must be used in making the final decision. In many of these applications, the flexibility of digital computers to accomplish other functions (in addition to process control) is the deciding factor.

TYPES OF DIGITAL COMPUTERS

We have been consistently talking about process control digital computers. This could lead to the question: Are there any other types? The answer is definitely yes; there are several basic types of digital computers. Each type, although they are all binary digital computers, has design features which optimize that particular design for specific applications. Competition between computer manufacturers has forced them to design around typical applications in order to compete successfully.

Differences among Types of Digital Computers

The basic design of the internal structure of digital computers is referred to as its architectural design, or its *architecture*. This term is used to describe the basic philosophy of data handling and processing around which a computer is designed and constructed.

For example, a computer may be designed to work with very large numbers and complicated formula solutions. Therefore, its architecture would be based upon optimization of its computational abilities; this would be a typical scientific computer design. At least one scientific computer has been designed in which over 100 simultaneous arithmetic calculations can be handled.

Another computer may be designed to process large volumes of information from many sources, primarily reassembling this information while performing relatively simple calculations. Therefore, its architecture would be optimized more around its input and output capabilities than its arithmetic capabilities. A typical business computer would have this type of architecture.

Many computers are designed to be *general-purpose*; that is, they have adequate arithmetic and input-output capability to be used in a wide variety of applications. The large majority of small computers in the mini- and micro-classes are of this general-purpose type.

Finally, there is the category of special-purpose computers. These computers have been specifically designed for particular applications (other than scientific or normal business applications). Frequently, the special-purpose computers are basically general-purpose computers that have been surrounded by specific peripheral equipment and programmed for the particular type of job they will be used on. The cost to design truly special-purpose computers is normally prohibitive; therefore, most of them are based around the general-purpose digital computers.

USE OF DIGITAL COMPUTER FOR PROCESS CONTROL

For process control applications, a computer is required which falls into the general-purpose category. Process control applications normally require only moderate arithmetic processing, as will be noted in Chapter 15; therefore, normally no extra scientific-type computational capability is included. The process control input/output requirements are extensive; however, they are of a different type than for business applications. Therefore, the process control computer is of neither the scientific nor the business type. Furthermore, normally both of those types have basic built-in capabilities far in excess of process control requirements, and they are simply too large and too expensive.

In the large majority of process control applications, smaller general-purpose digital computers have been successfully used and their interfaces have been modified to realize the necessary capabilities. In some instances, the central processing unit itself has been modified to make the computer more compatible with the process requirements. This degree of modification puts some of the processors into the special-purpose class, where a general-purpose computer was modified for certain applications. In the vast majority of process control systems, however, the basic computer has not been modified.

In the past several years, new computers have been introduced as being general purpose in nature but they have been designed such as to make them more adaptable to process control than general-purpose computers had previously been. The reason that computer manufacturers have done this should be obvious; the use of digital computers as process controllers has mushroomed to the extent where it has become a very profitable market for computer manufacturers to enter. This is based upon the volume of computers currently consumed in process control applications and the potential future market for process control computers.

HOW DIGITAL COMPUTERS WORK

Now that we have discussed just what digital computers are, how they can be used as controllers, how they compare with other types of controllers, what types of computers there are, and how they differ, we shall get into how they work in general. In Chapter 15 we will cover how they work in process control systems specifically.

The digital computer is basically a very complicated assembly of electronic circuits, which themselves (normally) perform very simple operations that can be represented mathematically in the binary number system or by Boolean equations. The computer performs simple binary and logical operations on binary-coded data in an extremely rapid, reliable, predictable, and sequential fashion. It does exactly what it is told to do (programmed to do) and nothing else. It cannot do anything besides the job it is programmed to accomplish; it cannot think for itself.

How do they accomplish their programmed tasks?

By interpreting various combinations of 1s and 0s as commands or orders to accomplish hardwired operations

By performing only one operation at a time

By rigidly following a sequential series of commands in order, unless commanded to alter that sequence

By performing simple tests and making relatively simple (programmed) decisions based upon the results of those tests

By performing operations in microseconds that would take human beings minutes (or even longer) to accomplish

By having very large memory banks in which to store and retrieve information when needed

By establishing bidirectional communications with "outside world" devices

As digital computers can be basically broken down into four functional units; the input/output unit, the arithmetic unit, the timing and control unit, and the memory unit. The combination of the first three units together is normally referred to as the central processing unit (CPU). The *input-output* (I/O) *unit* contains the digital logic necessary to interface the computer to external pieces of hardware equipment; such as analog-to-digital and digital-to-analog converters. It includes the logic necessary to generate and verify *synchronization* between the CPU and external devices so that data can be reliably transferred between them. It is the main avenue(s) of communications between the CPU and the outside world.

The *arithmetic unit* includes all the hardware used to modify data in the process of solving equations or performing other mathematical (or logical) operations on that data. It includes logic that will indicate the successful completion of each arithmetic operation performed. This logic can be tested by the timing and control unit in order to make decisions as to the procedure to be followed in either the event of successful arithmetic results or for unsuccessful results.

The *memory unit* can be visualized simply as a large warehouse full of post-office-type boxes, each one sequentially numbered. Digital information can be stored there (one "word" in each box) for later retrieval and use, as long as its location is remembered. The "words" stored in these post office boxes are all a specified number of binary bits in length. The number of binary bits handled by the computer (and normally its memory unit) simultaneously in a single operation is technically referred to as the *word length* of the computer. The vast majority of the computers currently used in process control have a word length of 16 bits; 32-bit computers are also commercially available.

There are several types of memory, each basically quite different in physical structure as well as in application, which are normally used with a process control computer. The magnetic-core-type memory has been the standby for primary computer memories because of its flexibility and the fact that it does not lose its information when power is lost. It requires a considerable amount of electrical power to operate and is the most expensive type of memory (based upon cost per bit stored). There are several basic designs of solid-state (semiconductor) memories which have become feasible for use in process control computers. These memories require considerably less electrical power to operate, are physically much smaller, and are considerably less expensive (dollars per bit). However, they all lose their storage capabilities when power is turned off (or lost due to a power casualty), and it may be a fatal disadvantage in many systems. New battery backed-up RAM semiconductor memories have overcome this problem. They have a built-in battery with a ten-year estimated data retention capability. They will be replacing core-type memories.

Some types of solid-state memories can be permanently (or at least relatively permanently) programmed so that neither power failure nor computer failure can alter the program. Various types of these *read-only memories* (ROMs) are finding application in certain types of process control systems. There are other types of solid-state memories which are finding application, particularly when dealing with microprocessors; however, these will suffice as examples for our purposes here. Suffice it to say that there is a need for large volumes of information (binary words) to be stored internally to the computer, where they are directly and immediately accessible to the computer's internal timing and control logic.

The *timing and control logic* comprise the fourth functional unit of any digital computer. It is the responsibility of this logic to synchronize and control each internal operation inside the computer. This logic is directly connected to each unit internal to the computer and it controls the operation of the CPU directly. It, in turn, receives its basic instructions from the programmer. Therefore, the individual *instructions* in a computer program are written to be interpreted by this timing and control logic, in order for it to cause some useful operation to be performed by the computer.

The four functional units are interconnected as illustrated in Fig. 14-1. Note that information can be transferred bidirectionally between both the arithmetic unit and memory, and external devices. This arrangement is typical of any digital computer and these avenues (or channels) of communication are termed *buses*. Note also that the timing and control unit is illustrated as being connected (also bidirectionally) with the other three units. In fact, it controls the operation of the other three functional units and is always understood to be there regardless of whether or not it is shown on the diagram.

Now let us follow the sequence of operations (executed by internal logic) caused by a typical instruction through a typical digital computer. It all starts out with a *computer instruction*. This is a binary word (or more than one word for some instructions) which is coded by a programmer (human type) for a specific computer. The actual coding varies between computers insofar as the significance of each bit is concerned; however, this coded computer instruction is meant to be interpreted by a portion of the computer's timing and control unit (called the *instruction register*). Before proceeding, we had better make sure that there is a clear understanding of just what a register is and how it is used.

A register, and there are many of these in computers, is a grouping of independent binary flip-flops. Each flip-flop has the capability of storing (memorizing) one binary bit of information which must be presented to it simultaneously with a command (also called a *strobe pulse* or a *trigger command*) to remember that bit. Without the occurrence of this strobe, the flip-flop ignores the binary bit at its input, or any changes in that bit, remembering only the single bit that was presented to it concurrent with its last strobe command. There are normally as many of these flip-flops assembled into a register (in a computer) as there are number of bits in that

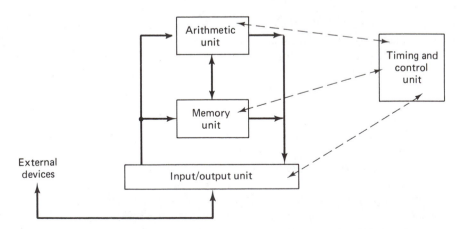

Figure 14-1 Four basic functional units of a digital computer.

computer's word. Therefore, for most process control computers, a register will consist of 16 independent flip-flops, all with a common strobe, and whose outputs are intended to be interpreted simultaneously as a complete word.

Continuing on with the operation of the computer, the programmer normally codes a series of instructions, not just a single instruction, each one of which will cause the computer to perform some elementary useful function (collectively called his "program"). The function could be data manipulation, data transfer, or testing for the existence of some condition internal or external to the computer. Once the programmer has determined the sequence of (computer) instructions required to perform the function he desires and has properly coded these instructions, the instruction sequence is written into the computer's main memory for future use.

The computer operator (probably the programmer) then loads a register in the timing and control unit (called the *program address register, program counter, sequence counter,* or other names, depending upon the specific computer being used) with the *address* (specific memory location or "P.O. box number") of where the *first* instruction of the program has been stored. He can then hit the start (or run) button on the computer's console and the computer will execute the program. But what happens inside the computer after the program itself has been *loaded* (written or entered) into the computer's memory, the address of the memory location where the *first* instruction of the program was stored has been loaded into the program address register, and the computer start (run) button has been pushed?

Certain sequences of operation are so routinely performed internal to the computer that computer manufacturers have relieved the programmer of the tedium of having to program their execution. The first operation after the start button has been pushed is one of these routine operations; it is called the *instruction fetch cycle*. In most computers, the following sequence of operations is performed.

First, the binary number in the program address register is transferred to the *memory address register* and the memory is commanded to READ the contents of that memory location. In most of the computers used in process control (mini- and microcomputers) the memory is a relatively self-contained unit, and it operates *asynchronously* from the computer. This means that it is not operated by the computer CPU's timing and control (T & C) logic but by its own timing and control logic. The memory's T & C logic "looks for" the address within memory (the P.O. box number) which is to be accessed and a signal to either retrieve (copy but not destroy) the information in that location (READ) or to replace the information that was previously in that location (destroying it) with new information (WRITE). When the memory is commanded by the computer's T & C unit to READ the first instruction, it (the memory unit) starts to run on its own internal timing and control (T & C) logic and completely ignores the operation of the rest of the computer until it has completed its cycle. This logic automatically uses the binary number in the memory address register (which was loaded with the address of the first program instruction) as the location in memory to work with (access). Since it (the memory's T & C logic) was given the command to READ the contents of that memory location, it *copies* the contents of that memory location into its transfer register (the *memory data register*). The original contents of that location have not been modified by this READ operation; the same information can be read indefinitely without modifying it. When the memory T & C unit has completed its READ cycle, it either sends a cycle-completed signal back to the computer's T & C unit or it simply stops; in either case, it does nothing else and waits for another READ or WRITE command from the computer's T & C unit.

When the computer's (CPU's) T & C unit is sure that the memory has completed the commanded READ of the first computer instruction, it "looks at" the memory data register, takes the binary word it finds there, and transfers this word to the *instruction register*. This completes the normal (and completely automatic) instruction-fetch cycle. It must be accomplished prior to the execution of *each* and every one of the instructions the programmer loaded into the computer's memory. One additional feature is normally included in this cycle: the program address register is normally changed to indicate or "point to" the memory address where the *next* program instruction should be fetched from once the current instruction (the one that has just been "fetched" but has not yet been *"executed"*) has been executed. This will normally be the very next sequential location in memory; therefore, the program address register is normally *incremented* (has a binary 1 added to it) prior to completion of the instruction fetch cycle. Figure 14-2 is a flow diagram illustrating and summarizing this process.

The computer is now (finally) ready to execute the programmer's first instruction. Up to this point, the programmer had no control over the operation of the computer; however, now (after the instruction fetch cycle has been completed) the computer's T & C unit "looks at" and decodes the binarily coded word in the instruction register and performs the operation it requires. As noted previously, the instruction can cause any of several possible types of operations to be performed. Some of the more common instructions:

Cause a binary word to be input to the computer or the computer's memory from some external device.

Cause a binary word to be output from the computer or the computer's memory to some external device.

Cause a binary word to be transferred from one register to another (both) internal to the computer.

Cause any of the following types of arithmetic operations to be performed on a binary word: ADDITION, NEGATION, SUBTRACTION (or MULTIPLICATION or DIVISION in some computers).

Cause a binary logical operation to be performed on a binary word, such as SHIFT LEFT, SHIFT RIGHT, ROTATE, or to logically compare (AND, OR, EXCLUSIVE OR) it with a second binary word.

Cause a binary word to be transferred from some register internal to the CPU to the computer's memory in any of several possible ways, or transferred from the memory to some CPU register (READ or WRITE).

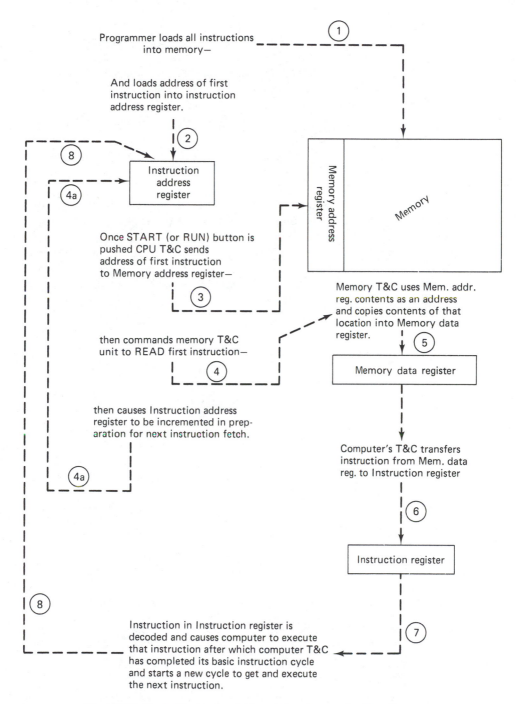

Figure 14-2 Typical instruction fetch and decode operation flow diagram.

Cause a test to be performed on a binary bit, on the algebraic sign of the result of an arithmetic operation, on the successful completion, or not, of an arithmetic operation (OVERFLOW), on the comparison of a binary word with zero (GREATER THAN, LESS THAN, or EQUAL TO) or another binary word, or on any of a number of various testable conditions that may exist within (or external to) the computer's CPU.

Cause the sequence of program execution to be conditionally altered based upon the result of one of those tests.

Control the actual state of the CPU itself (i.e., HALT the operation of the computer).

Figure 14-3 is a block diagram illustrating the functional pieces of hardware internal to a typical microcomputer, showing the paths of data transmission between them. This block diagram illustrates the architecture of this particular computer; other computers will have slightly different architectures and therefore different combinations of functional pieces of hardware and different interconnections. However, the previous (very brief) description of how a digital computer works applies generally to any of the computers.

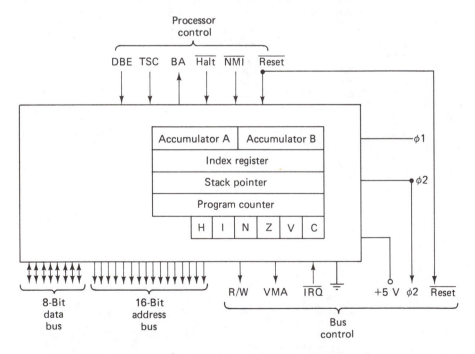

Figure 14-3 Architectural block diagram of the Motorola M6800 Microprocessor. (Courtesy of Motorola Semiconductor Products, Inc.)

Peripherals

Peripherals are pieces of hardware that are connected to the computer's CPU and work with it but are not essential to the basic operation of the CPU itself. Based on the text thus far, an analog-to-digital conversion system could be considered as a piece of peripheral hardware. In a system block diagram where the CPU is shown as a block, the other external equipment illustrated as being connected to and working with the computer is the peripheral equipment.

As illustrated in Fig. 14-4 and Fig. 14-5, there are many different types of peripherals commonly used in process control systems. Each one must be electronically interfaced to the CPU and performs a unique function in the system. The A/D and D/A conversion systems have already been discussed both as far as how they work and in their systems applications.

The combination of a dot matrix printer and a floppy disk unit are the most commonly used peripherals in small computer systems of the type used in process control systems, at least with mini and micro systems. Hard disk units are becoming increasingly popular due to their large storage and high-access-speed characteristics.

Cathode-ray display units (CRTs) have been rapidly increasing in their pop-

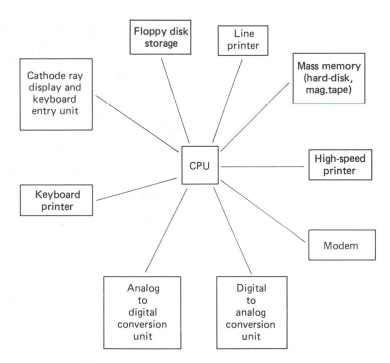

Figure 14-4 Computer system block diagram, illustrating commonly used process control peripherals.

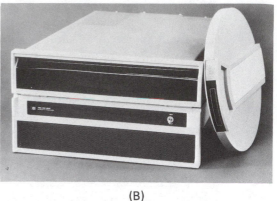

(B)

(A)

(C)

Figure 14-5 Typical process control digital computer peripheral devices: (A) terminal printer; (B) disc mass memory system; (C) cathode ray display, keyboard entry, and cassette tape storage terminal. [(A) Courtesy of Data General Corp.; (B) and (C) courtesy of Hewlett Packard.]

ularity as input/output (I/O) peripherals in process control systems. Different manufacturers provide a wide variety of features in their CRT units. All provide a television-type cathode-ray display tube, and most at least have the option of including a typewriter-type keyboard with the display. Many models have their own internal memories, so the device can be used without bothering the CPU (off-line) and then the computer can be alerted that the CRT's memory has information to be transferred (at computer speeds) when the operator is sure that the information is complete and correct. Many have "blink" capability to alert the operator to important pieces of information and many also have "color" capability.

High-speed paper tape readers and punches, high-speed typewriters, and high-speed line printers are advanced performance devices that duplicate (in many systems) the features available in the old style teletype unit. In many systems there is the need for both the older units and these advanced performance peripherals, each performing basic required system functions. Not infrequently, these peripherals are not physically located in the same room or even in the same building as the CPU.

Backup mass-memory devices are normally included in process control systems. The CPU's internal main memory has limited storage, and commonly there is the need to store portions of the control programs and data files in some backup mass-memory device. These programs and data files can easily be so large that the computers do not have the basic capability to work with internal memories of that size. Also, these internal memories are normally of the most expensive type on a dollar cost-per-bit stored basis. Therefore, less expensive mass memories would normally be used for much of the required storage. These mass memories include any of the various types of disc memories (fixed-head and movable-head discs, floppy discs, environmentally protected discs, cartridge-load discs, etc.), drum memories, and any of the many types of magnetic-tape memories (reel-to-reel, various cartridge designs, and cassette). Frequently, these mass-memory devices are provided a special channel of communications which is interfaced directly with the computer's main (internal) memory; this capability is called *direct memory access*, and it must be provided for in the basic architecture of the CPU as compared to an add-on option.

The final category of peripheral that we shall discuss here (and this does not mean there are no more types of peripherals, only the most commonly encountered types have been presented) is the *modem*. This peripheral provides an interface to (normally) telephone lines for long-distance communications. Some processes are spread over many miles, and there is the need for the CPU to communicate with remotely located I/O devices, remote sensors and servos, or even remote computers. The modem provides the interface electronics for this type of data transmission.

HOW DIGITAL COMPUTERS WORK WITH PROCESS CONTROL SYSTEMS

As you may have deduced from the preceding information, it is not easy, nor is it inexpensive, to apply digital computers to process control applications. However, it has proved to be possible and it is unquestionably economical to do so in many systems applications, despite the complexity and cost of the required system.

How it can be accomplished is the topic of the next chapter. Figures 14-6 and 14-7 illustrate commercial CPUs.

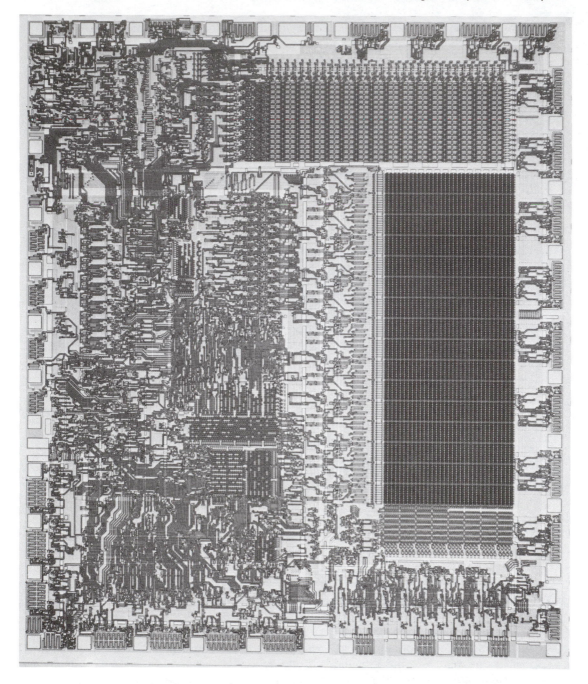

Figure 14-6 INTEL's 8748 complete microcomputer and memory in a single integrated-circuit chip. (Courtesy of INTEL Corp.)

Figure 14-7 Microprocessor chip in the foreground is the heart of the microcomputer system (on a printed-circuit board) in the center of the picture, and of the fully packaged computer in the background. (Courtesy of Data General Corp.)

QUESTIONS

1. In your own words, what is a digital computer?
2. What general criteria would be used to determine whether a digital computer should be considered as a controller for any specific system?

Define the following terms, as they relate to digital computers, in your own words.

3. Data.
4. Word.
5. Binary.
6. Logical operations.
7. Supervisory controller or computer.
8. Peripheral.
9. Central processing unit.
10. Register.
11. Flip-flop.
12. Mass memory.
13. Instruction fetch cycle.
14. Program.
15. Briefly compare analog and digital computers, insofar as process control applications are concerned.
16. What functions can a digital computer perform in a system that cannot be (economically) performed by or duplicated in an analog computer?
17. How can a single digital computer be used to replace an analog computer that controls many process control loops?
18. How do digital computers compare overall with other types of controllers for process control applications?
19. What are the basic types of digital computers?
20. Compare the basic types of digital computers from a hardware point of view.
21. In your own words, and as briefly as possible, describe how digital computers work.
22. Describe the four basic functional units of a digital computer and their interactions with each other.

What is (are) the function(s) of the following registers in a digital computer?

23. Program address register.
24. Memory address register.
25. Memory data register.
26. Instruction register.
27. Describe the instruction-fetch cycle in a typical digital computer.
28. List several of the more popular peripherals normally associated with process control digital computers.

GLOSSARY

Architecture   When this term is applied to digital computers it refers to the basic philosophy of data handling around which the computer is designed and constructed.

Arithmetic Logic Unit (ALU)   That portion of a digital computer central processing unit hardware and logic which specifically deals with the arithmetic and logical manipulation of data.

Assembler   A digital computer program (written in machine language) which performs the function of conversion (or translation) of a second computer program (written in symbols or mnemonics) into the specific binary codes which the computer must have in order to execute that program. There is a one-to-one correspondence between the mnemonic instructions and the machine language (binary codes) generated by the Assembler.

Assembly Language Programming   The process of writing a digital computer program using mnemonics (or symbols) to represent computer instructions and addresses. This program must then be translated into the binary sequences that the computer requires (the machine language codes) by an Assembler before it can be executed.

Asynchronous   Referring to a condition where two or more devices operate under the influence of their own individual timing networks. There is no intentional effort made to synchronize (or lock together in step) the individual timing networks.

Execute (a digital computer instruction)   A term used to describe the central processing unit sequence of operations which interpret a computer instruction and cause the computer's logic to perform the indicated operations as coded into that instruction.

General-Purpose Digital Computer   A digital computer which has been designed for mass production rather than a specific application. It has a relative balance between the arithmetic processing power, input/output capabilities, timing and control logic, and memory capacity; this balance makes it adaptable for a multitude of applications.

Input–Output (I/O) Unit   The portion of a digital computer central processing unit hardware and logic which specifically deals with the transmission of information between peripheral devices and the central processing unit.

Instruction (Computer)   A coded statement which is intended to control the internal operations and/or data manipulation within a digital computer when the computer "executes it". A group of instructions taken (and executed) in sequence is the computer program.

Instruction Fetch Cycle   That portion of the routine (hardware programmed) computer operation which locates, fetches, and stores the next program instruction to be executed in the instruction register, initiates the computer to execute the instruction (if necessary), and finally sets up the computer's logic so that the next sequential instruction will be fetched after this one has been executed.

Instruction Register   A hardware register which is part of the central processing unit timing of control logic, and whose function is to temporarily store the particular computer instruction which is currently being executed.

Machine Language Programming   The process of writing a digital computer program directly in the binary codes and addresses that the computer requires in order to execute that program.

Memory Address Register (MAR)   That portion of the computer memory unit which holds the address of the location (in the memory) which is currently being accessed.

Memory Data Register (MDR)   That portion of the computer memory unit which either (1) holds the word which has just been READ out of a memory location, or (2) holds the word which is to be stored (written) into a memory location (WRITE).

Memory Unit   That portion of a digital computer central processing unit which provides the primary storage for programs and data.

Mnemonic   An abbreviated term (usually three or four alphabetic characters) which is used to assist the programmer in his efforts to program a computer in assembly language.

Program Address Register   *See* Program Counter

Program Counter   A hardware register which is part of the central processing unit timing and control logic, and whose function is to (normally) hold the address (in the computer's main memory) where the *next* instruction to be executed is stored.

READ   The term used to describe the process of (nondestructively) copying information from a source.

Sequence Counter   *See* Program Counter

Symbolic Programming   *See* Assembly Language Programming

Synchronous   The slaving of two or more devices together in time so that they operate based upon a common "clock" signal; such that they are locked together in frequency and phase.

Timing and Control Unit   That portion of a digital computer central processing unit hardware and logic which controls the sequential operations within and external to the central processing unit. It includes the basic CPU timing and instruction decoding logic.

Word   In reference to digital computers this term refers to a group of binary bits which are handled (transferred, modified, and/or stored) as the primary unit of information.

Word Length (of a digital computer)   The number of bits which make up a word for a specific computer. This parameter is (normally) fixed by hardware design.

WRITE   The term used to describe the process of transferring information into a destination location (a register, device, or memory storge location) for storage. This process always results in destruction of the information previously stored in that location.

15

DIGITAL COMPUTERS AS PROCESS CONTROLLERS

INTRODUCTION

There are two basically opposite approaches to the study of applications of digital computers to process control systems. The first approach is to study a typical digital computer and then attempt to "bend" the process control system to work with the computer. This was necessarily the approach taken for the first designs of digital control systems because of the limited availability of digital computers. Understand, however practical the approach was at the time, that it certainly was not the optimum method of designing control systems.

The other basic approach is the "pie-in-the-sky" type of approach where first the needs of the process are defined, and then the digital computer is designed around these requirements. In modern control system design this is a very practical approach because of the wide variety of digital computers presently available on the market.

Competition between computer manufacturers has forced them to design computers around "typical" process control system requirements. The end results reflect the difference in technical opinions as to how the ideal process control computer should be designed (or what its architecture should be).

In order to determine what the architecture of a process control computer should be, it will first be necessary to investigate the sources and types of information required by the computer in order to control the process, then the processing required of this information must be determined, and finally the information required by the process (from the computer) must be determined.

Therefore, by taking the approach whereby the needs of the process are first defined, it will be possible to actually arrive at the functional specifications for a digital process control computer. This is the approach that will be followed here.

INFORMATION REQUIRED BY THE COMPUTER

Of course the actual information required by any process controller, whether it be digital or analog, depends upon the individual process itself. However, it is possible to categorize most of this information (in the form of electrical signals) by the type of electrical signal that the controller must work with.

The basic process variable being used as a source of input information might be humidity, pH, pressure, force, speed, motion, flow, or position. It will be assumed that these variables are being monitored by sensors having an electrical output. It is the electrical outputs from these sensors that are going to be categorized by their electrical values (i.e., variable resistance, millivolt potential, etc.). This was covered in Chapter 7, but will be reviewed in the following text to insure that the reader understands it in its proper context.

Generally these electrical outputs fall into two categories, analog and digital. The analog category can be further broken down into variable resistance types, millivolt potential types, larger voltage types, and variable current types (as in Chapter 7). Furthermore, as was noted in Chapter 13 sample-and-hold units and multiplexing will normally be used.

Analog Inputs

The variable-resistance-type sensor outputs are generally fed into some form of a Wheatstone bridge circuit and then to the A/D converter. Many temperature measurements, force measurements, and pressure measurements are commonly transduced to variable-resistance-type electrical signals.

The voltage-type sensor outputs are either fed into some form of a Wheatstone bridge (millivolt potentials) or through a level shifting, filtering, and impedance matching circuit (higher voltage potentials) and then to an A/D converter. Typically, some temperature, position, force, and pressure measurements are monitored by sensors having dc voltage-type outputs.

Current-type transducer outputs are normally fed to a small-value (sampling) resistor which converts the current to a small voltage signal, which in turn is handled as any other small potential. Ac-voltage-type signals are normally converted into dc voltage levels and handled as any other dc voltage signal; unless the information is contained in the frequency of the ac signal (as compared to its amplitude). In this case the ac signal may be converted to a square-wave signal and handled directly in that form or it may be converted to a dc voltage and handled as for any other dc signal.

Digital Inputs

This brings the presentation to the second category of electrical signals the process controller will be required to handle, *digital signals*. Digital signals, like analog signals, also fall into several types.

The first type includes those digital (on-off) signals which contain information in the frequency of a pulse train, similar to the ac signal previously discussed. Here the pulse train (serial information) is frequently converted to a parallel binary word by simply sampling the frequency for specific time intervals and counting (binary counter) the number of pulses during that time interval. Otherwise, the pulses can be passed directly to the processor. Digital shaft angle encoders and linear digital encoders frequently provide this type of computer input.

The second type of digital signals are those which contain information in their current 1 or 0 state, and the processor simply needs to know which of the two states the signal is in. Relay contacts, limit switches, safety devices, optical sensors, and switch positions are normally monitored in this manner. The voltages available from these switching sensors vary widely and frequently must be filtered and have their level shifted prior to being acceptable for the processor's use. Frequently, the processor also needs to know not only the present position (state) of a switch but also (and frequently even more important) the fact that the switching device has just changed position. Quite often this piece of information is important enough that circuitry will be required to override the processor's normal functions to demand that this change of state be immediately recognized and dealt with (technically referred to as *interrupt capability,* which will be further discussed later).

Frequently, a device such as the standard CRT terminal used with small computers has a series digital output. In this case a serial-to-parallel converter is required in the input interface. This consists of a binary shift register and its control logic. Essentially as each bit is received it is shifted into a register, bit by bit, until all bits in a word have been received. The output of the register then reads the same information, only in parallel.

The final type of digital signal that a digital processor will be required to work with is similar to the second type, in that the information is contained in the instantaneous value of the voltage (digital state). It has been included as a separate category due to the timing and control circuitry necessary to handle this information. This type of digital signal comes from bulk (external) memory devices such as magnetic tape, disc, or drums. Also this type of signal will be encountered when the digital processor works with keyboard-type inputs. The additional timing and control circuitry necessary to control the data transfers between the computer and these devices is very complex and normally is included in the computer interface electronics.

Summary

This generally includes all types of electrical signals that the processor will be required to work with in order to acquire the intelligence for it to make the decisions necessary to exercise control of the process. In summary, if the processor is to be a digital device, then all analog inputs must be converted to digital form prior to entry into the processor. The digital processor must also have the capability of dealing with series digital information, pulse-rate information, and with the transfers of information from other digital devices, including providing necessary control and timing for these transfers. Computer "interrupt" capability is normally required also. In all these examples, the conversion from the type of sensor output to digital voltages is accomplished in the computer input interface electronics. Therefore, the computer itself needs only to provide the capability of dealing with digital information in either parallel or serial formats. Figure 15-1 is a block diagram that illustrates (functionally) what might reasonably be expected to be included in the computer input interface electronics.

INFORMATION REQUIRED BY THE PROCESS

Now, turning attention to the information the process must receive back from the controller in order to be controlled, the computer output requirements will be defined. As with the inputs, there are the same two general categories, analog and digital.

Analog Outputs

The *analog output commands* (from the controller) will normally be used to drive some electromechanical device, such as servovalves and other electromechanical devices (which will be further discussed in Chapter 16). These outputs may require power amplification and level shifting in addition to simple D/A conversion. Other analog outputs will be used for direct input to analog control systems or computers which are "supervised" by the primary process controller.

Digital Outputs

Digital outputs from the controller may consist of serial (frequency-modulated) pulse trains, parallel digital words, or individual binary bits of information. Frequency-modulated pulse trains might be used to drive digital stepper motors (with proper power amplification) or to supply input to serial data devices such as matrix printers. Individual binary bits of information could be used to control the status of relays, interlocks, warning devices, or status indicators.

The parallel digital output could be required for computer output to auxiliary memory devices, digital peripherals such as line printers and cathode-ray-tube (CRT) displays. In these instances, as with input from this type of device, complicated timing and control signal generation circuitry will normally also be required.

Summary

This is a brief and very general description of the types of information required by the process itself (from the controller). Generally, the information will be used for either direct control of some process variable, storage in some digital peripheral

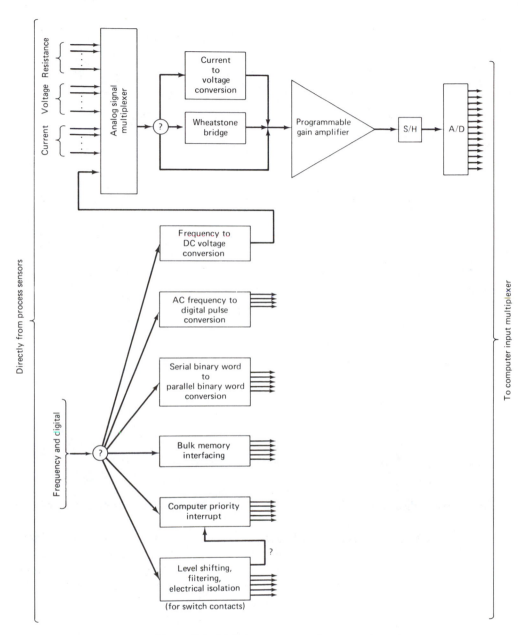

Figure 15-1 Digital process control computer input interface.

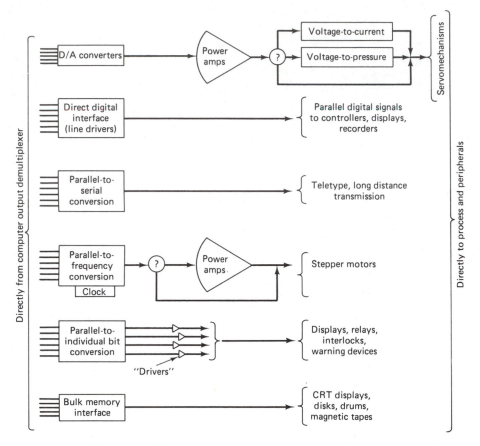

Figure 15-2 Digital process control computer output interface.

device, or indication of the current status of critical process variables for the operator.

In all cases, however, the computer itself is required only to output digital voltage levels, in either *serial* or *parallel* format. All other conversion required will be accomplished by the computer output interface electronics. Figure 15-2 is a block diagram which illustrates (functionally) what might reasonably be expected to be included in the computer output interface electronics.

COMPUTER INTERFACE ELECTRONICS

Based upon this brief presentation it must have become obvious that no computer can work directly with the process, especially if it is a digital computer. A considerable volume of electronic signal conditioning circuitry is necessary to provide *plug-*

in compatibility for the controller. Collectively, the electronics equipment that provides the necessary electronic signal conditioning is called the *interface electronics*. The total volume of interface electronics is normally broken down into the *input interface* and the *output interface* (referring to computer input and computer output). Each of these interfaces is subsequently broken down into *analog* and *digital interfaces*.

Now that the side of the input and output interfaces which will plug into the process have been defined, further attention will now be directed toward the side of the interfaces that will plug into a digital process control computer. Remember, in all cases, that the computer is required to work only with digital information which has voltage levels compatible with digital logic elements.

Computer Input Bus

There is obviously going to have to be an "avenue" for the computer to access all of the literally thousands of individual inputs. Many of these inputs will be multiplexed together through the analog-to-digital conversion system; however, that still leaves a great many inputs to be dealt with by the computer.

At this point it is necessary to point out a few of the realistic limitations of small digital computers (commonly referred to in technical literature as minicomputers and microcomputers). These computers, due to the use of integrated circuit electronics, are not small in capability or computing power as much as they are small from the standpoint of the physical size and dollar acquisition cost. Dollar cost is an extremely important specification, since there are so many different computers competing in this market.

Another reality that must be dealt with is the fact that most digital computers are constructed with only one set of logic with which to perform each of its functions. Therefore, it has the capability of doing only one operation or function at a time. These computers have the capability of accomplishing individual operations at such high rates of speed that they appear to be doing many operations simultaneously; but they are not.

With these realities in mind, even though it might seem that there is the need for many avenues of input to the computer, these many paths will require extensive digital logic and expense. Furthermore, since the computer can work with only one of the input devices at a time, it is practically possible to multiplex *all* the input devices to the same *computer input bus structure*. This bus structure acts as both a multiplexer for all digital computer input devices and as the single, common avenue of information to be transferred directly into the digital computer itself.

Since there is need for only a single input bus for the computer, the computer input interface electronics must provide all the signal conditioning necessary for each input device in order to be able to plug that device into the computer. The equipment necessary to accomplish this was discussed previously.

Let us take a few examples. The analog-to-digital conversion system has the capability of multiplexing all the analog signals to a common *data* line for each

A/D converter. This common digital data bus from the A/D converter(s) must have logic levels that are compatible with the computer input bus structure. Certainly, the actual bus structure specifications will have been economically determined in order to optimize the actual distribution of electronics between the interface and the computer itself. The parallel digital input information must have circuitry provided in the interface in order to temporarily hold this information so that it remains constant while it is actually present on the computer input bus. Typically the interface electronics are simply a flip-flop *buffer* register which accomplishes this function. The flip-flop register has the ability to ignore information present at its input until a memory command is received. It then stores the information present at its input when the memory command is given and retains this information until another memory command is received.

The interfacing between individual digital input signals (each containing one binary bit worth of information) such as relays is somewhat more complicated. As previously noted, each input may require level shifting, impedance matching and/ or filtering. The problem now is that in order for the computer to acquire information about only one of these devices, a full computer input cycle will be required. The computer cycles will be discussed later; however, all computers of the type being considered for process control are parallel devices, in that they do not ordinarily accomplish operations on a single digital bit of information at a time; they take many at a time, depending upon the word size of the computer.

Word size is the term that describes the number of binary bits that the computer handles in a single operation. Development of the specifications for computer word size is beyond the scope of this text; therefore, a word size of 16 binary bits will be assumed, since this is by far the most popular word size for computers of the size being discussed here. This is why word size was not considered when discussing analog inputs—since analog inputs are normally converted with only 10 or 12 bits of accuracy, which is well within the capability of most process control computers.

It should be noted that most of the microcomputers used in the past few years have an 8-bit word size. That word size is inadequate for most process control data applications; however, because these 8-bit microcomputers are so fast, so inexpensive, so available, and so reliable they have found wide acceptance. The word-size shortcoming has been overcome in those applications by storing the process control data in multiple precision format; that is, two sequential 8-bit words are mathematically combined into one 16-bit word and subsequently interpreted as a single 16-bit value. Thereby they have been, and still are, successfully used in numerous smaller processes.

Many modern microcomputers have a combination 8-bit and 16-bit architecture, wherein the computer typically uses a 16-bit word size internally but has only an 8-bit interface word size. These computers have realized a significant increase in processing power and are more directly suited to most process control applications. The most modern microcomputers are the full-16-bit-word-size machines of the type to be assumed in the balance of this chapter. Some of these 16-bit microcomputers actually perform mathematical operations on 32-bit words internally, but these have not yet (1986) found wide industrial process control application due to their cost and the expense and complexity of the circuitry required to interface to them.

To permit expanded use of the less expensive 8- and 16-bit architectures, support math coprocessing integrated circuits have been developed and are finding widespread use. These coprocessors are just that—additional special-purpose, dedicated-function computing chips designed to work with (but not replace) these other architectures. They typically accept several multiple precision data words from the main CPU and perform most commonly required mathematical operations on these values, interpreting them as up to 80-bit numbers, expressed in scientific (exponential)-notation format. This effectively overcomes the major shortcoming of the smaller-word-size microcomputers by providing them with a very powerful mathematical processing capability.

Since the 16-bit architectures seem to satisfy most process control applications, the following text material assumes that a 16-bit computer has been chosen as a benchmark. Simply restate references to 16-bit computers as 8-, 8/16-, or 32-bit computers and the text material remains valid in the new context.

The computer will have the capability of acquiring information about a group of 16 input signals of the type being discussed (i.e., relays) simultaneously. Therefore, the interface normally multiplexes 16 inputs to be handled by the computer as a group; again the flip-flop buffer register is used to synchronize all 16 independent signals while the computer is actually inputting this information.

Finally, there is a group of digital inputs that are considered to be either so important that when their values change, the computer must immediately stop routine processing and check their new values, or are so unimportant (or change value so infrequently) that it is not worth the time required for the computer to continually check their value. These signals are to have the capability to "interrupt" the computer (at some convenient point) during routine processing so that the computer will have their new status as soon as possible; normally within a few microseconds. The computer interface electronics must have provision to deal with each one of these inputs individually. These electronics are referred to as the *priority interrupt structure*. They not only accomplish the function of interrupting the computer during normal processing but provide a priority structure between each of these inputs in order to decide which is more important when several are attempting to interrupt simultaneously. An example of a high-priority interrupt would be the output from a special electronics circuit which monitors the computer power supply voltages. Whenever power is failing, the computer, owing to its speed, has the capability of safely shutting itself down so that normal processing can continue automatically once power is restored.

This briefly defines the requirements for the output of the computer's input interface electronics and simultaneously the requirements for the computer's input electronics, since they must both be (plug-in) compatible. In summary, the computer must have a 16-bit parallel digital input bus structure to which all input signals must be capable of being multiplexed. Furthermore, there must be provision for certain selected digital inputs to have computer interrupt capability, through some priority scheme. This is illustrated in block-diagram form in Fig. 15-3.

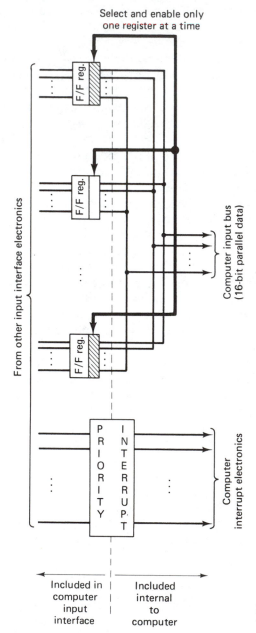

Figure 15-3 Computer input digital multiplexing scheme.

Computer Output Bus

Turning our attention to the computer output interface requirements, many (inverse) similarities to the input interface requirements will be noted.

The digital-to-analog converters require parallel digital input information plus memory. A digital parallel output bus structure with a buffer output register would meet all the D/A converter's requirements. Remember that there may be many D/A converters in a system; therefore, these buffer registers must be capable of being multiplexed at their input. The definition of flip-flop register offered previously describes how they have this capability by ignoring their inputs except when a memory command is being executed. At all other times the register maintains its output exactly equal to the input at the instant the memory command was given, regardless of changes at its input terminals.

This type of (computer) output interface will also be capable of serving any parallel digital output function, since the flip-flop register really does not care what its output is connected to. This leaves the pulse-train-type outputs and the individual (one-bit) binary outputs to be considered.

The one-bit of binary information outputs can be treated in a manner similar to the one-bit computer inputs. One 16-bit register can be used to handle 16 individual outputs simultaneously. Each flip-flop output would be connected directly to a single device. Frequently, this type of output is used to activate relays; therefore, relay driver amplifiers would also be required in the computer output interface. The only problem exists when one out of a group of 16 relays must have its status changed. The computer must command all 16 simultaneously; however, if its new command to any particular relay is exactly the same as the previous command, then the relay will not be affected. Therefore, even though only one out of 16 is being commanded to change its contacts, all 16 will be simultaneously commanded; only those commanded to *change* status will be affected.

The final type of computer output to be considered is the frequency-modulated-pulse-train-type output. The computer can handle this type of output as though it were the individual binary bit output as described in the previous paragraph, simply changing one bit output at the required rate, or else it can output a parallel binary word to a buffer register. This register would then be connected to a binary counter. These, together with a frequency source, would be used to generate a predetermined number of pulses at a constant frequency. This type of output would typically be used to drive digital stepper motors. Another possibility is to have the buffer register be connected to a shift register. The information in the shift register is then shifted (right or left) out of the register until each of the 16 bits has been shifted out in their proper order. This is a parallel-to-serial converter and is used for CRT terminal outputs among other applications.

Pulse-rate-modulated output signals could also be output through a digital-to-analog converter, the dc output of which is used as input to a voltage-to-frequency converter as described in the analog-to-digital converter section.

In summary, the actual digital computer output requirements are very nearly

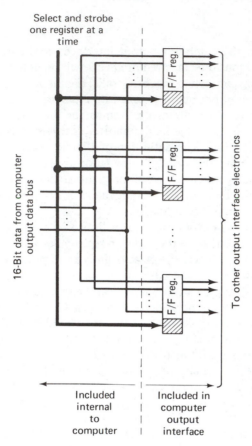

Figure 15-4 Computer output digital demultiplexing scheme.

the same as the computer input requirements. Both require the capability for parallel transfer of 16 bits of digital information. Both also require the capability of multiplexing all devices to these buses, all of which can be accomplished using flip-flop registers. This is illustrated in block-diagram form in Fig. 15-4.

DIGITAL COMPUTER INPUT–OUTPUT

Since the input and output data-transfer bus requirements are so nearly identical, and furthermore since the computer can do only one operation at a time, normally the same 16-bit bus is used for *both* input to and output from the computer. The common (or shared) computer input/output (I/O) bus arrangement has been in popular use for many years. Recently, however, several new computers have been introduced whose architecture is based upon separate input and output bus structures. They are specially designed so that both buses are simultaneously used, thereby increasing their input/output data rate capabilities.

Many factors must be taken into consideration when deciding whether a single I/O bus structure is preferred to dual bus structures. The overriding criterion is always economy. *Economy* refers not only to economy of hardware costs but also economy in programming costs, computer time economy, and economy in memory usage. Present indications are that the dual bus architecture is going to prove to be economical and highly efficient when used in process control applications.

This complicates the discussion of computer internal design and functioning. The dual bus architecture is coming into popular use, but most existing control systems, and the vast majority of existing digital minicomputers and microcomputers, use the shared (single) I/O bus architecture. The following discussion will be somewhat simplified if it is presented assuming a dual I/O bus structure, since the necessity for time multiplexing of the same bus for input and output is eliminated. Therefore, in the following discussion separate input and output buses will be assumed. In order to make the material applicable to a single bus system, simply realize that the same 16 wires (I/O bus) are alternatively used for input and for output—logic within the computer either connecting *line drivers* or *line receivers* to the computer end of these wires.

Having resolved that the computer will "look at" only digital flip-flop (buffer) registers attached to the 16-wire (bit) data input bus, and conversely will only output over a different 16-wire bus structure to the inputs of digital flip-flop buffer registers, the problem now is to figure out specifically which flip-flop registers, on which bus, each time. Figure 15-5A and B functionally illustrate digital computer I/O multiplexing.

I/O Addressing

Each of the input flip-flop registers is assigned a unique binary code number and is provided with digital logic which will decode *only* that exact code number. The inputs to this decoding logic are attached to a bus (functionally separate from the data buses) of normally six to eight wires called the *input address bus structure*. Whenever the decoding logic associated with any particular input register recognizes its unique code, it activates that specific computer input device's buffer register. This unique code is appropriately referred to as the *device address*. Whenever the computer requires information from any device it puts that device's code on the input address bus and the address decode logic enables (only) that addressed device to supply information to the computer. Schematically this is illustrated for two registers in Fig. 15-5A. Normally, there will be many more than two, but they will be similarly connected.

There is an exact duplicate of this device address bus structure and the address decode logic that is used for selectively enabling each particular output device (buffer register) to receive information from the computer. This is the (computer) output address bus and output device address decode logic. This additional bus system and decode logic provide the means for the computer to select any particular device to

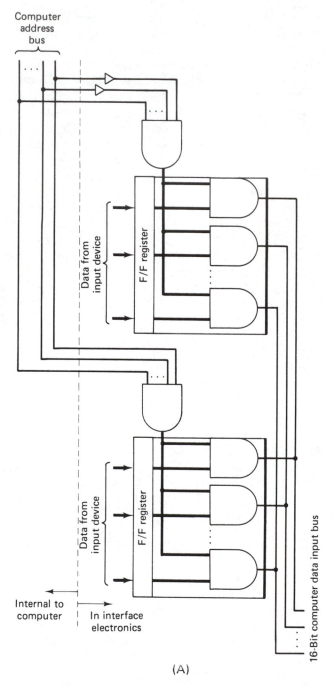

Figure 15-5 Device addressing in a digital computer: (A) input interface; (B) output interface.

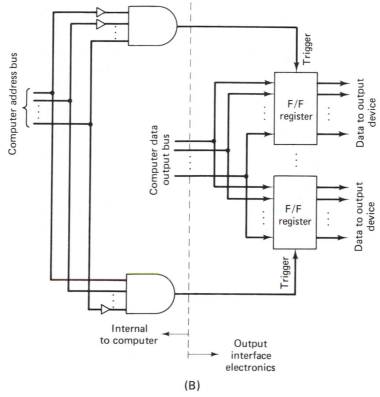

(B)

Figure 15-5 (*continued*)

transfer information to its buffer register and/or to alert the device to receive information. It is schematically illustrated in Fig. 15-5B.

Two additional bus structures have been mentioned for device addressing. Again, since the computer can do only one thing at a time, frequently these address bus structures are time-multiplexed to the data I/O bus wires, with the necessary timing and control so that devices know how and when to connect to the same 16 wires (the I/O bus). In this presentation two separate *I/O address* bus structures will be assumed, in addition to the two independent data bus structures.

COMPUTER PROCESSING OF DATA

At this point the computer must have the capability of addressing each of the possible devices connected to it for either input or output (or both if applicable), and transferring information to or from that device. Before proceeding to a discussion of the actual internal workings of the computer, a summary of the manipulations of these data required to exercise control should be developed.

Frequently, as the computer brings in the current (scaled) value of a process variable, it checks this value against the maximum and minimum values which this variable should not exceed. These limit values had been previously stored in the computer's memory for easy reference. This can be accomplished by a series of comparisons, which mathematically amount to *complementing,* incrementing, and adding the two values, while testing for positive or negative results. Should the value of the variable be "out of tolerance," this fact is normally output by the computer in the form of an alarm.

In addition to, or instead of, comparing data to minimum and maximum values, frequently the current value is compared to its past values to determine the time rate of change of the variable. This introduces the necessity for a *clock* internal to the computer. There are several means of keeping track of "real time" inside the computer, but at most this would require simple addition and either the ability to increment or decrement a number. A more sophisticated computer system might have this clock as a separate hardware function internal (or external) to the computer.

The method of handling relay contact status information (in groups of 16) introduces the requirement for being able to "logically compare" the values of digital words. This includes the necessity to perform logical AND and OR functions, both on 16-bit words and upon individual bits in that word (since each of the 16 bits represents a different relay in this example). This also implies the capability of testing the results of logical manipulations. The necessity for being able to work independently with each bit in the computer word may require the capability of shifting the word either right or left with a testable logic element, remembering the value of each bit, one at a time. Furthermore, the computer must be able to formulate and dissect 16-bit words, bit by bit, to input and output relay contact status information.

All of the aforementioned internal data manipulations require that previous values and/or limiting values are readily available in the computer's memory. Not only must these values be stored, but they must also be stored in logical and easily addressable locations. Normally, they will be stored in internal "tables." Numeric constants, trigonometric and other nonlinear curves, binary code conversion tables, and so on, may also be stored in the computer in table format. This introduces the requirement that the internal computer memory addressing logic include the capability to perform simple arithmetic on the memory addresses. This will require additional internal logic; however, the computer arithmetic unit will have the capability of performing the necessary arithmetic, and therefore no additional arithmetic logic will be required.

The final category of arithmetic processing of data includes the solution of both algebraic and logical equations. The solution of equations within binary digital computers forms the basis upon which the whole field of digital computer programming is based. Essentially the binary numbering system is not easily adaptable to the solution of equations; with the exception of binary logical equations. To perform even relatively simple algebraic manipulations, infinite series approximations, table-look-up techniques, successive approximations techniques, iterative equations solution techniques, equation *simulation,* and just plain mathematical "tricks" must be resorted to.

Most of these techniques require many relatively simple arithmetic manipulations which collectively simulate the solution of more complex mathematical processes, such as multiplication, division, integration, differentiation, raising numbers to powers, extracting roots of numbers, and so on. It is only the extremely high rate of execution of simple binary arithmetic that makes the binary digital computer at all acceptable for equation solution. The fact that the binary digital computer can perform literally hundreds of thousands of additions, subtractions, and shifts each second more than compensates for the difficulty in programming these calculations.

Therefore, solution of algebraic and calculus equations can be accomplished by the same binary arithmetic logic elements previously described. The further capability of hardwired digital multiplication hardware within the computer's arithmetic unit is frequently desirable, but more from the point of view of increasing computational speed than from increasing the basic computational ability of the computer. Hardware multiplication logic is expensive and frequently is not included in process control computers.

In summary, the computational requirements of the process control computer are based upon a binary adder, with the additional capabilities of shifting data words and dealing with each bit in a computer word on an individual basis. Also required is the logic necessary to test the results of binary addition (and therefore subtraction), test each bit in a computer data word, and finally to perform logical AND and OR comparisons of binary words and test the results of these logical operations.

DIGITAL PROCESS CONTROL COMPUTER DESIGN

The computer, as developed in the preceding discussions, has separate 16-bit data input and output buses, separate input and output address buses, priority interrupt structure, memory, an arithmetic unit, and of course all of the timing and control logic necessary to make all these other portions work together. Note that the approach taken to define the process control computer functional specifications based upon the requirements of the process itself has shown the need for the four basic functional units of *any* digital computer. These four functional units are the computer input/output unit, the computer arithmetic unit, the computer memory unit, and the computer timing and control unit.

The following discussion will deal with each of these four basic computer units from the point of view of how they control the flow of data and what operations they perform on the data. But first the definition of the term "data," as used herein, must be understood.

Data refers to any 16-bit digital word for which no other designation (such as address, instruction, etc.) exists. It includes any and all process variable information, codes representing these variables' values, digital information being transferred to or from any external device, and so on. Any 16-bit binary word which is transmitted to or from the computer via the data input and output bus structures qualifies under this definition, regardless of what the information actually represents or how it is going to be ultimately used.

Now, turning to the digital computer itself, a quick look at Fig. 15-6 will reveal the foregoing functional description of the process control digital computer in block-diagram form. The priority interrupt "request" originates in the computer input interface electronics and must have direct access to the computer's timing and control circuitry. Addresses for both the input and output computer interfaces originate within the computer timing and control circuitry.

Computer input information (data) must pass directly to either the arithmetic unit *or* directly into the computer memory for future use. Computer output information (data) must come from either the computer's memory or the arithmetic unit. Finally, there must be provision for bidirectional data flow between the computer's memory and arithmetic units.

Several general concepts illustrated in Fig. 15-6 are worthy of special note.

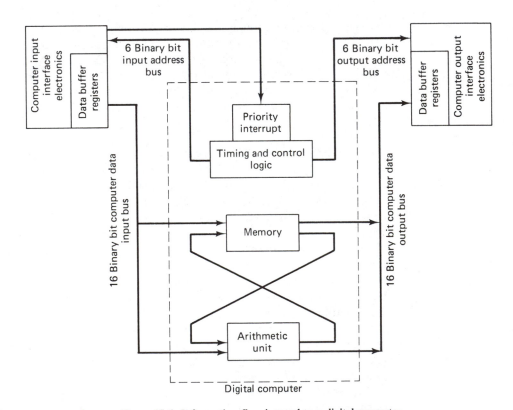

Figure 15-6 Information flow internal to a digital computer.

First, the flow of any binary information between two (or more) functional units is simply shown as a single line with an arrow at one end of the line indicating where the information is destined to go. Bidirectional data flow is indicated by a single line with arrows at *both* ends. Second, a single information flow line can actually represent any number of wires. For example, the computer data buses are all shown as single lines whereas they are actually 16 wires (one per binary bit of data). Third, note that the terms "input" and "output" *always* refer to the computer as the center of the system, unless otherwise noted. Fourth, note that even though the computer timing and control unit *must* have control over every operation and device the computer works with, the lines indicating that information is coming from this unit and going to every other unit are generally "understood" to be present but are omitted from the schematic. Finally, note that data is transferred from place to place within the computer itself and that this data flow is noted exactly the same way as is flow external to the computer.

In a very general sense at this point, this discussion has developed (functionally) the basic hardware requirements required to exercise control over an automated process. To further develop the internal workings of a digital computer, it will be necessary to develop a working knowledge of the basics of computer machine language programming.

COMPUTER PROGRAMMING

Taking Fig. 15-6 and attempting to implement the functional blocks with digital logic will reveal that essentially a digital computer is a very complex assembly of circuits which themselves are capable of performing only very simple operations.

Take, for example, the arithmetic unit. Figure 15-7 is a block diagram of a typical computer arithmetic unit that has the capability of performing binary additions only. Note the gating networks, temporary storage registers, and data flow paths (again no timing and control information flow is illustrated). Each of these functional blocks is capable of performing relatively simple operations on 16-bit binary data words.

In order to add two binary numbers together, the following process must be followed. First, get one of the numbers, either from computer memory or from the computer data input bus, and store this number in one register. Second, get the other number to be added from either source and store it in the other register. Third, "gate" the information in these registers to the binary adder. Fourth, take the output from the binary adder and store it somewhere. Fifth, check the overflow indicator to verify that the output from the binary adder is a meaningful quantity (the addition has not overflowed the capacity of the computer).

This happens to be one of the most frequently used sequences of operations in the computer, and in order to execute it, there are many individual steps. In order to "program" the computer to perform this simple process it must be specifically

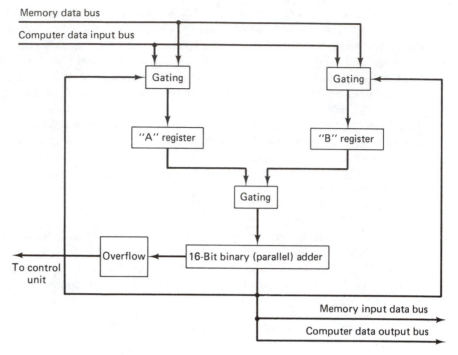

Figure 15-7 Block diagram of a typical computer arithmetic unit capable of binary addition only.

told to do each operation—in sequence. In fact, this is exactly how all computers must ultimately be programmed.

Note that the computer program is the actual mechanism through which the operation of the computer is controlled. Therefore, decoding and execution of the individual steps of the computer program is the function of the computer's control logic. Programming a digital computer therefore takes on the connotation of controlling each operation that takes place internal to the computer.

Machine Language Programming

Ultimately, this definition is accurate; however, computer manufacturers have somewhat reduced the quantity of programming required by combining several frequently used combinations of operations and therefore requiring only one program step (or command) to initiate a sequence of simpler and more basic operations. This is in addition to the (automatic) instruction fetch cycle described in Chapter 14, which only retrieves the instruction and places it in the instruction register. For example, a specific program *instruction* (or *command*) might cause a 16-bit data word to be retrieved from the computer's memory and stored in register A. Therefore, one computer instruction has actually caused several internal commands to be sequentially executed. First, the memory was "cycled" and the desired data word was retrieved (several operations itself which were described as the instruction fetch cycle). Second, the necessary gating was set up in order for this data word to be transmitted to the A register. Third, the proper command (called a *strobe pulse*) was generated and the data word was stored in this register. In actuality, there were probably more operations performed than this.

Programming the computer at this level is commonly termed *machine language programming* or *symbolic programming*. This is the lowest level of detail that any computer programmer must (or can) work at. The programmer *codes* each of the individual commands to the computer logic, which perform the elementary control of information flow from one point to another within the computer (or between a specific piece of hardware internal to the computer and an external device). Note, however, that each of his program instructions actually generates a sequence of even more basic and elementary operations to be executed by the computer logic elements. However, the computer programmer has no control over these more basic operations; they are hardwire-programmed into the computer's control logic, are "triggered" by his instructions, and controlled sequentially by an internal timing mechanism. The term *microprogrammed* is typically used to refer to these instruction sequences.

Probably the most basic and routine computer operation that must be performed is the cycling of the computer's memory unit. The instructions of the program itself are always stored somewhere within the computer's memory, and as each program instruction is executed, the computer must simultaneously cause the next program step to be retrieved from the computer's memory (the instruction fetch cycle), regardless of what other internal operations are required by that particular instruction.

Computer Cycle

This gives rise to a need for a routine sequence of events to be performed during the execution of almost any computer instruction. This sequence of events is controlled by the computer's timing and control logic automatically every time any computer instruction is executed, and it need not be specifically programmed by the computer (human) programmer, as previously noted.

This very routine sequence of events requires a period of time which, together with the amount of time required to execute the most elementary program instruction, is termed a *computer cycle* or *machine cycle*. This is the basic instruction execution time required by the computer and is the specification which the computer manufacturer includes in all the advertising for any particular computer. Thus a 1.76-μs computer cycle is the total amount of time required for that computer to do all its routine "housekeeping" (obtaining the next instruction, routing data through the various functional hardware, checking for priority interrupts, etc.), plus the time required to execute a basic *one-cycle computer instruction*. Note that some of the more complex computer instructions may require more time than one cycle for their execution. This is specially true for microprocessors, which typically require from three to seven (or more) cycles for instruction execution. Data transfers between the computer and its peripherals (input/output instructions) typically require several *machine cycles* for their execution.

It is important never to lose sight of the fact that even though it takes one computer cycle to execute the most basic computer instruction, in actuality there are many sequential operations taking place internal to the computer during that time interval (one of which is the instruction fetch cycle).

Compiler Programming

In any process control application, the actual hardware cost of installing a digital computer to automate the process represents only a fraction of the total "automation" costs. The largest single cost item is normally the programming expense. The process of actually making the computer "do something" once it is all plugged in to the process is the responsibility of the computer programmer, not the technician. However, the technician must be somewhat familiar with the problems that the programmer could possibly encounter so that when something does not work, he will be in a position to decide whether it is a programming error or a hardware failure.

Generally, the programmer will be writing the process control program using machine language as previously described. However, in order to simplify his job and thereby reduce the programming expenses, other computer program languages have been developed. These other languages have been developed specifically to make it easier and quicker to program the computer. Therefore, they are designed to eliminate the requirement that the programmer tell the computer every single basic operation to perform.

Take the simple addition example that was previously used. It took several machine language program steps (instructions) to cause two numbers to be added together. It would be much simpler if the programmer could simply write an instruction similar to "ADD X PLUS Y" and have the computer itself worry about how to accomplish this (and also what and where X and Y are).

Languages such as this exist and are frequently used; however, from the preceding discussion it should be obvious that the computer cannot "understand" such languages. Its hardware timing and control unit still requires each of the individual binary-coded commands or instructions that were required before. Therefore, what is needed is a "translation" for this simplified programming language to actual instructions which the computer can understand and execute.

The translation is accomplished by a computer program which has been specifically designed to accept each of the simplified instructions and convert each one to the series of machine language instructions that the computer requires in order to accomplish the same result. Furthermore, this "translation program" keeps track of where each data value, each program instruction, and so on, is going to be stored in the computer. It therefore relieves the programmer from the necessity of doing all the internal housekeeping chores in his program and frees him to worry only about the function the program is going to perform. This *computer language transition program* is called a *compiler.* The simplified programming language is called a *compiler-level language,* some of the most popular being FORTRAN and COBOL.

The most obvious question is then: Why would anyone bother to write computer programs in machine language with these compiler languages available, and so much easier to use? The answer to this question, as with most other questions concerning the process control computer, is based upon economics. Economics frequently dictate that compiler language programming is too expensive, for any of several reasons. To go into the criterion for arriving at this decision would be too lengthy and involved for the purposes of this text; however, be aware that machine language programming is very widely used.

Furthermore, as a technician you will be involved with repairing the computer hardware, which actually responds *only* to the machine language instruction, regardless of which type of language is used for the program. Therefore, you *must* learn the machine language programming for your specific computer. The reason that the term ''specific computer'' was used is that *every single* brand of computer on the market uses a different machine language. Of course, there are numerous similarities between them, but in general they differ widely.

Symbolic Programming

Getting back to machine language computer programming, it should be noted that there are two methods of programming in machine language (for the same computer). First, of course, is by setting up a binary code, each bit of which (or combinations of several bits) has specific meaning to the computer's timing and control unit, and which we have decided the technician must learn.

The second method of machine language programming is frequently called *symbolic programming.* In this case, rather than programming a string of 1s and 0s, the computer manufacturer has provided a program similar in theory to the compiler. Each machine language binary code instruction is assigned (normally) a three-or-four-letter mnemonic which is a relatively easy code to remember. The computer machine language program is then written using these mnemonics, and a computer program translates these mnemonics into the string of binary bits which the computer requires. This program is called an *assembler program* and sometimes the process is called *assembly language programming.* Figure 15-8 is a summary of the symbolic instruction set mnemonics for one particular industrial microcomputer.

The primary difference between the assembler and the compiler is in the ratio of machine language (binary code) instructions generated for each of the assembly language instructions as compared to the number of machine language instructions generated for each of the compiler-level instructions. For a compiler, a single compiler language instruction will normally require many machine language instructions

ABA	Add accumulators	CLR	Clear	PUL	Pull data
ADC	Add with carry	CLV	Clear overflow		
ADD	Add	CMP	Compare	ROL	Rotate left
AND	Logical AND	COM	Complement	ROR	Rotate right
ASL	Arithmetic shift left	CPX	Compare index register	RTI	Return from interrupt
ASR	Arithmetic shift right			RTS	Return from subroutine
		DAA	Decimal adjust		
BCC	Branch if carry clear	DEC	Decrement	SBA	Subtract accumulators
BCS	Branch if carry set	DES	Decrement stack pointer	SBC	Subtract with carry
BEQ	Branch if equal to zero	DEX	Decrement index register	SEC	Set carry
BGE	Branch if greater or equal zero			SEI	Set interrupt mask
BGT	Branch if greater than zero	EOR	Exclusive OR	SEV	Set overflow
BHI	Branch if higher			STA	Store accumulator
BIT	Bit test	INC	Increment	STS	Store stack register
BLE	Branch if less or equal	INS	Increment stack pointer	STX	Store index register
BLS	Branch if lower or same	INX	Increment index register	SUB	Subtract
BLT	Branch if less than zero			SWI	Software interrupt
BMI	Branch if minus	JMP	Jump		
BNE	Branch if not equal to zero	JSR	Jump to subroutine	TAB	Transfer accumulators
BPL	Branch if plus			TAP	Transfer accumulators to condition code reg.
BRA	Branch always	LDA	Load accumulator		
BSR	Branch to subroutine	LDS	Load stack pointer	TBA	Transfer accumulators
BVC	Branch if overflow clear	LDX	Load index register	TPA	Transfer condition code reg. to accumulator
BVS	Branch if overflow set	LSR	Logical shift right		
				TST	Test
CBA	Compare accumulators	NEG	Negate	TSX	Transfer stack pointer to index register
CLC	Clear carry	NOP	No operation		
CLI	Clear interrupt mask			TXS	Transfer index register to stack pointer
		ORA	Inclusive OR accumulator		
				WAI	Wait for interrupt
		PSH	Push data		

Figure 15-8 Typical microprocessor's assembly language (symbolic) instruction set. (Courtesy of Motorola Semiconductor Products, Inc.)

in order to perform the same functions. In assembly language (symbolic) programming, there is a one-to-one correspondence between the number of symbolic instructions and the number of binary-coded instructions required to accomplish the same function.

Normally, the process control computer programmer will be writing the control program in the symbolic machine language. As the program is written and tried on the system, there inevitably will be times when "something" is not working. There are two general possibilities: either the program is not written correctly or else there is a hardware malfunction. Since the programmer knows little about the internal workings of the computer hardware, it is normally left up to the technician to work with the programmers to try and isolate the problem—another very good reason for the technician to become familiar with programming.

A technician's most valuable troubleshooting tool is the computer itself. If he can program it to do very simple and basic operations, he can normally locate the hardware fault by noting which of these program instructions the fault affects. When these programs are entered, it is normally by directly entering the program into the computer's memory through a series of switches. The switches are only two-position devices; therefore, the program must be entered as a binary code, which therefore must be learned.

As you are troubleshooting a computer, frequently you must "look at" the contents of a register or memory location. The information contained therein is, of course, only a string of binary 1s and 0s. Therefore, you must be able to break this code in order to decide whether the string of bits is correct or not. This, again, requires that the technician be able to interpret the effect of computer instructions based upon their binary codes.

SUMMARY

The computer that has been developed from this presentation must have capability for handling 16 bits in parallel, internally and for both input and output of data. It must have memory and the timing and control logic necessary to cause the computer to operate and execute instructions which will be required by the control program. It must have data input and output bus structures plus input and output address buses. These buses may be multiplexed together. A priority interrupt structure will be required to handle emergencies. The arithmetic unit must have the capability of adding two 16-bit data words, negating data words (converting between positive and negative values), shifting and incrementing data, performing logical comparisons on data words, and finally must have the capability of indicating arithmetic overflow. A testable logic element must be included to indicate the results of logical operations. There must also be some means whereby the computer can keep track of real time.

The computer must have both a machine language and a symbolic programming language. Therefore, the manufacturer will supply an assembly program. Ordinarily, the manufacturer will also supply other computer programs in addition to the assembler. These might include some version of a FORTRAN compiler, *diagnostic programs* (programs designed to test various portions of the computer, such as the memory), programs designed to "load" other computer programs into the computer's memory, programs designed to "copy" the contents of memory onto peripheral devices, and so on. The total of these programs and the program that links them all together is called the *operating system*.

Figure 15-9 Digital computer process control system in a refinery application. (Courtesy of Bristol Division of ACCO.)

CONCLUSIONS

The foregoing discussion has hopefully provided a very brief and general overview of the functional requirements of a process control digital computer. The functional requirements of the computer, and its input and output interface electronics, were developed assuming that a computer could be purchased which would meet these specifications. The resulting specifications can be adequately met by many different brands of computers currently available. The computer that best suits a given application must be decided on the basis of other criteria in addition to these functional specifications. Some of these other criteria are acquisition cost, ease of programming, backup support available from the computer manufacturer, economic stability of the computer manufacturer, options and peripheral equipment available, memory and I/O addressing limitations of the computer, and a multitude of other criteria.

Figures 15-9 and 15-10 illustrate modern process control computers in actual industrial applications. These computer systems operate based on principles discussed in this chapter and include all the subsystems and software also discussed herein.

Figure 15-10 Modern process control system operator's console. The (color) display on the left CRT is showing a loop diagram; the center CRT is displaying the control parameters and current status of many loops; and the right CRT is displaying the trend of activity of three selected process variables over (up to) the past 16 hours. The cassette tape unit on the right maintains this past history. The buttons and switches on the console table provide control of the displays and of all the process conditions. (Courtesy of Taylor Instrument Process Control Division/SYBRON Corp.)

QUESTIONS

1. What does the term *computer architecture* mean?
2. Briefly summarize the types of signals required for (by) a process control digital computer and the necessary processing of these signals.
3. Briefly summarize the types of signals required by the process from a digital control computer and the necessary processing of these signals.
4. Explain the concept of digital multiplexing.
5. What is the purpose for, and the effects of, computer interrupts?
6. What purpose does device addressing serve?
7. Basically what hardware arithmetic capabilities are required by process control computers?
8. What is the computer interface, and why is it required?
9. What pieces of functional equipment would you expect to find in a computer's input interface?
10. What pieces of functional equipment would you expect to find in a computer's output interface?
11. What is a real-time clock used for?
12. What is the net effect of programming a digital computer?
13. What is machine language programming?
14. Where are program instructions located when a computer is running and executing a program?
15. Explain what a computer cycle is.
16. What is compiler-level programming, and how does it differ from machine language programming?
17. What is symbolic programming, and how does it differ from machine language and compiler-level programming?
18. List the hardware and software capabilities (or functions) that any process control digital computer must have.
19. Why is an understanding of machine language programming necessary in order to understand digital computer internal operations?
20. Explain, in your own words, what is illustrated in Fig. 15-6.
21. Basically why is a general-purpose digital computer more economical for large process control systems than hardwired logic to accomplish the same function?
22. Which of the following lists of equipment would you expect to find only in the *interface* to a digital process control computer (either interface)?

    A. Core memory
       A/D converter
       Serial-to-parallel converter
       I/O buffer register
       Magnetic tape controller

    B. Instruction register
       D/A converter
       Parallel-to-serial converter
       I/O buffer register
       Paper tape punch controller

C. D/A converter
    Serial-to-parallel converter
    I/O buffer register
    Magnetic drum unit controller

D. Memory buffer register
    Serial-to-parallel converter
    I/O buffer register
    Magnetic tape unit controller

23. Supervisory (or setpoint) control requires:
    A. A digital computer monitoring the system performance.
    B. Individual analog controllers actually exercising automatic control over each loop.
    C. Analog controllers capable of receiving their setpoint from a digital computer.
    D. Signal transmission channels between the digital computer and the analog controller.
    E. All of the above are required.

For questions 24 through 27, select the single best definition for each term.

24. Priority interrupt.

25. Compiler.

26. Assembler program.

27. Computer cycle.

    A. A computer program designed to "translate" a "foreign" programming language into "machine" language.
    B. The length of time required to cycle the computer's memory unit.
    C. A computer program designed to translate "mnemonics" into "machine" language.
    D. Hardware which causes the computer to deal immediately (not routinely) with emergency conditions.
    E. The length of time required by the computer to execute its simplest instruction.

28. Data (information) transfer between A/D (and D/A) converters and the computer is accomplished by use of functional devices called
    A. Discs or drums.
    B. Modems.
    C. Teletype units.
    D. Flip-flop registers.
    E. Directly addressed memory reference instructions.

29. What would an interrupt be used for?

30. The most basic difference between modern mini- and microcomputers is the physical size of the central processing unit (CPU). True or false?

31. In a computerized control system, normally the analog-to-digital converter(s) operate(s) asynchronously. This means that they operate on a time base which is _____ the computer's time base, and are _____ to the computer in time by use of the computer's priority-interrupt structure.
    A. Proportional to; not slaved.
    B. A function of; not synchronized.
    C. Independent of; synchronized.
    D. A multiple of; slaved.
    E. Dependent upon; not synchronized.

32. Direct digital control (DDC) refers to a system in which _____.

33. Why are "interface electronics" required when applying any digital computer to process control?

34. Basically, why is a general-purpose digital computer more economical for large process control systems than hardwired logic to accomplish the same functions?

35. The term used when selecting one out of many possible devices to access a logical circuit (or wire) is _____.

36. Which of the following pieces of hardware would you expect to find in the interface electronics portion of a digital system?
    A. Program register and core memory.
    B. Core memory and A/D converter.
    C. Parallel-to-serial and serial-to-parallel digital converters.
    D. D/A converter and instruction register.
    E. All of the pieces of equipment listed above may be included in the computer's interface electronics.

For questions 37 through 40, select the single *best* definition for each term.

37. Address.
38. Computer cycle.
39. Word size.
40. Data.

A. Any combination of binary bits being handled by the computer.
B. Hardware that causes the computer to deal immediately (not routinely) with emergency conditions.
C. The length of time required by the computer to execute its simplest instruction.
D. A binary word coded to select one of several possible choices.
E. The maximum number of bits that can be simultaneously handled by the computer.

41. What is the purpose of the instruction fetch cycle?

Questions 42 through 45 refer to the following list of operations which must be performed in order to enter information into a typical microcomputer system's memory. List them *in the order in which they must be performed.*

42. The first operation must:
43. The second operation must:
44. The third operation must:
45. The fourth operation must:

A. Key the information to be stored into the display.
B. Push the address key.
C. Push the data key.
D. Key the 16-bit memory address into the display.
E. Push the " + " key.

46. _____ is a term that describes the number of binary bits that the computer normally handles in a single operation.

47. What are computer diagnostic programs used for?
    A. Troubleshooting of peripheral hardware devices.
    B. Troubleshooting of internal portions of the computer's timing and control logic.
    C. Troubleshooting of the computer's memory system.
    D. Troubleshooting of miscellaneous computer hardware such as adders, general-purpose registers, etc.
    E. Computer diagnostics are used for all of the purposes above.

48. What is the net effect of programming a digital computer?
    A. Generating a lot of paper with confusing sequences of ones and zeros all over it.
    B. Sequentially controlling each internal operation of the computer.
    C. Randomly controlling each internal operation of the computer.
    D. Randomly controlling *only* access to the computer's working registers.
    E. Sequentially controlling *only* access to the computer's memory.

49. What is a computer "instruction"?

50. The hardwired series of operations which the computer must perform to move each instruction itself into the instruction register is collectively called
    A. The memory cycle.
    B. The computer cycle.
    C. The instruction fetch cycle.
    D. The input/output cycle.
    E. The interrupt cycle.

51. How many effective data operations can a microcomputer perform at once?

52. Most commonly available microprocessors have a _____-bit data bus.

53. Most commonly available microprocessors do all their operations on how many bits (simultaneously)?

54. All commonly available microprocessors normally do what type of arithmetic operations?

55. The major difference between hardwired logic and computers is _____.

56. What types of arithmetic-logic circuits are normally included in a microprocessor?

57. In a microprocessor the major difference betwen an accumulator and any other data register is _____.

58. The process of determining which instructions to use, and in what order, in order to solve a specific problem is called _____.

59. When a series of instructions are assembled in sequence to perform some useful operation (or solve some specific problem), they are collectively called _____.

60. Execution of a computer program normally proceeds _____ through the instructions.

61. What general category of instructions provide the decision-making intelligence for a microcomputer?

62. The instruction register in a microcomputer controls _____.

63. The basic purpose of the instruction register is to _____.

64. Which piece of equipment is included in the computer to specifically keep track of which memory locations contain program as compared to other types of information?

65. How does the computer tell whether a binary word is an instruction or data?

66. The single purpose of the program counter is _____.

67. When doing signed binary arithmetic the polarity of the result of an arithmetic operation would be determined by _____.

68. When doing binary arithmetic (unsigned) the fact that an overflow occurred would be determined by
   A. Testing the C bit.
   B. Testing the V bit.
   C. Testing the I bit.
   D. Testing the Z bit.
   E. Testing the N bit.

GLOSSARY

Address   A unique identification code assigned to each memory location and device with which a digital computer must communicate.

Address Bus   The set of electrical conductors which are used to transmit the address codes for each device attached to a computer.

Buffer Register   A flip-flop register which is used to isolate a source of information from its destination. An intermediate temporary storage register.

Bus   A set of electrical conductors used to transmit information from one place to another, both internally and between the computer and external devices.

Byte   A group of binary digits which are smaller in number than a binary word, but are intended to be interpreted, processed, and/or transferred as a unit. Commonly 16-bit words are processed as two 8-bit bytes by computer peripheral devices.

Clock   The hardware in a digital computer which generates the periodic timing pulses that control its internal processing.

Compiler   A digital computer program similar in concept and effect to an assembler program; however, instead of mnemonics, which represent machine language instructions on a one-to-one basis, the programmer codes "English" type statements (such as SKIP IF..., READ, $X = 4.3 \ Ye^{-z}$, etc.). The primary characteristic of the compiler is that for each program instruction written, the compiler must generate more than one machine language instruction which will accomplish the same function.

Complementing   The binary logical process of converting all the ones in a binary word to zeros, and all the zeros in the original word to ones; each in its corresponding relative position.

Data   As used herein, this term refers to information which is generated, processed, and/or transferred by a digital computer.

Device Address   A unique identification code assigned to each device with which a digital computer must communicate.

Input–Output (I/O) Addressing   The hardware process of sending out a device address over the computer address bus and synchronizing information transfer.

Interrupt   A temporary pause in the routine processing by a digital computer caused by a source which has been wired in such a manner that it is capable of causing this condition so that it can be immediately recognized and dealt with.

One's Complement   *See* Complementing.

Parallel   The simultaneous transfer and/or procesing of *all* bits in a word (or byte) with each on a separate (pair) of wires or using separate circuits.

Priority Interrupt   A computer interrupt which has been ranked in its importance as compared to other Interrupts should more than one occur simultaneously.

Serial   The sequential transfer and/or processing of each individual bit in a word (or byte) over a single (pair) of wires or using a single circuit; one bit at a time.

Simulate   To represent the dynamic behavior of one system using a second system. Typically, various processes are simulated by computers so that the effect of different controller designs can be studied without actually tying up the process.

16

ACTUATORS

INTRODUCTION

All the sensing, signal conditioning, and controlling discussed up to this point normally has as its final systems objective the appropriate activation of some process *actuator (servomechanism)*.

The *actuator,* by definition, is a physical piece of hardware designed to accept a *signal* at its input and convert this signal to an appropriate mechanical motion. The input signal (normally) comes from a controller, sometimes via a transmitter, and is normally electrical or pneumatic in nature. The output is some form of mechanical motion (linear, rotary, eccentric, or reciprocating), and it is this mechanical motion which somehow directly alters process variables, frequently via valves and the like.

This *actuator* is selected and physically placed in the process in such a way as to have the desired affect on the variable that is being controlled by the control system. Automatic control system theory has as its goal the calculation of the command signal necessary to activate the actuator(s) in the process in order to obtain satisfactory overall control. The actuators are the "end of the line" insofar as automatic control system hardware is concerned. They are directly (physically) connected to the process, and all hardware from this point on is specifically determined by the process requirements themselves, not by control system requirements.

The actuator is the direct opposite of a sensor. Whereas the sensor converts process variable information into a form compatible for control systems' use (without affecting the process variable), the actuator takes a signal from the control system and causes changes in the process variables. They are both transducers, ow-

394

ing to the fact that they change information available in one energy system to another energy system.

It is appropriate to point out that a bourdon tube (and some other devices discussed in Chapter 3) could be either a sensor or an actuator, depending upon its use in the system. If the bourdon tube is converting pressure information into mechanical motion such that the resulting motion is going to be used as input information to the process controller, then it is a sensing element. On the other hand, if it is receiving a pneumatic signal from a controller and the resulting mechanical motion is being used to actuate a valve that is controlling flow through some portion of the controlled process, it is an actuator. Its designation is therefore as much a function of its use in the system as it is a function of its physical construction or its basic principles of operation.

For the purposes of the following discussion, we shall divide actuators basically into electrical and nonelectrical categories. Both categories will be subdivided by the type of output motion they produce, linear or rotary, and by whether they accept analog or digital input signals.

ELECTROMECHANICAL ACTUATORS

Linear Motion—Solenoids

The most common electrical actuator in use is the *solenoid* (Fig. 16-1). It accepts an electrical signal at its input and converts this signal to an electromagnetic field. A piece of magnetic material (core) is physically located in this field and is caused to move by the field.

The solenoid can be either analog or digital in operation. If the amount of electrical power it is excited with varies in an analog fashion, so will the resulting field and mechanical motion vary in an analog fashion. A common application for an analog solenoid would be in proportional control of braking mechanisms for linear or rotating motions (i.e., recreational vehicles with electrical brakes). Another example of linear analog solenoid operation is the activation coil in a normal audio speaker.

More commonly, however, the solenoid is used as a digital device, where either maximum power is applied or no power is applied. Common relays are an excellent example (Fig. 16-2). The *relay* consists of a solenoid that operates a set of electrical contacts; they are either opened or they are closed (i.e., digital operation). Solenoids are also used to operate many other devices; they either push (push a defective part from a conveyor on an assembly line) or pull (activate safety interlocks) against spring pressure. The spring returns the movable piece to its deenergized position. Solenoids are also commonly referred to as electromagnetic *force motors.*

There are not many other linear motion electromagnetic actuators; most electromagnetic actuators develop rotary motion. Not infrequently the rotary actuators are connected to *rack and pinion,* worm-screw, or other types of rotary motion-to-

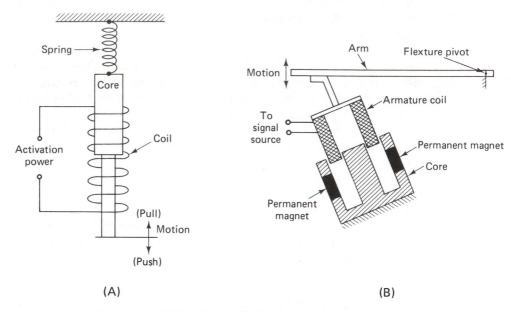

Figure 16-1 (A) Digital solenoid; (B) analog "solenoid" or force motor.

linear motion converters (Fig. 16-3). Therefore, many of the more powerful linear electromagnetic actuators are actually rotary electromagnetic actuators with rotary-to-linear motion conversion elements included.

Rotary Motion—Solenoids

There are pieces of equipment that effectively operate as rotary solenoids. The *D'Arsonval meter movement* is by far the most common rotary analog solenoid or torque motor (Fig. 16-4). It accepts an analog electrical signal in and produces proportional rotary motion at its output. For many reasons the D'Arsonval movements are normally not very powerful; however, by sacrificing some of the meter's excellent characteristics (such as low friction and sensitivity), more powerful analog torque motors can be, and are, made.

Rotary digital solenoids are also used throughout industry, although they are not as common as their linear-motion counterparts.

The most common rotary servomechanisms are the dc and ac motors. Whereas the rotary solenoids and torque motors are normally restricted to a very small range of rotary motion, dc and ac motors are continuous-operation devices, with no angular-motion restrictions. These motors are much more complicated than solenoids in theory and operation, and frequently motor starters are necessary for their successful operation. The following discussion will deal not only with the theory of operation but also with applications, limitations, and (where necessary) motor starters and controllers.

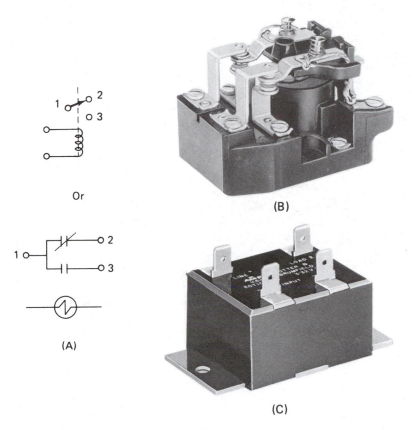

Figure 16-2 (A) Relay symbols; (B) solenoid-operated relay; (C) solid-state relay. [(B) and (C) Courtesy of AMF Potter-Brumfield.]

Direct-Current Motors

Direct-current motors operate on the principles of attraction of unlike magnetic poles (and the repulsion of like magnetic poles), the fact that current flowing through a conductor creates a magnetic field around that conductor, and the fact that a conductor moving through a magnetic field has a current induced in it. Referring to Fig. 16-5, a single-turn winding is located in a permanent magnetic field. This winding is on a form which is free to rotate in that magnetic field. The ends of the windings are connected to the *split-ring commutator,* which is also free to rotate *with* the windings. This entire rotating assembly is termed either the *armature* or the *rotor.*

The *brushes* establish electrical contact between the power source for the motor and the winding on the armature. They remain stationary while the commutator rotates beneath them; therefore, this split-ring commutator and brush assembly provides a rotary switching function.

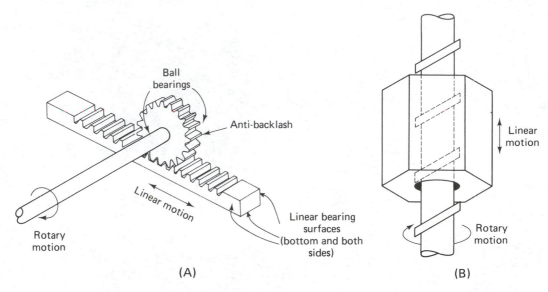

Figure 16-3 Rotary-to-linear motion converters: (A) rack and pinion; (B) worm screw.

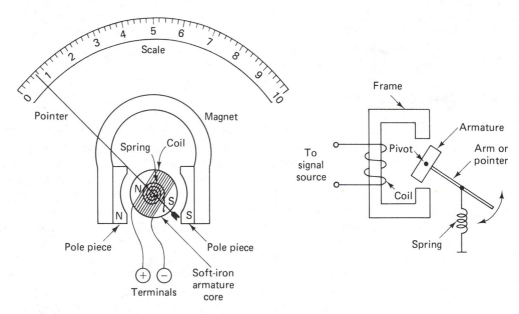

Figure 16-4 Analog torque motors.

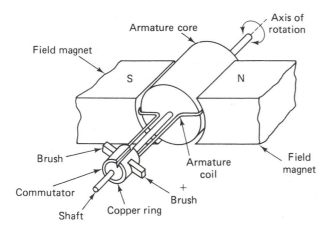

Armature core

Axis of rotation

Field magnet

S

N

Brush

Armature coil

Field magnet

Commutator

+

Brush

Shaft

Copper ring

Figure 16-5 Direct-current motor.

Now that the physical construction is clear, let us see how this dc motor operates. If the armature winding (in Fig. 16-5) is energized with the indicated polarity, current will flow through it, from the negative brush through the winding and back to the positive brush. Any conductor in which a current is flowing will have a magnetic field created around it. In this example, assume that the electromagnetic field is such as to create a north magnetic field on the rotor next to the north magnetic pole of the field magnet. There will simultaneously be a south magnetic pole on the armature next to the south field magnet.

In this condition the like magnetic poles are next to each other, which results in the creation of forces of repulsion between them. As long as the mechanical structure of this example motor does not have all these forces acting in the same plane, a rotating torque will be developed. Let us assume clockwise motion for this example.

As the armature begins to turn, the turning torque is due to repulsion of like magnetic poles. However, as it rotates 90°, not only is the torque due to forces of repulsion of like poles, but it is also due to forces of attraction between unlike magnetic poles. It is more these forces of attraction that will generate the torque necessary for the armature to complete 180° of rotation.

After rotating 180°, the motor would have unlike poles next to each other and it would stop . . . except! Except that as the armature was rotating, the commutator also rotated, and now the opposite polarity power is connected to the armature. This causes the armature current to reverse, which results in a reversal of the armature's magnetic polarity. Therefore, the armature will now have like magnetic poles adjacent to the field magnets, and the process will repeat itself, alternating the direction of current flowing through the armature windings each 180° of armature rotation.

The foregoing discussion explains how a dc motor will start. However, there is another very important naturally occurring phenomenon that occurs simultaneously, or our dc motor would not work for long—it would melt. To understand why this last comment is true, stop and think of what the dc motor (refer again to Fig. 16-5) looks like (electrically) to the source of power it is connected to. The motor armature looks like an electrical short-circuit! Therefore, when power is first applied to the motor, very heavy currents flow and there is nothing in the circuit to oppose or control this current flow except the inductance of the windings and low resistance of the brushes and armature windings.

This current flow does, in fact, occur when starting dc motors and, depending upon the motor, it can be destructive. Therefore, in larger dc motor applications, motor starters are required. *Motor starters* consist essentially of a bank of series resistors that are electromechanically switched into series with the armature before power is applied. Figure 16-6 is a schematic of a simple dc motor starter. After applying power the resistors are shorted out sequentially (by a timing mechanism) until they are all out of the circuit. Therefore, this starter is used only for initially starting the motor; why isn't it necessary to restrict the current flowing in the armature winding when the motor is running?

The answer is that it *is* necessary to restrict armature current at all times. However, nature takes over and does it automatically once the armature starts to rotate. The armature has many turns of wire on it and as it rotates a situation exists wherein conductors are being moved through a magnetic field. Therefore, there is a current induced in these conductors. This is the tendency of a dc motor to act like a dc generator when running. Furthermore, it can be shown that this induced current will be in a direction opposite to the direction of the current that is being supplied to the motor; therefore, it is a countercurrent and results in a countervoltage being generated (counter electromotive force). It is this counter emf that controls the amount of current flowing in the armature windings *while* the armature is rotating. Obviously, there will be no counter emf when the armature is stationary (therefore,

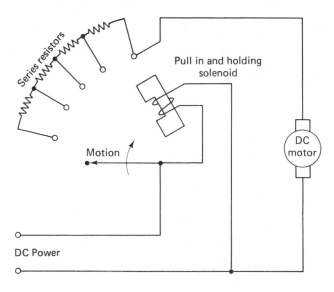

Figure 16-6 Direct-current motor starter.

the need for the motor starter), and the magnitude of the counter emf will vary from zero up to a maximum as the motor armature comes up to operating speed.

If no power was being taken from the motor and *if* there were no internal friction or electrical losses, the tendency to act like a generator (the counter emf) would exactly balance the power being supplied to the motor and zero armature current would flow. In reality, there is friction and there are losses; therefore, there must be some power supplied to the motor to overcome these losses. The counter emf must thus be somewhat less than the applied voltage and some armature current will flow—exactly enough to provide the power necessary to overcome these losses. This counter emf is a naturally occurring negative feedback.

In order to predict the effect of load changes on dc motors, one only needs to consider the effect of the counter emf. For example, if a motor is running at its no-load speed and power is subsequently taken from the motor via a clutch mechanism, what will happen to the speed of the motor? If more power is going to be taken from the motor, common sense dictates that more power must be supplied to the motor. In order to cause more armature current to flow, the counter emf must be reduced, and therefore the operating speed of the motor must also be reduced.

The reverse situation would exist if the motor were running under a load and the load was removed. In this situation the power being supplied to the motor would cause it to speed up, thereby increasing the magnitude of the counter emf and reducing the armature current flow until the motor stabilizes at a higher speed. This brief discussion also explains how the speed of a motor can be controlled. Either the magnitude of the applied voltage can be varied or the value of the counter emf can be controlled. In practice, both effects are used, and some techniques of accomplishing speed control will be covered later in this chapter.

That is basically how a dc permanent magnet motor works. In practice, the armature would have many turns of wire (not one, as in Fig. 16-5) and would also have more than two segments on the split-ring commutator. In order to get more power from the motor, either the field magnets can have their magnetic strength increased or the electromagnetic strength of the armature can be increased.

Figure 16-7 is a cutaway view of a commercial dc motor illustrating the commutator and brush assembly previously discussed, as well as the rotor and stator windings which will be discussed next.

Increasing the electromagnetic strength of the armature field requires many turns of heavier wire and subsequently higher currents to be passed through the brush/commutator assembly; this results in power loss and heating of this assembly. Furthermore, as the current is reversed in the (inductive) armature circuit, the effects of inductive kickback cause arcing at the brushes, which also results in heating of the assembly. In practice, larger dc motors use many sets of brushes connected in parallel to handle this power dissipation. Therefore, it may be more desirable to increase the strength of the field rather than increase the armature current. Practically speaking, in many cases, increasing *both* is required.

One way to increase field strength is by using more powerful permanent mag-

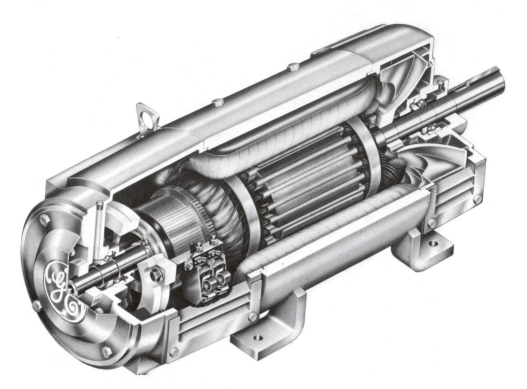

Figure 16-7 Cutaway of commercial dc motor illustrating the split-ring commutator and brush assembly. (Courtesy of General Electric Co.)

nets, but there is definitely an upper limit to the magnetic strength of available permanent magnets. The only other choice is then an electromagnetic field (or combination of the two). The symbols for both permanent magnetic field dc motors and electromagnetic field dc motors are shown in Fig. 16-8A.

Once the complexity of an electromagnetic field is introduced, one must consider how it can be connected into the circuit and the effects on the motor of each type of connection. Fig. 16-8B illustrates the coil connected in series with the armature. In this configuration, all the armature current must pass through the field coil; therefore, it would consist of relatively few turns of fairly heavy gage wire. The operating characteristics of this (series-wound) motor can be explained by investigating the effects of motor currents flowing from zero speed to full speed.

As power is first applied to the motor, very heavy currents flow through both the armature and the field. Therefore, both fields are at their maximum strength, and very large starting torques are developed. As this motor comes up to speed, the counter emf reduces the current through both the field and the armature until just enough power is supplied to the motor to maintain it under its load. As the load changes, so does the armature speed and the counter emf and therefore the current supplied to the motor. However, this changing current affects not only the magnetic

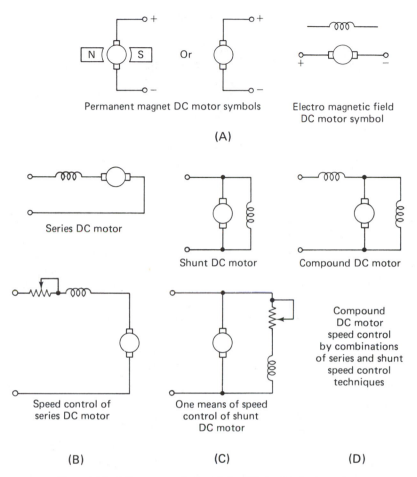

Permanent magnet DC motor symbols   Electro magnetic field
DC motor symbol

(A)

Series DC motor

Shunt DC motor   Compound DC motor

Speed control of
series DC motor

One means of speed
control of shunt
DC motor

Compound
DC motor
speed control
by combinations
of series and shunt
speed control
techniques

(B)   (C)   (D)

Figure 16-8 DC motor symbols and simplified control schematics.

strength of the armature but also the field strength; and speed control is very poor. Therefore, series-wound dc motors are used in applications either where they operate essentially under constant load (i.e., electric fans in automobiles) or in applications where high starting torque is required and constant-speed control is of secondary importance (i.e., starting motors for automobile engines).

A second method of connecting the electromagnetic field in a dc motor is the shunt configuration illustrated in Fig. 16-8C. In this configuration the field is connected across the supplied power line and therefore will remain at a fairly constant strength. It must consist of many turns of fairly small gage wire (higher impedance), so it will have a restricted but relatively constant amount of current flowing through it at all times. This configuration results in a field strength that is relatively independent of armature current, quite similar to the permanent magnet motor. The overall result is much improved steady-state speed control with motor load variations; however, starting torque is much less than with the series-wound motor.

In order to realize the benefits of both the series- and the shunt-wound motors both types of field windings are included in some motors, called *compound motors* (Fig. 16-8D). By varying the numbers of turns and sizes of the wire in each of the windings, motors with reasonably good starting torque and reasonably good steady-state speed control can be designed.

Dc motor speed control can be achieved in all the motor designs by varying the applied voltage. Higher voltages will permit higher-speed operation, because the armature will have to rotate faster in order to generate the proper counter emf, and conversely with lower voltages. In a dc motor with a shunt field winding, a variable resistor in series with that winding will control the strength of the magnetic field and thereby the motor speed.

The stronger the field, the more lines of magnetic flux that are cut by the armature at any given speed. Therefore, as the strength of the field is increased, the armature does not have to rotate as fast to generate the proper counter emf, and the motor slows down. The converse is also true; as the field strength is decreased by the motor, the armature must speed up to generate the proper counter emf.

The limit of high speed for a dc motor is either where the field is too weak to permit the motor to develop enough power to continue running—in which case it may burn out due to high armature currents—or the motor will speed up to self-destruction. In either case it is obvious that some protection must be built into the motor in the event that the magnetic field is lost. If the motor is permanently connected to a load, this is normally sufficient to reduce overspeed possibilities, and a thermal switch built into the motor will protect for overcurrent situations. For overspeed, a centrifugal switch can be mounted on the armature which interrupts the circuit at a predetermined speed.

Direction of rotation for a dc motor is controlled by controlling the polarity of dc voltage applied to the motor. Bidirectional motor controllers would have a switch installed configured as in Fig. 16-9.

Note that in a dc motor the field is of constant magnetic polarity and that the magnetic polarity of the armature is varied. This is exactly the opposite, as we will soon see, as for ac motors, our next category of analog rotary motion actuators.

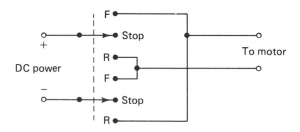

Figure 16-9 Direction control for dc motors.

Alternating-Current Motors

Alternating-current motors operate basically on the same principles as do dc motors, the interaction of magnetic fields. In both types of motors, one magnetic field rotates and one is stationary. In both types there are permanent polarity magnetic fields and alternating polarity magnetic fields. The difference between the two types is primarily in how the magnetic fields of the armatures (rotor) and fields (*stator*) are produced. There are several different types of ac motors just as there are several different types of dc motors. We shall start our discussion with the simplest types.

Figure 16-10 illustrates one type of ac motor, an *induction motor.* In the single-phase ac induction motor illustrated, the field winding (*stator winding*) is connected directly to a source of ac power. The armature winding is illustrated as a single turn of very heavy gage wire which is wound around the armature and simply shorted together at its ends. This results in a transformer-like situation, where the stator winding acts as a primary of the transformer and the single turn on the armature acts as the secondary.

When ac currents flow in the stator winding (primary), currents are induced in the armature winding (secondary). Since the armature winding is only a single turn of very heavy wire, very large currents will be induced in it. These currents cause the armature to have a strong electromagnetic field. Note that there are no electrical connections necessary to the armature for the ac induction motor (which, in any one of its several forms, is the most popular type of ac motor in common use).

When investigating ac circuits, it is convenient to investigate the circuit at different times during a single cycle of the ac. Let us first investigate the reaction of the motor in Fig. 16-10 to stator currents flowing in the direction indicated by the arrows. The stator is wound such that, with currents flowing in this direction, a north magnetic field will be developed at the left stator pole piece, and a south magnetic field will be developed at the right pole piece. The winding on the armature will be in a changing magnetic field, and we can prove that it will have a current

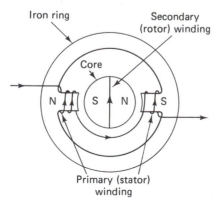

Figure 16-10 Schematic of an ac single-phase induction motor.

induced in it in a direction such as to cause an electromagnetic field to appear with a south pole on the left and a north pole on the right.

What will the motor do in this situation? Nothing, absolutely nothing. Unlike magnetic poles are adjacent to each other, attracting each other, and the rotor will remain stationary. On the other half cycle of the supplied ac power, the magnetic poles will reverse on both the stator and the armature, the net result being unlike magnetic poles adjacent to each other, and again *no* motion. Therefore, we have just proved that single-phase ac induction motors do not work—right? Wrong. They do work, but they will not *start* by themselves. If the rotor is given a spin, the motor illustrated in Fig. 16-10 will run well; however, some additional means of getting it started is necessary.

One means of designing an ac motor so it will start by itself is by connecting the motor to two-phase ac, with two independent stator windings. Two-phase ac power is developed by two ac generators whose shafts are bolted together such that their electrical outputs are 90° out of phase with each other. Figure 16-11 illustrates one cycle of the ac current output from each of the two ac generators. The schematics in Fig. 16-12 illustrate a two-phase ac motor which we will assume is electrically connected to the two generators. Stator winding 1 is connected to generator 1, and winding 2 to generator 2. The armature is not going to be considered at present.

Fig. 16-12A illustrates the direction of current flow for time period *a* in Fig. 16-11. Generator 1 has a maximum current output in one direction and the output from generator 2 is zero at this instant in time. The stator winding 1 is wound such that, with the current flowing as indicated, a north magnetic pole will appear at the left and a south pole at the right. Therefore, the net magnetic polarity (since there is no current flowing in winding 2) will be left to right, as illustrated.

At time *b* in Fig. 16-11 (90° of generator rotation later) there will be a maximum current output from generator 2, and no output from generator 1. Therefore, stator winding 2 will have a maximum magnetic field, and it is wound such that the field will be north on top and south on the bottom (Fig. 16-12B). Note that the net magnetic effect in the center of the motor has been a 90° clockwise rotation of the magnetic field.

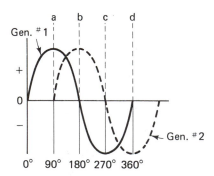

Figure 16-11 Current output waveform from two-phase ac generator.

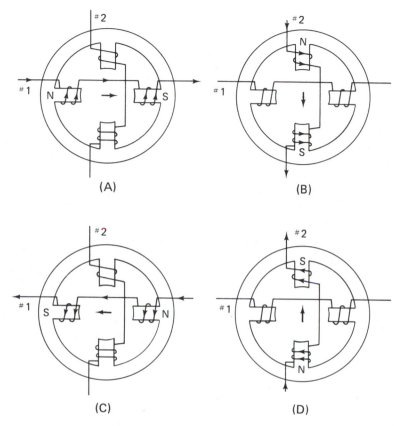

Figure 16-12 Two-phase motor schematic illustrating rotating magnetic field phenomenon.

At time periods c and d the currents in the stator windings will flow in the opposite direction (as compared to currents flowing during time periods a and b, respectively); first with winding 1 having maximum current and winding 2 with no current (Fig. 16-12C) and then with winding 2 having maximum current and winding 1 no current (Fig. 16-12D). The magnetic fields generated by the stator windings will also be in the opposite directions as compared to Figs. 16-12A and 16-12B.

Note the magnetic field in the center of the stator; it apparently rotates 90° clockwise for each time period. Since Fig. 16-12A-D represents the effect for one complete rotation of the ac generators, and since the motor's magnetic field has made one complete rotation during the same period of time, the *field* in the motor is said to rotate synchronously with the generator's. This is exactly the opposite of a dc motor, where the field remains steady and the rotor alternates magnetic polarity.

If we took a permanent magnet rotor and put it into this rotating magnetic field, it would start and it would follow the field exactly. Therefore, the permanent magnet rotor would rotate in synchronism with the rotating field and we have a *synchronous* ac motor. The speed of rotation of a synchronous motor is dependent only on the frequency of rotation of the stator's magnetic field, which is dependent only on the frequency of the ac power supplied to it.

Since the specifications for modern commercial ac power distribution systems are so rigid, this type of motor will have such precise speed control that it can be used for timing mechanisms. For normal timing mechanisms, however, two-phase power is neither available nor necessary. The second phase of this two-phase synchronous motor causes the motor to have more power (which is not necessary for the large majority of timing motor applications) and to start. There are other techniques available to get a permanent magnetic ac synchronous motor started, the most popular being the *shaded-pole* technique.

If we take the single-phase motor shown in Fig. 16-10, and replace the rotor with a permanent magnet rotor, we have the typical ac synchronous motor talked about in earlier chapters, and used exclusively in common household clocks and timing mechanisms. The only difference (as compared to Fig. 16-10) is the way the stator poles are configured. The stator poles have an additional single-turn heavy-gage-wire *shading coil* on them. Refer to Fig. 16-13 for the effect of this coil.

As noted in Fig. 16-13A, on the positive increasing portion of the applied stator coil ac power, the magnetic field in the one-pole piece illustrated will be building up in one direction. The important thing is that the field will be changing (because the current in the field coil will be changing). Since the single turn shading

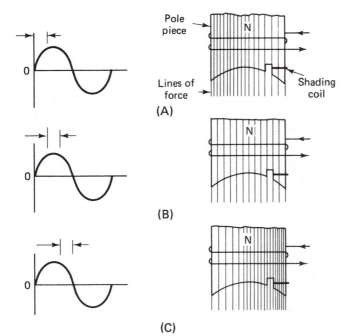

Figure 16-13 Diagram illustrating the phenomenon of pole shading.

coil is physically located in this changing magnetic field it will have a current induced in it. This induced current will be in such a direction as to cause the shading coil to have an electromagnetic polarity *opposite* to that of the field that is causing it. Therefore, the magnetic fields of the pole and of the shading coil will cancel each other out to some extent, resulting in an uneven magnetic flux distribution over the pole piece, the stronger magnetic field being on the left.

During the peak of the applied ac power, the magnetic field in that pole piece is not changing appreciably. Therefore, the shading coil will not have much of a field of its own, and there will be a relatively even distribution of flux across the pole piece (as illustrated in Fig. 16-13B).

On the falling (decreasing) portion of the positive half-cycle of the ac power, the field across the pole piece will be decreasing in magnitude. As it is changing, the shading coil will again be in a changing magnetic field, but this time the field is collapsing, not building up. Therefore, the current in the shading coil, and therefore its magnetic polarity, will be opposite to the way they were in Fig. 16-13A. This will result in the shading coil's magnetic field aiding the pole's magnetic field, and the sum of the two results in a stronger magnetic field to the right (in Fig. 16-13C).

As can be noted, on comparing the magnetic field intensity distribution from Fig. 16-13A-C, apparently the strongest magnetic field has shifted from the left edge of the pole piece to the right. This small motion of the magnetic field is sufficient to start a single-phase motor's permanent magnet armature rotating, even under light loads. In fact, this technique is used throughout industry in basic timing mechanisms.

That takes care of single-phase permanent magnetic motors. To make more powerful synchronous motors, either multiphase power can be applied (as previously noted) or the rotor can be converted to a dc electromagnet, where dc current is applied through brush and slip-ring assemblies.

Now let us insert an electromagnetic ac rotor (without any electrical connections to the rotor) into the two-phase motor of Fig. 16-12. As the magnetic field created by the stator windings rotates, an electromagnetic armature will have to rotate at some lesser speed. If it rotated in synchronism with the stator field, the armature windings would never be cut by any lines of magnetic flux; therefore, the armature would have no magnetic field of its own. In practice, the armature will rotate at approximately half the speed of the rotating stator field. Therefore, the armature windings will be continuously cut (in the same direction) by the rotating stator field and the armature will be an induced electromagnet that will always have the same electromagnetic polarity.

This motor will start and run, but it requires two-phase ac power which is not normally conveniently available. It is called a two-phase *induction motor,* since the rotor is an induced electromagnet.

Now back to our original single-phase ac induction motor (Fig. 16-10). Remember that the only problem with this motor was that it would not start by itself. Pole shading will not help because of the electromagnetic armature. Therefore, another scheme is required. One of the most commercially successful schemes is illustrated in Fig. 16-14.

Figure 16-14 illustrates an ac induction motor (electromagnetic rotor) operating on a single phase but having two windings. The main field winding is always connected directly to the applied ac power. The second field winding is connected in series with a switch and a capacitor. When the motor is initially started, the switch is closed and the rotor sees essentially a two-phase motor because the series capacitor shifts the phase of the current in the starting winding as compared to the main winding. Therefore, the motor will start much like a two-phase motor. When the armature reaches approximately 75% of its full speed, the switch opens and the motor continues to operate on the main winding alone.

Therefore, this second winding, capacitor, and switch are only used for starting the motor; they are disconnected from the circuit while the motor is running. The switch mechanism is physically mounted on the armature and is centrifugally operated. This means that the electrical connections to the centrifugal switch must be through brushes. Figure 16-15 illustrates schematically a brush and slip-ring assembly along with the centrifugal switch. This motor is called a *capacitor-start split-phase induction motor* and is the type you will find in the larger household electrical appliances. Its direction of rotation can be changed by reversing the connections on one of the two windings.

There is one additional type of ac motor you will commonly find used in smaller appliances and hand tools. It is operated directly on ac, although it is a series-wound dc motor with split-ring commutator and brushes. When ac power is applied to this motor (refer to Fig. 16-8B), the polarities of both the field and the rotor will change, but will change together; therefore, the motor will work. The sure sign that this is a series-wound dc motor being operated on ac (and called a *universal motor*) is the split-ring commutator.

For variable-speed ac hand tools and appliances, the universal motor is used along with silicon-controlled rectifiers (SCR) or *triacs*. As you may have deduced, ac induction motors are essentially constant-speed devices, their speed being determined by their physical construction and the frequency of the applied power. For speed control, ac motors normally require the use of auxiliary equipment such as gear trains or clutch mechanisms. Therefore, the majority of speed-control applications are handled by dc motors. Also, ac motors, due to the inductance of their windings, inherently limit the current supplied to them and normally do not require starters.

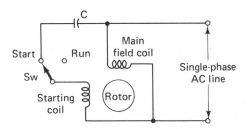

Figure 16-14 Circuit schematic for a split-phase, ac, capacitor-start induction motor.

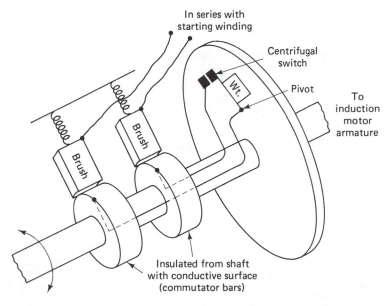

In series with
starting winding

Centrifugal
switch

Wt.

Pivot

To
induction
motor
armature

Brush

Brush

Brush

Insulated from shaft
with conductive surface
(commutator bars)

Figure 16-15 Commutator assembly.

There are other types of ac motors, but the types discussed are the most common. Some examples of others include the ac servomotor, the ac synchromotor, and the amplidyne. Each has its advantages and its applications. Figure 16-16 illustrates several commercial fractional horsepower ac motors.

Stepper Motors

The final category of rotating electrical actuator to be discussed here is the digital *stepper motor*. As the name implies, this motor responds directly to a digital code, which makes its application in digital control systems a natural one. It is one of the very few direct digital actuators currently available.

There are several different basic principles upon which the designs of *digital stepper motors* (simply referred to as *steppers* from this point on) can be based. Only two will be discussed in this text, as they are very commonly used and relatively easy to understand.

The first type is the *permanent-magnet* (PM) *stepper*. Figure 16-17 is a simplified schematic of a typical PM stepper. Note that the rotor is a permanent magnet, this one being only two poles; normally, the rotor contains a great number of magnetic poles. The stator consists of four sets of windings, grouped into two pairs, each pair of which is wound around a common magnetic pole piece. The two windings common to each pole piece are mutually reverse wound so that energizing one of them gives the pole one magnetic polarity, and energizing the other results in that same pole piece having the opposite magnetic polarity. A simple mechanical switch-

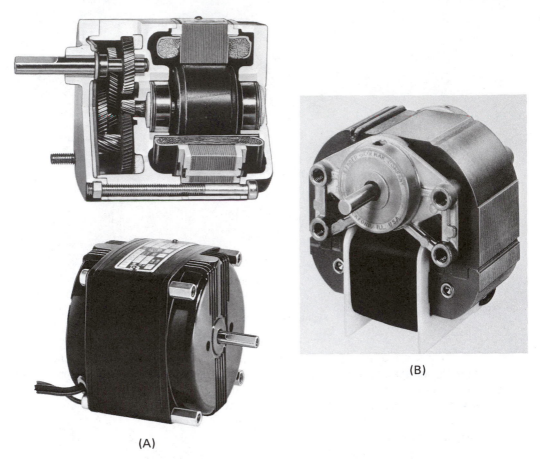

Figure 16-16 (A) Industrial ac fractional-horsepower servo gearmotor; (B) fractional-horse-power shaded-pole synchronous ac motor. [(A) Courtesy of Bodine Electric Co.; (B) courtesy of Barber-Coleman Co.]

ing arrangement will be used in this example to explain the motor's operation. In practice, the drive circuits would all be solid-state electronics.

Referring to Fig. 16-17A, with switches X and Y in the indicated positions, coils F and H will be energized. These coils are wound such that the pole pieces will have the magnetic polarities as indicated. The permanent magnet rotor has no choice but to assume the position where the south pole is in position 1. The entire torque generated by the two electromagnetic poles and the permanent magnet rotor can be used to move the rotor to this position while moving a load attached to the rotor.

Furthermore, any tendency for the load (i.e., due to its inertia) to try to move the rotor from this position will be resisted by the same amount of torque that resulted in its assuming this position. This tendency for the PM stepper to hold its

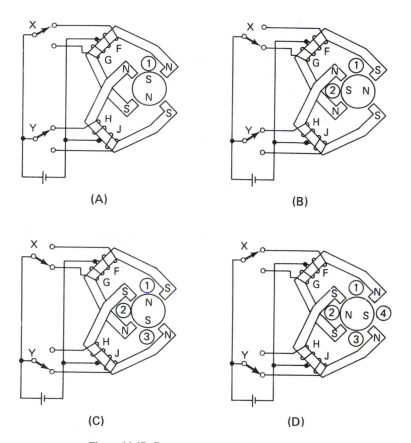

(A)   (B)

(C)   (D)

Figure 16-17 Permanent-magnet stepper motor.

position is referred to as *detent torque* and is a very valuable feature not inherent in basic ac or dc motors.

Now referring to Fig. 16-17B notice that switch *X* has changed position (switch *Y* is in the same position it was previously). Coil *G* is now energized and it has the effect of causing the magnetic polarity of its pole piece to reverse. Therefore, the south pole of the rotor is now forced into position 2 and is held there by the motor's detent torque. The rotor has stepped 90° counterclockwise.

By now changing the position of the *Y* switch (leaving *X* as it was), we can cause the rotor to take another 90° step in a counterclockwise direction (Fig. 16-17C). Subsequently changing the position of the *X* switch (only) will result in still another counterclockwise step (Fig. 16-17D). Note that the sequence of operation of the switches was very important.

At any time the sequence could have been reversed, which would cause the rotor to step 90° in a clockwise direction. For example, refer to Fig. 16-17C. With the south pole of the rotor in position 3, energizing the X switch will cause the rotor to step counterclockwise to position 4. However, energizing the Y switch instead would cause the magnetic situation depicted in Fig. 16-17B, and the rotor would take a 90° step clockwise back to position 2. Therefore, as long as the basic sequence of energizing windings is maintained, the rotor can be stepped bidirectionally.

If the sequence is not adhered to, then the operation of the rotor may become unpredictable. For example, with the rotor in position 1 (Fig. 16-17A), causing both the X and Y switches to change positions simultaneously will result in the rotor not moving at all, moving clockwise to position 3, or moving counterclockwise to position 3. The sequential operation is even more important with motors having more poles on the rotor and/or more field coils.

Another very important characteristic of the stepper motor is the absolute accuracy of the position of the rotor after any number of steps. The absolute position accuracy for each position of the rotor is dependent only upon the mechanical construction of the motor itself, and this parameter is specified by the manufacturer. The absolute position accuracy after the completion of any number of steps (even after bidirectional operation) is dependent only upon the position accuracy of the last step. In other words, the errors do not add up (are not cumulative) as long as the rate of the step commands does not exceed the ability of the motor to keep up.

The motor has inherently got negative feedback built in. When the stepper is commanded to a position, the system can be reasonably sure that the motor has responded and can calculate the position of the rotor without the need for an additional position feedback sensor in many applications.

The circuitry for generating the specific sequence necessary in order to properly sequentially energize windings of stepper motors is normally hardwired into individual controllers for each motor. Functionally, these controllers would be constructed as per the block diagram in Fig. 16-18. This releases the system from the responsibility of remembering the sequence to be generated next; all the system has to do is issue step commands and tell this interface (controller) logic which direction for each step.

By using a digital pulse train derived from a precision oscillator circuit as the source of step commands, the rotating speed of the stepper motor can be controlled to within the tolerance of the oscillator. This technique, combined with the absolute position accuracy inherent in stepper motors, makes them ideally suited for applications where a load must be accelerated according to some specifications, maintained under precise speed control while running, and then decelerated, again according to specifications. Precision digital magnetic-tape-drive units must be controlled in a manner similar to this.

PM steppers definitely have advantages and applications, but they also have their disadvantages. They normally have windings energized continuously, which uses up a significant amount of power. They have limited step angles, although small step angles can be obtained through proper reduction gearing.

The second type of digital stepping motor that we will consider is the *variable-*

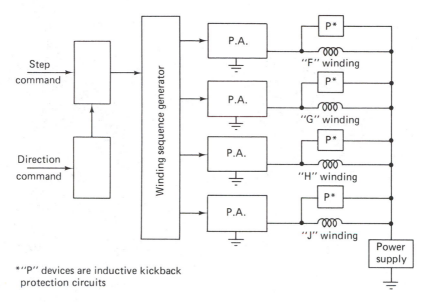

*"P" devices are inductive kickback
protection circuits

Figure 16-18 Control logic for PM stepper.

reluctance (VR) *stepper.* Figure 16-19 illustrates a typical variable-reluctance stepper. It has 12 stator coils energized in groups of four (called *phases*). The illustration shows the complete set (of four) coils for one phase in each schematic; the coils for the other two phases are omitted for clarity. The rotor is made of a magnetic material (it is not a permanent magnet) and has a series of teeth machined in it.

The illustration shows a rotor with eight teeth and Fig. 16-19A illustrates the rotor in the position it would assume during (or after) phase A windings (only) have been energized. The teeth on the rotor will align themselves with the set of energized poles that are nearest any of the rotor teeth, which is the position of least magnetic reluctance between the rotor and the stator field. Subsequently deenergizing the phase A winding and energizing the phase B windings will cause the rotor teeth nearest the phase B windings to be pulled into position directly in line with their magnetic fields, as illustrated in Fig. 16-19B. This causes the rotor to move (15° for the motor illustrated) one step in a counterclockwise direction.

Energizing phase C windings (Fig. 16-19C) will cause the rotor to step another 15° counterclockwise. However, if the phase A windings have been reenergized instead of the phase C windings, the rotor would have stepped in a clockwise direction back to the position indicated in Fig. 16-19A.

Therefore, as with the permanent-magnet stepper, the VR stepper requires its windings to be energized in the proper sequence for predictable operation. Also, it can be made to step bidirectionally. The VR stepper can be made with smaller stepping angles than the PM type; however, it does not exhibit the same "detent" ability with all windings deenergized as do PM-type steppers.

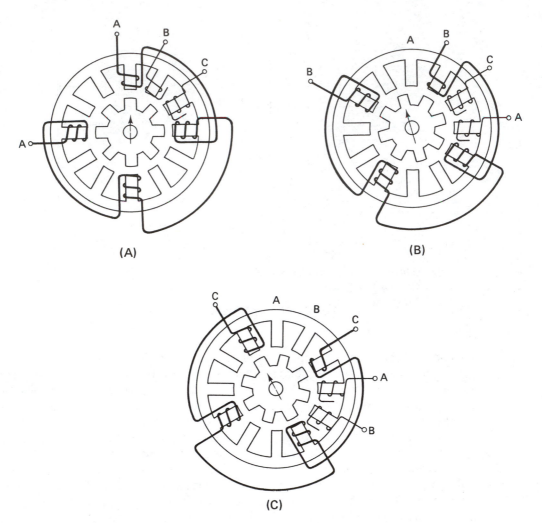

Figure 16-19 Three-phase variable-reluctance digital stepper motor schematic: (A) rotor position for "A" winding energized; (B) rotor position for "B" winding energized; (C) rotor position for "C" winding energized.

Other types of digital stepping motors are currently available, but the majority are either the PM or VR types. Also, the drive electronics for these stepper motors can vary widely in the number of windings that are simultaneously energized, in power capabilities, and in other features. These devices definitely have their place in the field of process control; however, they have design limitations which have made them economically disadvantaged as compared to other (more standard) drive mechanisms in the majority of applications. Figure 16-20 illustrates several commercial stepper motors.

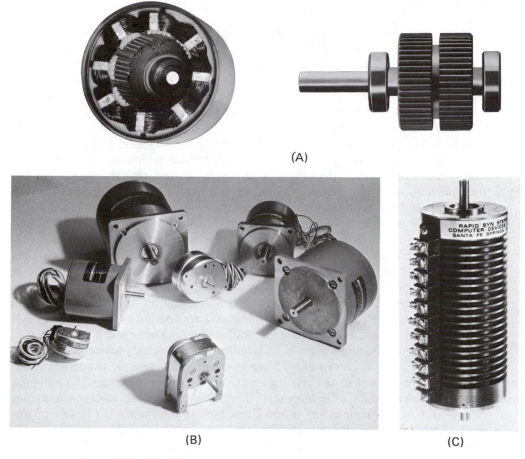

(A)

(B)   (C)

Figure 16-20 (A) Disassembled permanent-magnet stepper motor, showing internal construction; (B) group of commercial stepper motors; (C) commercial stepper motor. [(A) and (B) Courtesy of Sigma Instruments, Inc.; (C) courtesy of Computer Devices of California.]

HYDRAULIC-PNEUMATIC ACTUATORS

There is a relatively limited selection of nonelectrical actuators available in the commercial marketplace. Actually, however, this limited selection is quite adequate for most applications. There are excellent linear position actuators which can be made in any size desired and there are also several very excellent rotary actuators. Therefore, there is no big requirement for new conceptual designs in nonelectrical actuators. Although new designs have appeared over the past years, basically the techniques primarily being used in systems have been around for many years with only minor design changes.

Linear Actuators

There are essentially two basic linear actuator concepts used. The first is the fluid (hydraulic or pneumatic) operated cylinder; the second is the diaphragm (force) motor that was referred to earlier (Chapter 12).

The hydraulic cylinder (Fig. 16-21) consists of a hollow cylindrical tube that is divided into two parts by a movable piston. The cylinder itself is sealed at both ends except for a single connection for tubing to be attached at each end. The piston is free to slide back and forth in the cylindrical tube. It has a power-takeoff shaft attached to one side, and there are seals around both the piston and the shaft.

These cylinders are designed to be operated either pneumatically or hydraulically (or both); however, the principles of operation are identical regardless of the fluid medium used. A pressure admitted to either end of the cylinder will develop a differential force acting across the piston that will cause the piston (and the attached power-takeoff shaft) to move.

If the piston is designed for activation by pressure applied to either end, it is termed a *double-acting piston*. Fluid pressure causes motion in both directions. One technique for control of the pressures applied to this type of piston is illustrated in Fig. 16-22. The control device is the common slide valve (or spool valve, as it is also commonly called), which is moved in a linear motion to open and close the valve ports. Figure 16-23 illustrates an industrial application of double-acting cylinders; other common examples are in construction backhoes and bulldozers.

In many applications there is no requirement for double-acting pistons. The only requirement is to exert a linear force in one direction and the piston will either return to the other end of the cylinder due to an internally (or externally) mounted spring, or else a weight will return the piston. One of the most common examples of a weight-return single-acting piston is the typical automobile lift in service stations. Figure 16-24 illustrates spring-return single-acting cylinders, again using a spool valve to control the pressure applied to the piston.

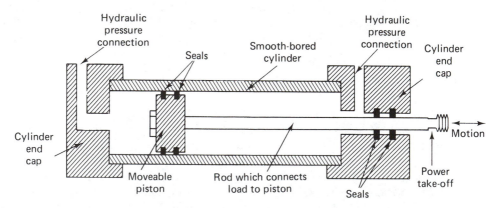

Figure 16-21 Fluid cylinder.

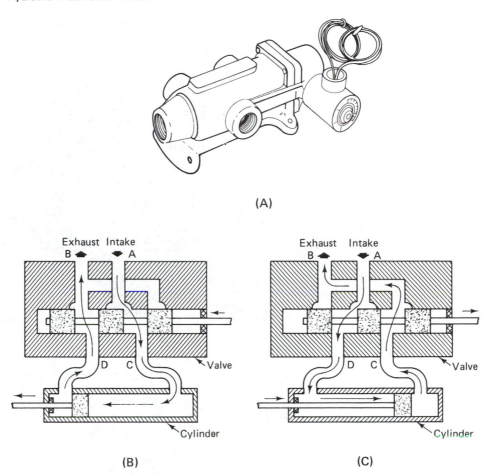

(A)

(B)   (C)

Figure 16-22 Spool valve control for double-acting hydraulic cylinder.

Another very common linear-motion actuator is the *diaphragm force motor.* It is an extremely simple device and is schematically illustrated in Fig. 16-25. Pressure (normally pneumatic) is admitted to one side of the diaphragm, which causes it to move against spring pressure. Typically, a flapper-nozzle (baffle-nozzle) assembly is used to control the pressure admitted to the diaphragm. The reader is referred back to Chapter 12 for further explanation of the action of the flapper-nozzle and diaphragm motor.

Other linear actuators exist and are used in control systems; however, the ones described here are by far the most common types. Also, it is not uncommon to find a linear fluid actuator connected to a simple lever-arm arrangement in order to obtain rotary motion. The cylinder arrangements used on dump trucks and backhoes are good examples of hydraulic linear actuators used to obtain rotary motion.

Figure 16-23 Double-acting cylinder controlled ball valve. (Courtesy of Kamyr Valves, Inc. & Neles Oy.)

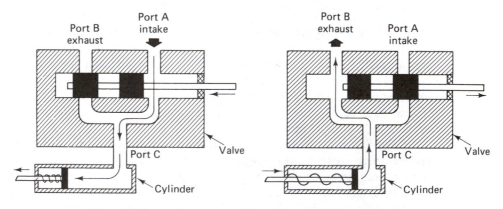

Figure 16-24 Spool valve control for single-acting hydraulic cylinder.

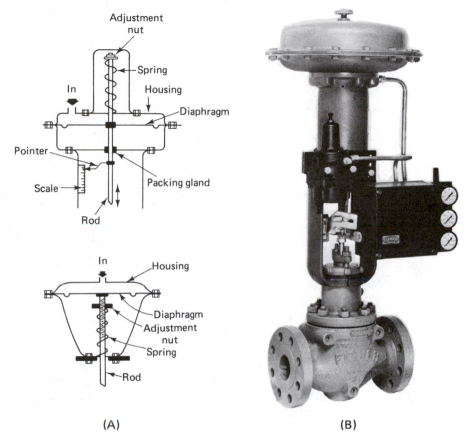

Figure 16-25 (A) Pneumatic diaphragm-motors; (B) diaphragm motor controlled valve. [(B) Courtesy of Fisher Controls Co.]

Rotary Actuators

Although several types of rotary fluid (normally hydraulic) actuators are used industrially, they are not used as commonly as linear actuators. Probably the simplest of the rotary actuators is illustrated in Fig. 16-26. It consists of a sealed chamber divided into two parts by the rectangular-shaped piston. A differential pressure acting across the piston will produce a rotary motion of the output shaft. The motion of the output shaft is limited to much less than 360°; however, tremendous torques can be obtained from this type of motor.

There are several other types of rotary hydraulic actuators which have continuous rotation capabilities. The axial and radial piston hydraulic motors are two excellent examples; however, they have highly specialized applications and the operation of these devices is beyond the scope of this text.

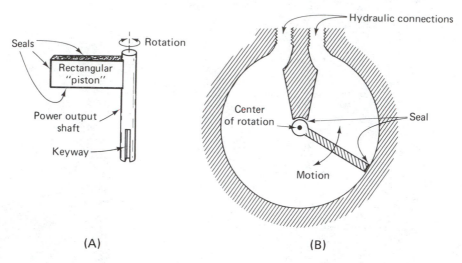

Figure 16-26 Rotary hydraulic actuator: (A) "piston"; (B) actuator.

The spool valves that control the pressure admitted to linear or rotary fluid-operated cylinders are themselves controlled by a linear-motion servomechanism. In many instances the spool valves are controlled by electrical solenoids, both the digital and the analog types. Therefore, the cylinders can act either in digital or analog systems, depending upon the actuation used to move the spool valve. In many other applications, diaphragm force motors are used to operate the spool valves, again for either digital or analog control.

QUESTIONS

1. What is an actuator?
2. What is a servomechanism?
3. Where do actuators functionally fit into the automation of process control systems?
4. How do actuators differ from sensors? How are they similar to each other?
5. Explain the phenomenon of counter electromotive force in dc motors and its effect upon the speed of a dc motor.
6. Why are motor starters required on larger dc motors?
7. List several means of controlling the speed of dc motors.
8. How can the direction of rotation of a dc motor be controlled?
9. Explain the rotating magnetic field in an ac motor and its effect on the speed of the ac motor.
10. Why is it desirable to use pole shading in single-phase ac motors?

11. Explain the sequence of operation of a split-phase capacitor-start single-phase induction motor.

12. Basically, what is a universal ac motor? How can you recognize one simply by looking?

13. How do you control the speed of an ac motor?

14. What is the difference between synchronous ac motors and ac induction motors?

15. How can the direction of rotation of a single-phase ac induction motor be controlled?

16. Compare briefly the theory of operation of dc motors with the theory of operation of ac motors.

17. Upon what does the accuracy of the final rest position of the rotor of a digital stepping motor depend?

18. Generally, what is the principle of operation of all digital stepping motors?

19. Compare the permanent-magnet and variable-reluctance stepping motors.

20. Explain the operation of the double-acting cylinder as controlled by a spool valve (illustrated in Fig. 16-22).

21. How does the theory of operation of the pneumatic diaphragm motor in Fig. 16-25 differ from the explanation of operation of the pneumatic actuator pictured in Chapter 12?

22. Make a table, listing in one vertical column each of the actuators covered in this chapter. Then make four adjacent columns, labeling them: Variable controlled, Principle of operation, Advantage, and Disadvantage. Attempt to fill every blank space in the table.

23. Which connections would be made in order for these relays to perform an OR function (i.e., continuity between points *X* and *Y* when circuit 1 or 2 is energized)?

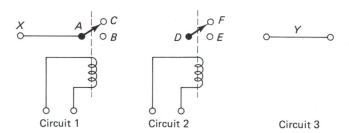

Circuit 1   Circuit 2   Circuit 3

A. $C \rightarrow D$ and $F \rightarrow Y$.
B. $B \rightarrow D$ and $E \rightarrow Y$.
C. $C \rightarrow Y$, $A \rightarrow D$, and $F \rightarrow Y$.
D. $B \rightarrow Y$, $A \rightarrow D$, and $E \rightarrow Y$.
E. $B \rightarrow C$, $F \rightarrow E$, and $E \rightarrow Y$.

24. Which of the following functions can be performed by an ordinary electromechanical relay?
A. Electrical isolation between two circuits.
B. Power amplification.
C. Memory.
D. Ac-to-dc or dc-to-ac conversion.
E. All of the answers above are correct.

25. The following relay symbols—in order from left to right—represent what type of contact configurations?

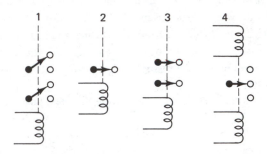

A. DPST, DPDT, SPDT, triple PST.
B. DPDT, SPST, DPST, SP triple throw.
C. DPDT, DPDT, SPST, SPDT.
D. SPDT, DPST, DPDT, SP triple throw.

26. Schematics for relays always illustrate the condition of the relay contacts
A. With the coil energized.
B. With the control circuit energized.
C. With the coil not energized.
D. With the controlled circuit energized.

The next three questions refer to the following figure:

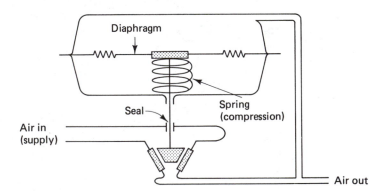

27. If the diaphragm fails, what will the output be?

28. What function will this system perform?

29. What is the device that actuates the valve called?
A. A pneumatic cylinder.
B. A spring motor.
C. A diaphragm motor.
D. A hydraulic motor.

30. A certain hydraulic device is to be used to lift heavy weights straight up and then set them down again. What type of hydraulic device would be the *most economical* for this purpose?
  A. A rotary actuator.
  B. A double-acting cylinder.
  C. A diaphragm motor.
  D. A single-acting cylinder.

31. Which of the following statements describe conditions employed to make a "small" ac motor start *and* continue to run?
  A. A phase difference between the ac exciting the different pairs of windings.
  B. A revolving (rotating) magnetic field.
  C. A set of starting windings or shaded-pole windings.
  D. A permanent-magnet (or electromagnet) armature.
  E. All of the statements above describe phenomena that can be used to make an ac motor operate.

32. The most basic difference between actuators and servomechanisms is _____.

33. The accuracy of the final rest position of the rotor of a digital stepping motor after a number of steps depends
  A. On the number of steps it has taken.
  B. On the applied voltage.
  C. On the mechanical positioning of the poles.
  D. On the digital logic that determines which windings are energized.
  E. On all of the above.

34. Which of the following is a characteristic of permanent-magnet-type digital stepper motors?
  A. Excellent starting torque.
  B. Extremely precise position control of the rotor.
  C. Detent action (holding ability when not rotating).
  D. Less overshoot than other types of motors.
  E. All of the above are distinct characteristics of stepper motors.

35. A universal motor is a dc series-wound motor which has been rated for use with either ac or dc excitation power. True or false?

36. 110-V ac small hand power tools, which operate at variable speeds, are powered by _____.

37. The secret to understanding and predicting the behavior of a dc motor lies in an understanding of
  A. Rotating fields.
  B. "Pole shading".
  C. The sequential switching of its windings.
  D. The effects of counter emf.
  E. It is necessary to consider all of the effects above.

38. A series field winding improves what characteristic of a dc motor?

39. What is the primary advantage of having a shunt winding on a dc motor?

40. Wound field dc motors are lumped into three general classes: _____,
_____, and _____.

41. What limits the current through the armature of a dc motor?

42. How can you cause a "small" dc motor to run bidirectionally?

43. Why is it necessary to use "pole shading" in single-phase ac motors?

44. Essentially what is the difference in the use of permanent-magnet rotors and electromagnet rotors in ac motors?

45. What is counter emf in a dc motor?

46. Which of the following statements is true when comparing differences between ac and
dc motors?
   A. In dc motors, the field polarity remains constant and the rotor field rotates.
   B. In ac motors, the field polarity rotates and the rotor field maintains a constant mag-
      netic polarity.
   C. The speed of an ac motor is determined by applied frequency.
   D. The speed of a dc motor is determined by applied voltages.
   E. All of the statements above are valid comparisons.

47. In application where precise speed control is required over a wide range of speeds, which
type of motor would be preferred? (No variable-frequency power supply is available.)
   A. An ac synchronous motor.
   B. A dc compound wound motor.
   C. An ac induction motor.
   D. A dc torque motor.

48. Digital stepper motors operate
   A. By sequentially energizing various of its many field windings.
   B. By shading each of its several poles.
   C. By utilizing an electromagnetic rotor.
   D. By applying ac excitation power to its field coils.
   E. By using all of the techniques above simultaneously.

GLOSSARY

AC Motor   An electromechanical rotating machine (transducer) whose electrical input is
alternating current.

Actuator   A transducer whose output is mechanical motion.

DC Motor   An electromechanical rotating machine (transducer) whose input is direct current.

Induction Motor   An ac motor having a wound electromagnetic rotor. The motor stator
windings act as the primary of a transformer while the windings on the rotor act as a
secondary. The rotating stator field rotates significantly faster than the rotor itself, thereby
inducing an electromagnetic field in the rotor which transmits power to the load. The
induction motor rotor must rotate at a slower speed than the frequency of the ac excitation
source would cause in a synchronous motor.

Rack and Pinion   A mechanical rotary/linear motion transducer physically made up of (1) the rack, which is a bar having a gear teeth cut into one side of it; and (2) a pinion, which is a gear having matching teeth cut around its periphery. When they are properly meshed (including bearings) either can be used as the input and the other will yield the transduced mechanical motion output.

Servomechanism   An actuator which has been included in its own automatic feedback control system.

Shaded-Pole Motor   A single-phase permanent magnet ac motor which has one or more auxiliary, secondary, shorted stator windings whose effect is to create an apparent motion to the magnetic field across each pole piece. This shift is used to cause the motor to start.

Solenoid   An electromagnetic device which has an electrical input and causes the mechanical motion of a movable core piece located in its magnetic field.

Stepper Motor   An electromagnetic motor so designed that its output shaft moves in a series of (equal) angular displacements when its several stator windings are sequentially energized.

Synchronous Motor   An ac motor whose (normally permanent magnet) rotor rotates at a speed which is fixed by the frequency of the ac excitation source.

17

TROUBLESHOOTING SYSTEM FAILURES

INTRODUCTION

Troubleshooting: if there is one word that describes the responsibilities of a technician, this is it. It is the capability and training in the area of *troubleshooting* the ailments of electronic equipment that distinguishes the technician from the engineer. Make no mistake, though, neither one is an island unto himself; they must work in pairs in order to realize success.

The engineer has been educated from the point of view that he will eventually be employed in the field of original design. Therefore his courses are very theoretical and highly dependent upon higher mathematics, which is the only available means of describing systems so that they can be theoretically analyzed and designed. It is through mathematics that the engineer can express the complicated interrelationships among the components of any system (the system is normally partly electronic and partly hydraulic, mechanical, pneumatic, and/or thermodynamic), and thereby predict the overall system performance of a new system design with any degree of accuracy.

The technician, on the other hand, receives a more practical formal education. It is not intended that he should receive an education that would make him competent to perform other than very basic original equipment design. Rather, he is trained to have enough of an understanding of electronic components to analyze a circuit and pinpoint the probable cause of failure of that circuit so that it can be repaired. To accomplish this function, it is not necessary for the technician to be burdened with the requirement for understanding the theory and mathematics that the engineer needs. Frequently, the study of so much theory has a tendency to confuse rather than to clarify.

428

Therefore, the engineering team consists of those who can perform the original theoretical design and those who can reduce the paper design to a practical operating circuit. Very seldom will a paper design work when first wired up and plugged in. There is a period between the point the circuit is designed and the point that it operates satisfactorily to meet the original design specifications. Neither the engineer nor the technician can handle this interim phase alone; it is a joint, shoulder-to-shoulder effort if success is to be realized.

Success will, of course, result only if all parties in the design and maintenance/repair team can fulfill the roles assigned to them. Of concern here is the role that the design, maintenance, and repair technician fills. Regardless of the position or job description an electronics technician is hired to fill, sooner or later he will be required to troubleshoot and repair systems. Otherwise, the technician is wasting his education in a position that does not require the amount of formal education he has acquired.

The purpose of this text is to give the student a basic background in automatic control systems, their component parts, and the basic theory behind their operation. However, to stop at this point would be only telling half of the story. The rest of the story lies in the *documentation* provided by the system supplier. The purpose for this documentation is to enable the user of the system to learn and understand its operating capabilities and limitations, to apply the system to fulfill his requirements, and finally to maintain, repair, and possibly even alter the equipment himself.

Therefore, this chapter will be aimed at giving the student an idea of the types of documentation typically supplied with a process control system and the use made of this documentation by the technician to understand, use, maintain, and repair the system will also be covered. General guidelines as to the method of diagnosing and tracking down equipment failures will be covered where appropriate.

Throughout this presentation, do not lose sight of the fact that every system will have different documentation supplied with it. Some documentation will not be clearly written, some will be downright inaccurate, and most documentation will be inadequate. Therefore, you, as the technician responsible for a system and its operation, must, in most cases, develop the documentation you want but which has not been supplied by the system manufacturer. You will almost always have to make corrections to the supplied documentation and add missing information to this documentation.

A short digression seems to be appropriate here to answer an obvious question: Why is this documentation normally so poor?

The answer is basically a question of economics. Once the original design engineers have completed their portion of a system, make all the alterations necessary to make it work by itself, and then redesign it to work with the rest of the control system, they amend the original documentation to include the changes that were made. But keep in mind that this is usually not the only project they are working on, so they may forget some changes; also, many "bench" changes are never documented to begin with. Furthermore, the process of reengineering even an operational system never ceases. The time and money required to continually update schematics, drawings, and so on, is substantial. At some point in the process, a project manager "freezes" all documentation and it is mass-produced. After that time, to include a change would require that all the existing copies be either updated or replaced.

This is a very brief, and not all-inclusive, general explanation of the derivation of inaccurate documentation. Adequacy of documentation supplied with a system is another problem, but is still based upon economics. It requires much time and expense to develop documentation for a system, if for no other reason, because of the complexity of most modern systems, especially those which include an electronic computer as the controller. The expenses are classified by management as overhead expenses, in other words, expenses that cannot be tied to any specific salable item (as another example of overhead, consider secretarial expenses associated with a project). Therefore, they want the "minimum adequate" documentation to be produced for a system, in order to keep these overhead costs to a minimum.

Whether justified or not, these are the general lines of reasoning used to justify the documentation (or lack of documentation, if you prefer) supplied with automatic control systems. They are normally all based upon economics, however, not the desire to make your life as a technician more difficult.

There is one further item I would like to discuss before delving into the secrets of successful system troubleshooting. Process control systems are very expensive. Because of the expense involved, the decision to install automation on any specific process is normally made at the level of the board of directors of the company. The justification is frequently based upon maximum utilization of the system. This means that if the system is inoperative, some very influential personnel within the company become upset. This, in turn, means that their subordinates become even more upset, and the only two things available to vent their anxieties on are the system and the technician. Since the system is already "sick," it is obviously no fun to take out one's anger on it, and that leaves the technician. Now you have not only the problem of repairing the equipment but also the problem of putting up with these managerial personnel who may have no concept of what you are up against but want "their" system operational—ten minutes ago at the latest. I have no solution to this problem but remember throughout the ordeal that system downtime may often cost hundreds or even thousands of dollars *per minute,* so the managerial personnel have a real reason to be upset.

Now, on with the introduction of basic troubleshooting of automatic process control systems.

PRELIMINARY STEPS

This may come as a surprise to some of you, but the *last* steps in troubleshooting a large system are those in which you use oscilloscopes, meters, and other tools of the trade as you know them. If you stop to think about it, this statement does make sense. Up to this point in your formal education you have been faced with rather small systems. "Small systems," as used in this context, are those which are housed within one electronic enclosure. There was never any question as to where to begin. You were familiar with the possible modes of operation of this equipment, so you could decide which of its possible functions were not operating. Furthermore, you had a general idea of the various "functions" performed by portions of the internal circuitry in order to achieve the output that was not operating. With a circuit schematic and a few meters, the problem could be located.

Now envision yourself in the control room of a typical industrial process control system (see Fig. 17-1). There are many electronic enclosures, all filled with various types of electronic equipment, including a computer (digital or analog or both). Outside the control room is a process having electrical, hydraulic, pneumatic, thermodynamic, and mechanical components. The system is down. What do you do? Where do you start? A somewhat more involved problem than you are used to, but the troubleshooting procedure is roughly the same as for the small system.

First Step: Understand System Operation

For any system you must first know what it is capable of doing. You must know how it is being used and how it should operate in the process you are dealing with. For a larger system with many different outputs, possibly not one person in the company thoroughly understands all facets of how the system operates. The original system supplier should have provided operating manuals for the various portions of the system. These manuals should explain the various functions that the system is capable of performing and go into great detail as to how it accomplishes these functions.

Frequently, the original system supplier has purchased the bulk of the subsystems from other suppliers and merely ties the subsystems together into a complete system. When he purchased these subsystems (a digital central processor, analog-to-digital conversion systems, electrical to pneumatic and hydraulic converters, etc.), he merely bound together the operating manuals, schematics, blue prints, and so on, supplied by the subsystem manufacturers. He then was left with the necessity of providing documentation on the interface equipment he had designed, on the arrangement and interconnections between the subsystems, on the operation of the system as a whole, and on the computer program developed to make the system operate.

Therefore, there will not be only one manual to read to get the necessary background information; there may well be many manuals, all written by different people and all going into various amounts of detail on how the applicable equipment operates. So this particular phase of preparing yourself to be able to troubleshoot the system is difficult and time-consuming, but if it is not done well, there is little hope of troubleshooting the system successfully.

Figure 17-1 Two views of the control room for the pulverized coal-fired boiler at the Columbus and Southern Ohio Electric Company's Conesville Station. This is a typical process control room, with a multitude of displays and controls. This particular system has the added complexity of being a direct digital control (DDC) system. (Courtesy of Bailey Meter Co.)

Second Step: Pinpoint the Problem

In a large control system, any of literally hundreds of functions could fail and cause the system to be down. Your problem will be to determine, as specifically as possible, what function(s) are the ones that have failed. Obviously, this would be impossible without a good understanding of how the system should operate (step 1).

To help in this second phase of troubleshooting, there should be available (from the original system supplier) a series of information flow block diagrams. These block diagrams illustrate the flow of information from the process sensors and command inputs through the various major portions of the system and finally show where the outputs go.

There should be several levels of breakdown for these block diagrams. Figures 17-2 to 17-5 illustrate this for a typical process control system using a general-purpose digital computer for the controller. Figure 17-2 shows a very general block diagram for the entire process control system, of which the computer is simply one block. As an example, let us assume a fault that has as its symptom the loss of all meaningful analog information to the process servos.

Referring to Fig. 17-2, it is obvious that the digital output module probably is not at fault. However, the problem could be that the analog input or digital input modules are malfunctioning and no meaningful information is being supplied to the computer, and therefore the computer's outputs are meaningless. From the knowledge gained in step 1, you would have learned that the computer program is written to make sure that these inputs remain within certain limits and to sound an alarm if any input goes beyond the preset limits. For our example, assume that no such alarm was sounded. Therefore, assuming that the alarm itself is operating (a simple test will confirm this), the input modules can be assumed to be working.

By the way, the analog input module is located in two electronic cabinets and the digital input module in a third. Therefore, you have eliminated three cabinets of electronics to check for the problem. Again referring to Fig. 17-2, this narrows the problem down to either the analog output module, or the digital computer. There is usually provision made to easily insert test patterns into the input side of the analog output module to facilitate rapid troubleshooting of this module. Assume that we insert a couple of test patterns and the analog output module checks out. This seems to indicate that the probelm is located in the digital computer. Furthermore, since the digital outputs are normal, the fault must lie within a portion of the computer that is concerned only with the analog outputs.

As you can see, we have done a lot of troubleshooting with simply a general block diagram of the system. Stop and remember the last piece of electronic equipment you worked on; didn't you have the block diagrams of that system in your head, and didn't you follow essentially the same procedure?

Now that we have narrowed the problem down to the digital computer (also called the controller or the central processing unit, CPU) let us get a block diagram of the CPU itself so that we can follow the information flow through it and further narrow the specific area of failure.

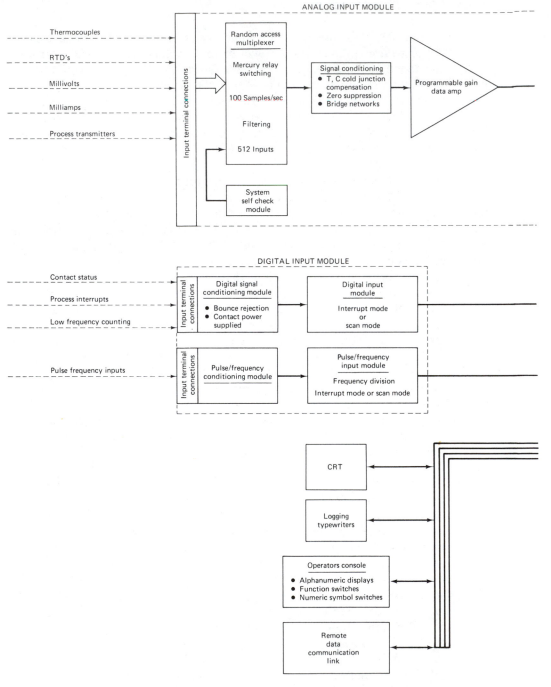

Figure 17-2 Typical direct digital control system block diagram.

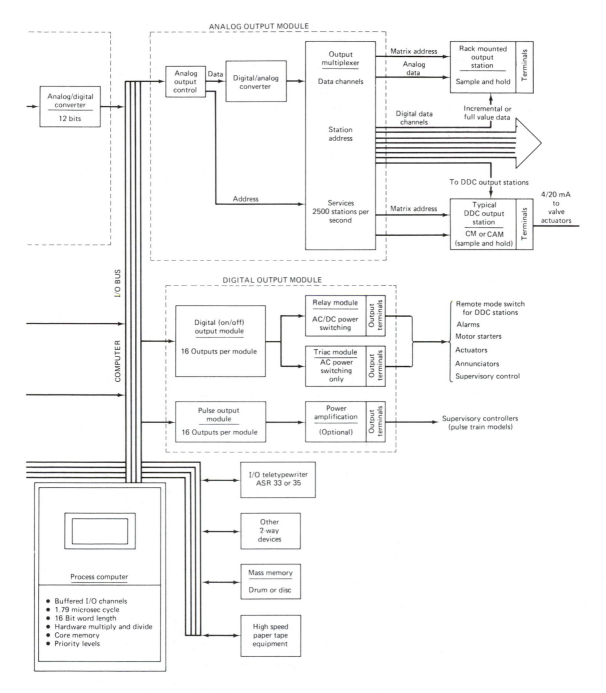

Figure 17-2 (*Continued*)

Figure 17-3 is a block diagram of the CPU with its input and output buffer electronics. There should also be block diagrams that further break down each of the major functional blocks shown in Fig. 17-3. Referring to Fig. 17-3, it is noted that all inputs to the CPU and all output from the CPU use the single I/O bus. Therefore, if the fault was with the I/O bus, there would be no valid output from the computer. In our example the digital output is working. Therefore, the problem could be either in one portion of the CPU that only the analog output section uses (such as a particular computer command or instruction), or it could be in the analog output buffer area in memory.

Manual operation of the computer will confirm that the analog output buffer is operating properly. At this point we have narrowed the problem down to the internal workings of the CPU—but more than this, the problem is narrowed down to some internal circuitry which is evidently used *only* by the analog output module. Furthermore, the problem has been narrowed down to one electronic enclosure of the 10 or 20 that are located in the control room.

You should be getting the idea now of how to proceed. The process consists of going from each block diagram to more detailed block diagrams; each time *eliminating* as many functional pieces of hardware as possible. Our next step in this example is to go to a block diagram that breaks the CPU down into more detail. This block diagram is shown in Fig. 17-4.

Referring to Fig. 17-4, it can be shown that most of the individual pieces of hardware are used for nearly all operations of the CPU. So we seem to be at a blind alley. A further breakdown of the CPU would reveal no further information. Note that by this time you would have troubleshot your problem down to one of the primary elements in the data flow path. Also desirable is a diagram which includes the following information: schematic drawing numbers, control signal mnemonic designations, and the card and pin numbers where an oscilloscope can be connected to observe these signals.

The next level of breakdown is the manufacturer's individual card schematics (drawings). Also there may be included timing diagrams illustrating the relative timing of the control signals in order to accomplish a function. A typical timing diagram for this system is illustrated in Fig. 17-5.

You should also be warned at this point that Figs. 17-3, 17-4, and 17-5 were not supplied by the computer manufacturer. They were developed by the person in charge of training technicians to maintain this computer system (the examples used herein are of an operating direct digital control system). Therefore, when you are confronted with a large and complicated system to maintain, you may need to develop these block diagrams, information flow schematics, control signal schematics, and timing diagrams for yourself. The process of developing these items will aid immensely in your understanding of the system.

Now, getting back to the example problem: without even opening a cabinet door, the problem has been troubleshot to a particular portion of the internal workings of the CPU which is used (evidently) only by the analog output module.

By properly preparing yourself during step 1, becoming familiar with the op-

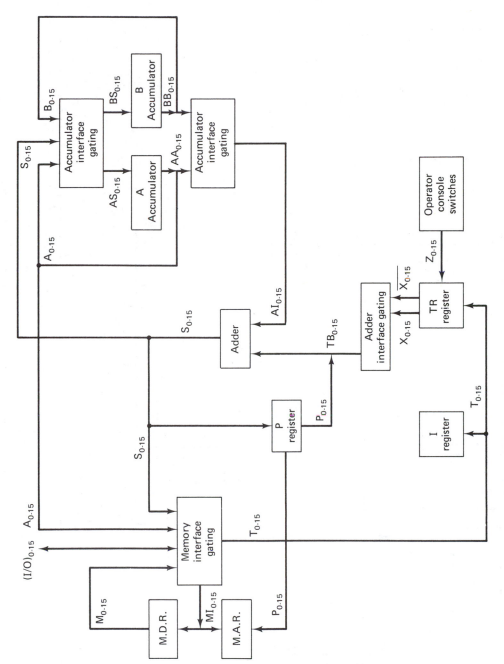

Figure 17-3 Typical digital computer data flow block diagram.

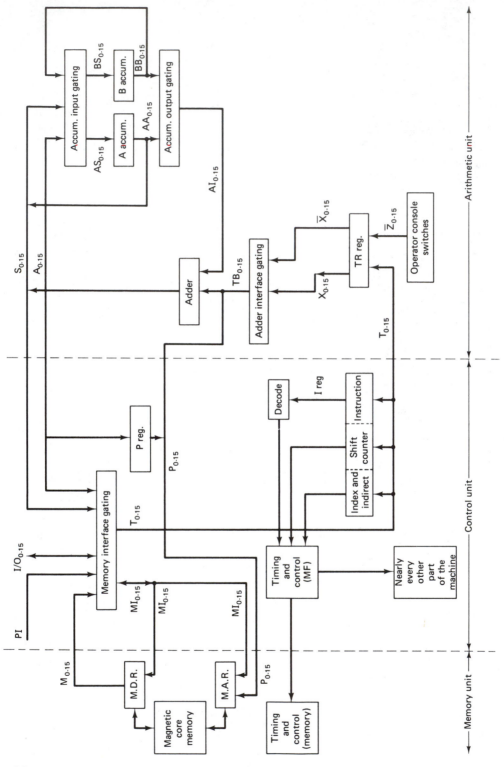

Figure 17-4 Typical digital computer information flow block diagram.

438

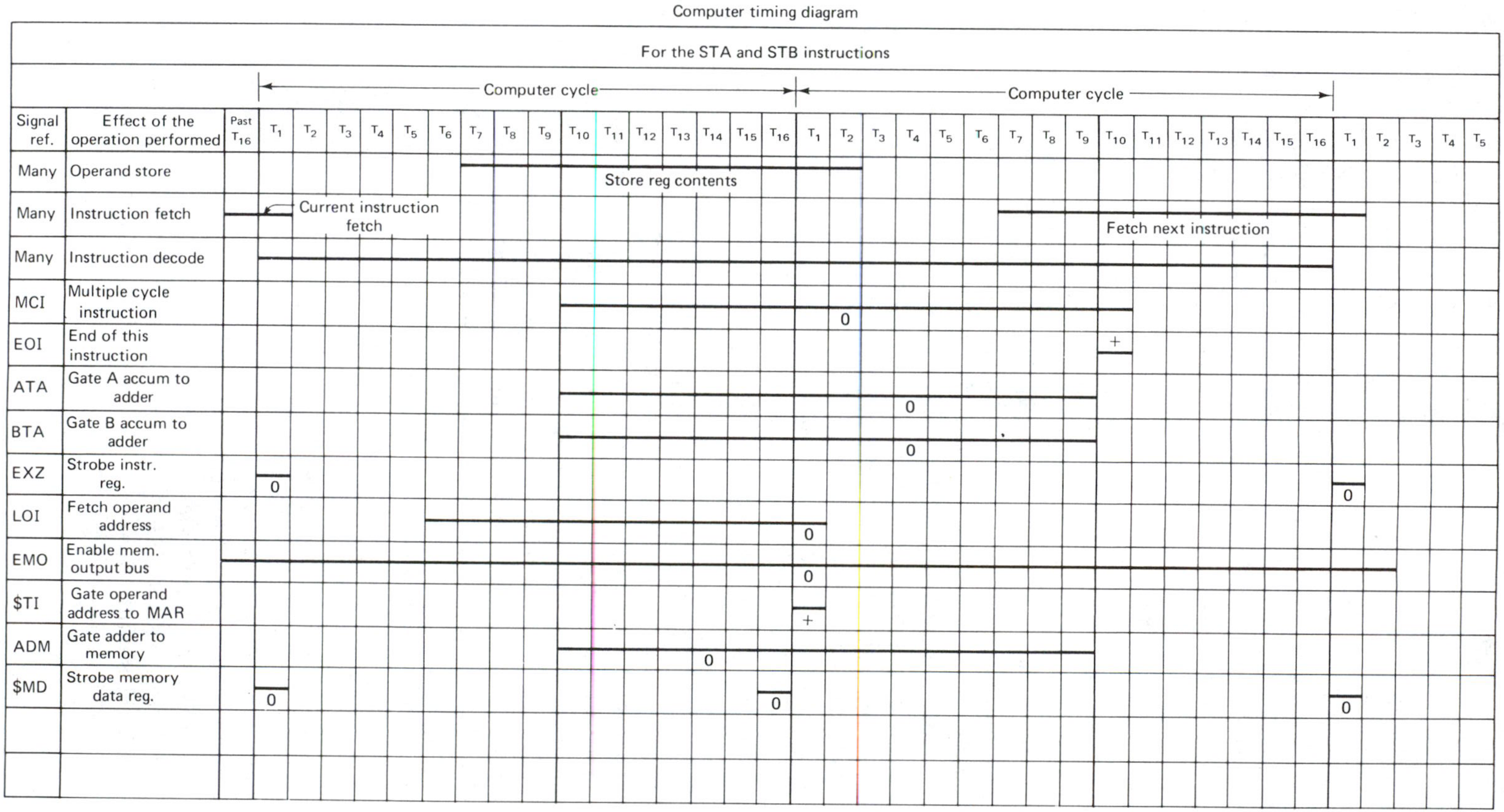

Figure 17-5 Typical digital computer timing diagram.

eration of the CPU, you would be able to eliminate all the functional blocks in Fig. 17-4, with the exception of portions of the following blocks: *mainframe* (MF) timing and control, and the instruction and decode portion.

Figure 17-4 is a data flow schematic and does not illustrate these two functions. The reasoning behind their elimination from Fig. 17-4 is that they are well documented on one or two of the manufacturer's schematic drawings. The next step is to go to these drawings.

Final Step: Use Test Equipment

Now is the time to get out the oscilloscope(s) and other applicable electronic test equipment. If you stop to think about it, where would you have hooked the scope up to before this, and what would you have looked for? There really is no other way to manually troubleshoot the system; there is simply too much of it. (There *are* automatic ways, which will be discussed later.)

Both the mainframe timing and control module and the instruction and decode module are used whenever the computer is running. All the mainframe timing signals and all the instruction registers are used to perform most other computer functions, which, in our example, are operating properly; therefore, the problem probably lies with the mainframe control or instruction decode module.

By using the computer schematics for these two modules (which are physically located on a total of five logic boards), the oscilloscope can be used to determine what is wrong.

Digital computers execute instructions which have the effect of routing data through the various modules of the computer (see Figs. 17-3 and 17-4). As the data pass through these modules, they are operated upon; for example, they are added to (or subtracted from) other data already in the adder, incremented, shifted, multiplied (or divided), and so on, as they pass through these functional modules.

These instructions are decoded in the instruction decode module, which then (with control signals from the mainframe control module) sets up the proper data paths. There may be anywhere from 30 to hundreds of possible instructions and variations of instructions. At this point it might seem as though each type of instruction must be checked to determine exactly what has failed.

A discussion with one of the computer programmers on the system would reveal that the Multiply instruction is used only by the analog data manipulation programs and in no other output programs. Therefore, it would seem that, of the approximately 50 instructions the CPU is capable of executing, the Multiply instruction should be checked first, because except for the analog output module, the rest of the system appears to be operational.

Finally, the oscilloscope is used. A check of the instruction decode module would, in this example, reveal that a logic gate had failed, one that is only used in decoding the Multiply instruction. A check with the manufacturer's board drawings will reveal the defective logic element.

OTHER TROUBLESHOOTING AIDS

The system used in the example above is typical of most medium-sized computer control systems. Other, more complicated systems may have several computers tied in together. The method of attacking the problem of troubleshooting is similar to our example, except that there are more complicated block diagrams to contend with.

To help reduce the time required to locate a malfunction in these computer systems, frequently the manufacturer will supply a series of computer programs termed *diagnostic routines*. These are relatively small computer programs designed to run specific functional modules in the system through all of their modes of operation (at computer speed), with some method of giving the technician an indication of what test (or tests) failed. These programs can be so important that most computer technicians develop their own "diagnostics" in addition to those supplied by the system manufacturer.

SUMMARY

There are, of course, many shortcuts and "pet" techniques that individual technicians use, but in general they are modifications of the general technique outlined previously. There *must* be a logical approach taken to eliminate operational portions of the system, thereby reducing the possible sources of failure to only a few possibilities before actually beginning the process of troubleshooting with electronic test equipment. This equipment is not used until the possible faults are reduced to only a few possible electronic signals (to the circuit board or even component level). Then, as the final step in the troubleshooting procedure, the electronic cabinets are opened and the circuitry attacked with oscilloscopes and meters.

THE TECHNICIAN'S NIGHTMARES

There are two additional classes of automatic control system faults which have purposely not been mentioned in the foregoing presentation. The first, and easier-to-solve problem is a computer software fault. A digital (or analog) computer will do exactly what it is programmed to do. If the process of hardware troubleshooting results in no hardware faults, the possibility of a programming error is very possible. In most cases the computer electronic technician must be sufficiently familiar with the computer programming language in order to discuss failure symptoms with the programmers in order to help determine whether the fault is hardware or software. In this case, the two must work together to find the problem; and it may be more logical to do this before proceeding with the process of hardware troubleshooting. If the computer program being used has been successfully used previously, suspect a hardware fault; if it is a relatively new program, suspect the program of being the problem.

The second, and definitely the most difficult type of electronic hardware failure to troubleshoot is the intermittent failure. Unfortunately, most well-designed electronic equipment will not have complete hardware failures; the components simply drift far enough out of their specified tolerances to the point where they normally will function in the circuit but sometimes cannot quite perform their function and cause a circuit failure.

Essentially, this problem cannot be troubleshot effectively unless the fault is obvious when it happens. Unfortunately, electronic component intermittent failures within the system (within the CPU especially) can leave no indication of their location. Sometimes in order to locate the failure, the technician simply has to replace boards one at a time until the failure no longer occurs.

Other types of intermittent failures include bad solder joints, intermittent board connectors, RFI pickup from outside sources, and variations in the supply power to the control system electronics. They are very difficult to trace and require every bit of cunning available to finally locate.

QUESTIONS

1. Why should you expect systems documentation to be somewhat inaccurate?
2. Why should you expect systems documentation to be somewhat inadequate?
3. Explain the three steps to troubleshoot a system.
4. Which of the three steps to troubleshooting a system is the most difficult to accomplish, and why?
5. List as many ways as you can think of to help you accomplish step 1 of the troubleshooting procedure—understanding the system in all its possible modes of operation.
6. When troubleshooting systems having digital computers in them, what is normally your best troubleshooting aid (tool)? Why?
7. Basically how does the approach to troubleshooting a large, complicated system differ from that needed to troubleshoot a simple piece of equipment such as a volt-ohmmeter?
8. To make sure that you have automatic control systems in their proper perspective, imagine that you are the manager of a medium-sized manufacturing company which produces a product made largely by manual labor. A manufacturer's representative calls on you and explains that his company has developed an automated machine that can potentially reduce by 60% the quantity of manual labor required in your company.

There is a catch, however; the cost of this equipment is roughly equal to the annual net earnings of your company. You have done a lot of investigation into the salesman's claims and are convinced that the equipment would probably be a good long-term investment.

Now you have to sell the idea to the board of directors and the stockholders. You want to present a good case for the long-term potential economic benefits; however, you also want them to realize the potential problems (your job may depend upon it if problems arise during installation or startup).

Make two lists, one of potential long-term benefits and another of long- and short-term potential problems as notes for your presentation to the board of directors. Note that not all of the potential arguments have been discussed in the text.

9. What is the proper order of the following items for a logical approach to the process of troubleshooting large electronic systems?
   A. Use electronic test equipment to locate the fault.
   B. Use system block diagrams to eliminate operational equipment.
   C. Understand system operation.
   D. Start replacing and/or interchanging PC boards.
   E. Use the manufacturer's drawings to tie the problem down to a couple of possibilities.

10. How are manufacturers' system drawings and schematics normally indexed or cross-referenced?
   A. From the source of each signal to each place it is used in the system.
   B. From major control signals back to their source.
   C. From page to page.
   D. From front to rear.
   E. From top to bottom.

11. When preparing yourself for troubleshooting of large industrial electronics systems, what is the most difficult area of preparation?
   A. Eliminating obviously operational functions.
   B. Getting an understanding of the many capabilities and operations of which the system is capable.
   C. Understanding the proper application and operation of the test equipment.
   D. Actually applying the test instruments to the problem.

12. Which of the following is (are) major differences between engineers and technicians?
   A. Engineers are more "theoretically" educated than are technicians.
   B. Technicians are educated along more "practical" lines than are engineers.
   C. Engineers are educated more for original design work than are technicians.
   D. Technicians are educated more for troubleshooting and repair than are engineers.
   E. All of the above are generally true differences.

13. When troubleshooting an electronic system containing a digital computer, what would probably be the piece of test equipment you would *first* use?
   A. An oscilloscope.
   B. The computer itself.
   C. An ungrounded VOM.
   D. A curve tracer.
   E. A digital logic probe.

14. Basically, how does the approach to troubleshooting large computer control systems differ from troubleshooting a simple electronic device such as a VOM?
   A. It takes a degreed engineer to troubleshoot large control systems.
   B. Basically there is no difference in the approach.
   C. In troubleshooting large systems the first step is to inspect visually for "smoked" components.
   D. There is no similarity to the basic approaches.

GLOSSARY

Diagnostic Program   A computer program which has been specifically written to exercise certain computer functions or peripheral devices in order to locate hardware faults.

Documentation   As used herein, this term refers to the supporting reference material supplied with process control system hardware and software (computer programs). It includes any combination of the following: narrative information explaining the operation, construction, and application of the equipment; schematic diagrams; blueprints; logic diagrams; wiring diagrams; test and calibration procedures; and troubleshooting information.

Mainframe   Another technical term used to refer to a digital computer's central processing unit.

Module   As used herein, this term refers to a functionally identifiable portion of the system, whether or not it is individually packaged as such.

Troubleshooting   The process of diagnosing, locating, and repairing malfunctions.

ROBOTS

WHAT ARE ROBOTS?

Ther term *robot* evidently originated as a descriptive term applied to a human-like, mechanical, automated being who was a character in a play written by Karel Čapek, titled *R.U.R.* (Rossum's Universal Robots). The term (or the concept at least) has inspired many playwrights and story writers ever since, resulting in numerous published works and motion pictures in which robots appear in a human-like form and either supplement or replace human beings in various applications. They are typically afforded human-like (or even superhuman) intelligence and sensory abilities.

We have been so conditioned to equate the term *robot* with these human-like devices that even popular dictionaries use that analogy. The terms *android* and *humanoid* have also been used to refer to science-fiction human-like automated devices.

The robots referred to in a modern industrial automation context positively do not resemble a human being in physical construction, mobility, or intelligence. State-of-the-art sensing devices and end effectors are certainly not as capable as their human counterparts in many ways. Then just what are we referring to when we discuss industrial robots?

They are simply machines, typically run by digital computers, which have been programmed by human beings to perform a wide variety of tasks. The literature covering the field of industrial robots, hereafter referred to simply as robots, has about as many specific definitions as there are authors. The definitions collectively include the following concepts:

Do routine manual work

Substitute for human beings

Do work for human beings

Operate automatically

Perform tasks in a seemingly human way

Can be reprogrammed to perform different tasks

Using the same license as they have used, we will try to develop a definition of the concept of industrial robots which will be used throughout the rest of this chapter.

First of all, they are simply machines; they can be as simple as a pneumatic cylinder programmed to perform a simple push/pull operation, or as complex as automated devices which basically simulate all of the motions of a human arm (from the shoulder to the wrist) and even specialized functions of the hand.

They are always machines, designed and made by human beings, their automated functions decided upon by human beings, and the computers that control them are programmed by human beings. They are used to perform the following general types of functions:

1. Relatively simple, mundane, repeated operations
2. Relatively complex, repeated tasks
3. Tasks which must be performed in environments that are unsuitable, dangerous, or even impossible for human beings.
4. Tasks which are basically unpleasant for human beings
5. Tasks which can be accomplished by human workers, but adequate quantities of skilled workers are not economically available
6. Tasks which are basically impossible to accomplish by human beings

In addition, these automated systems (industrial robots) are typically designed to perform those tasks listed above

1. With higher accuracy and improved quality
2. With higher repeatability and more consistent quality
3. At higher production rates
4. With less scrap product generated
5. At higher energy efficiency
6. With less downtime
7. With high reliability
8. At reduced manufacturing expense

In an industrial context, the bottom line must always be that robots must be economically beneficial. World industry runs on profits, producing goods and services that must compete in the world marketplace. Obviously, robotic devices fulfill this basic requirement or else they would still be relegated to science fiction and R&D laboratories.

We started out to derive a definition of robot, but so far we have investigated only the justifications for using industrial robots, although our definition must be general enough to include them, too. Other concepts that must also be included in our definition are the following:

1. They are flexible automated systems which can typically be reprogrammed and, with relatively minor physical modifications (or even none at all), can be used to accomplish many tasks.
2. They must supplement a human work force and not be unacceptably dangerous or obnoxious for human beings to work with.

Taking all of these matters into consideration, the definition I have arrived at is as follows:

In an industrial automation context, *robot* refers to a programmable automated mechanical system designed to supplement or replace a human work force and to perform manual work in a seemingly human way. A robot is typically capable of performing multiple functions or tasks with minor (or no) hardware modifications, and with economic and/or environmental benefits.

This definition seems to include the intent of other published definitions, in a more general form, and expands those definitions to include human compatibility references and economic benefits. This is what we mean by the term *robot* when we refer to it in the balance of this text. Figure 18-1 is a photo of a typical industrial robot that fits this definition.

ROBOTS AND PROCESS CONTROL SYSTEMS

What is the relationship of robots to the automated industrial process control systems discussed throughout the rest of this book?

Robots are a specific subset of automated process control systems. They have evolved from dedicated-function automated control systems as a class of general-purpose automated systems; typically designed to be flexible enough to be included as subsystems in larger automated systems or even as complete automated systems individually. They normally have the ability to be operated on their own, whereas automated process control systems are typically designed to operate only as a complete system.

They are normally marketed as standard catalog devices, having specified characteristics which make them economically adaptable to more than one specific production application. For example, several companies manufacture robots which

(A)   (B)

Figure 18-1 Typical industrial multifunctional, multiaxis, general-purpose robots. [(A) Courtesy of TOKICO AMERICA, INC.; (B) courtesy of Cincinnati Milacron, Industrial Robot Division.]

have been designed as painting robots or as welding robots. However, those robots have not been designed for a specific type of production line. The same model of painting robots are capable of being programmed to paint different portions of a automobile chassis on an automobile body assembly line as well home appliance cabinets on a completely different type of assembly line. The same model of welding robots may be used in automotive, marine, and commercial appliance industries without mechanical modifications.

Since the mechanical portion of the robot does not change from application to application, how can they be so flexible? It is the applications programs in the computers which control the robots that are changed between applications. Also, the end effectors may vary between applications; they will be discussed specifically later in this chapter, but they are the devices which are held on the ends of the robot arms, the "hands" or the tools for the robot. Therefore, the same hardware, both basic robot and the computer hardware, has been designed to be so flexible that it can be used in multiple applications within the same industry (even the same production line) and also in multiple industries. Only the applications programs executed by the computers and the end effectors vary.

The manufacturers of industrial robots provide a software library of (robot) hardware-specific control functions such that the applications programmer need not reprogram basic functions such as system initialization, joint motion motor control algorithms, operator input and output control functions, data acquisition functions, and basic system operational control functions.

Some robot manufacturers have developed their own high-level (compiler-type) robot control languages. To reduce the software effort required to get their robots up and running in any process, these languages include such instructions as the following: move from coordinate point 1 to coordinate point 2, move up, move down, move home, open, and close.

Other manufacturers have included a *teach mode* in their basic robot systems. Basically, this mode can allow an experienced operator to perform a complicated task, such as spray painting cylindrical surfaces or welding along a complicated path, using the robot arm itself as an input device. As the operator goes through his procedure (moving the robot arm as he goes) the computer monitors and records all motions. When the operator has completed the desired procedure the computer is signaled that the teach sequence is complete and the entire sequence is then stored in the computer's memory.

The computer now has the ability to cause the robot to mimic the motions that were performed by the skilled operator, thereby reproducing the desired end result. The robot can continue to cycle through this sequence as desired until "taught" to execute a new sequence.

There is another basic type of teach mode commonly used throughout industry to provide the applications programs for robots. It is the *teach pendant* technique and is used for robots where no provision is made (or is physically possible) for the operator to physically force the robot to learn a teach sequence.

In this mode a control box is connected to the robot's control computer by an umbilical. The operator has control of the robot's motions through a pushbutton keyboard on the teach pendant. The operator sets the robot into motion to execute each step of the desired sequence, signaling the computer when each step is completed. The computer again "remembers" the sequence and is able to recreate the exact motion sequence on demand after the teach sequence is complete.

Both of these teach techniques are in addition to the ability of a computer programmer to write a control sequence program for the robot to follow. There may be significant differences in the purchase cost for the different programming modes. The particular robot application may dictate the most cost-effective technique. The robot system illustrated in Fig. 18-4 includes a teach pendant as well as its own robot control program language.

It should be quite obvious by now that any number of identical robots can be taught (or programmed) to execute different tasks and that any specific robot can be retaught as often as necessary. Therein lies their economic attraction to industry.

Also therein lies the most basic difference between robots and the automated process control systems discussed previously in this book. Nonrobot industrial control systems are designed for a specific limited class of functions, normally within a single industry, whereas the automated process control systems referred to by the term *robot* are general purpose in nature, designed to cross applications as well as industry boundaries. It would hardly be economically feasible to modify the Louisiana Power and Light control system pictured on the frontispiece of this book to replace the refinery computer control system pictured in Fig. 15-9, even though they may share many similar hardware items. The same basic robot can be used in cross-industry applications with relatively minor and economical modifications. It should further be noted that all of the measurement and control theory and devices discussed previously in this book are completely applicable to robots and robot control.

Robots basically evolved from the dedicated process control systems in an effort to provide more flexible, more adaptable, more economic means of automatic control. Robots are typically included as subsystems within larger automated process control systems. Several figures in the following text illustrate robots contained within larger systems which automatically provide workpieces to the robot and dispose of workpieces completed by the robot, possibly passing them to another robot for further processing. Figure 18-2 illustrates such a system where several robots sequentially process the workpiece to a completed product.

Figure 18-2 Welding robots included as subsystems within an automotive assembly production system. (Courtesy of Cincinnati Milacron, Industrial Robot Division.)

ROBOT CLASSIFICATIONS

Robots may be classified in many ways; one classification scheme follows.

1. *Manual:* manipulated directly by a human being to extend reach, amplify strength, reach into hostile environments, etc. Also referred to as teleoperator systems. Examples include remote manipulators used on underwater research vehicles, in nuclear environs, and on space vehicles.

2. *Fixed sequence:* follows a preprogrammed sequence of events without much flexibility for easy modification. Devices controlled by hardwired logic, such as fluidic logic, are fixed sequence.

3. *Sequence:* operates a sequence, as with fixed sequence, but has the capability to make alternative decisions in accordance with changes in existing conditions. The sequence is relatively easy to modify. Examples include robots that are programmed and controlled by computers.

4. *Playback:* includes robots programmed in a teach mode. Examples include spray painting and welding robot applications described previously as being programmed via a teach mode in which a skilled operator leads the robot through a sequence which it remembers and later reproduces.

5. *Numerically controlled:* commands are executed in accordance with a numerical program which defines the position, motion, or sequence of operation. The type of robot which is programmed by a teach pendant, in which end points are remembered in the form of numerical coordinates, would fall into this class.

6. *Intelligent:* has the ability to alter its programmed sequence based upon its own sensing capabilities. This class includes robots having sensory feedback, such as vision systems, and automatically adjust their operations based upon feedback from those devices.

Robots can also be classified by the type of drive system used. Common drives include hydraulic, pneumatic, electrical, and wire cables. They can be further classified by the number of movable joints, load capacity, work envelope size, maximum part size, maximum reach, type of programming, level of intelligence, architectural form, and number of degrees of freedom, which is the subject of the next section.

As a further means of classifying robots, we can identify three distinct generations in the evolution of industrial robots. The first generation includes numerically controlled machine tools. They are characterized by programmed movement sequences and are only capable of those programmed repetitive sequences and operations.

The second generation of robots are characterized by rudimentary sensing capability, including basic visual, force, tactile (or touch), and so on, sensors. These robots can control their operation based upon feedback from those sensors. The majority of robots pictured in this chapter are second-generation robots.

The third generation of robots, which are more or less still under development, have the capability of artificial intelligence. They are developed around a much improved visual sensory capability which includes a digital simulation of the robot's environment, visual perception of the robot's interaction with that environment, pattern recognition for workpiece orientation, and extended use of other sensors and control elements.

DEGREES OF FREEDOM

Robots are also commonly classified by the number of *degrees of freedom* of which they are capable. "Degrees of freedom" generally refers to axes of motion. Any free object in space has only six degrees of freedom (Fig. 18-3), three translational and three rotational.

The translational motions can be provided to a robot by using translational tables (X-Y and Z motions). The robot may be fixed to these movable tables, which creates mechanical stability and accuracy problems for some robots, or the workpiece may be manipulated by an X-Y (and even Z if necessary) table within the robot's work envelope. See the items numbered 3 and 23 in Fig. 18-4 as examples of translational tables.

Figure 18-5 illustrates a gantry-type robot, which is mounted on an overhead X-Y-Z track system. This configuration is typically used to extend the work envelope to the required size for larger parts while maintaining the degrees of freedom and a high degree of positional accuracy.

Typically, arm-type robots are mounted upon stationary bases and use combinations of rotational motions to achieve the X-Y-Z capability. This method of mounting provides a mechanically more stable reference base.

Figure 18-6 illustrates a jointed-arm-type robot. Note that six degrees of free-

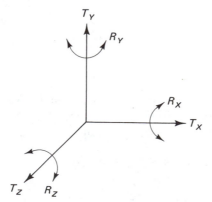

Figure 18-3 Degrees of freedom of an object in free space.

1. Rhino® XR Mark II Robotic Arm
2. Mark II Controller
3. Linear Slide Base
4. Software Manuals Rasp 1.04
5. Rasp 3.04
6. IBM Software
7. TRS-80
8. Language Card Manual
9. Teach Pendant Manual
10. Experimental Motor Kit
11. Input Output Card
12. RHINO-Com Language Card
13. CCS 7710A Card
14. "Hands-On-Introduction to ROBOTICS"
15. Teach Pendant
16. Moto-Dremel® Holder
17. Vacuum Fingers
18. Aluminum Base
19. Rotary Indexing Carousel
20. Magnetic Pickup
21. Triple Finger Attachment
22. Rhino-Rite Hand
23. X-Y Table
24. Extra Long Finger Attachments
25. Narrow Fingers
26. Clamshell Attachment
27. Shovel Attachment
28. Chain Belt Conveyor

Figure 18-4 Arm-type robot with a wide variety of accessories. Included are programming languages, teach pendant, workpiece manipulation tables, conveyor, and a variety of end effectors. (Courtesy of RHINO Robots, Inc.)

dom are also illustrated; however, they are all rotational. Translational motion of the end effector is achieved by coordinated movement of multiple rotary motions. This seems to be a very complex way to achieve simple translational motion, but mechanically this is the preferred mounting technique for this type of robot, and we have to live with the additional complexities. In fact, once the computer software has been developed to coordinate these motions, and is supplied as high-level software functions to robot computer applications programmers, it is very simple, effective, and efficient to use.

The additional degrees of freedom required by a specific process are provided by multiple-axis and effectors. Figure 18-6 illustrates an end effector having three degrees of freedom: pitch, roll, and yaw. Using coordinated combinations of motions, this type of robot can reach any free side of a workpiece within its work envelope.

Figure 18-5 Industrial overhead gantry-type robot. The vertical head in the center of the photograph can move forward and back along the side rails, left and right along the cross-head rails, and up and down relative to the cross head. The end effectors would be mounted on the bottom of the vertical element. (Courtesy of GCA Corporation, Industrial Systems Group.)

Note that specialized robots can have multiple joints along the arm (basically multiple elbows or multiple shoulders) which permit access to the back side of workpieces which are not accessible from either the top or bottom or from either side. Some painting and welding robots have these extra joints. These extra joints may also permit the robot to access areas behind its back by various over-the-head maneuvers.

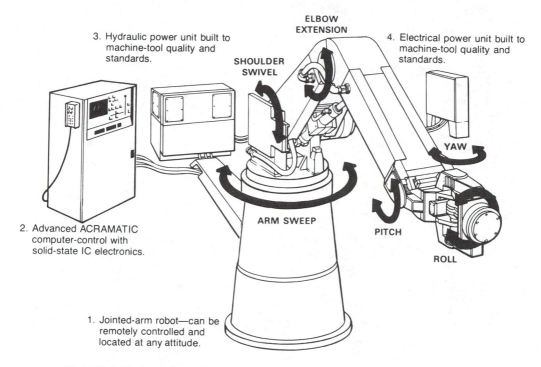

3. Hydraulic power unit built to machine-tool quality and standards.

ELBOW EXTENSION

SHOULDER SWIVEL

4. Electrical power unit built to machine-tool quality and standards.

YAW

2. Advanced ACRAMATIC computer-control with solid-state IC electronics.

ARM SWEEP

PITCH

ROLL

1. Jointed-arm robot—can be remotely controlled and located at any attitude.

Figure 18-6 Typical industrial stationary, base-mounted, jointed-arm robot, illustrating the six degrees of freedom normally available with this type of robot. (Courtesy of Cincinnati Milacron, Industrial Robot Division.)

FACTORIES OF THE FUTURE

It is appropriate to discuss here a relatively new concept called the factory of the future. This factory theoretically will be designed to accept raw materials at its input and automatically process these raw materials into a finished product which will be delivered at its output—plus a lot more.

Figure 18-7 is one manufacturer's concept of what this factory of the future might look like. Referring to that figure, raw materials are introduced at the left, where they are preprocessed as necessary to be passed on to the component-part assembly area. This procedure includes quality control at appropriate points.

Those component parts are passed on either directly to the shipping area to be shipped as components, or to the component assembly area, where they are combined with additional materials and are further processed into subassemblies. Note that the types of robots required may change as the size and character of the parts and subassemblies change.

Finally, the subassemblies are passed either to shipping, to be shipped in that form, or to the product assembly center, where they are again combined with ad-

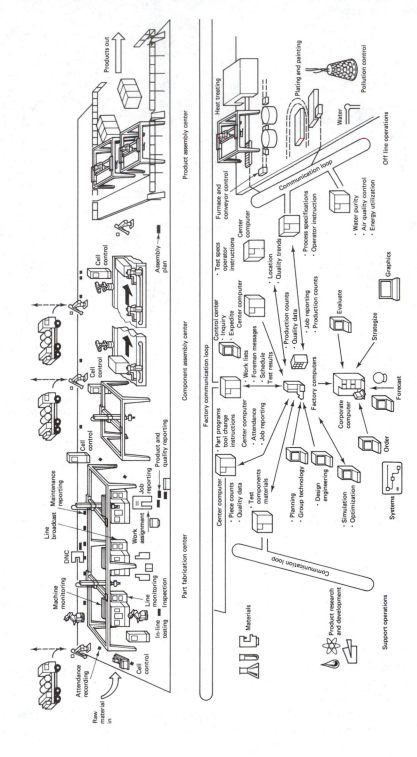

Figure 18-7 Factory-of-the-future concept. (Courtesy of GCA Corporation, Industrial Systems Group.)

456

ditional materials as necessary and are fabricated into finished product. Again, the types of robots may change in size and function, and quality control continues to be enforced at appropriate points in the process.

There are several very important concepts included in this diagram which need to be discussed further. First, the entire concept is based upon the availability of different types of robots; also, several of the same type of robots may be required to perform completely different functions. We have discussed them adequately previously. Second, there is a need for a product transport and delivery system in order to maintain the flow of product through this factory, and also to feed the robot work cells and to dispose of the finished assemblies as the robots complete their work. Delivery and transport handling systems are discussed in the next section.

Another concept implied in that figure is the flexible manufacturing system concept. Each of the robots and work cells are to be designed such that they are so flexible as to be quickly, easily, and economically (and possibly even automatically) modified or reconfigured so as to change the product being manufactured. Figure 18-15 is an excellent example of a work cell of the type envisioned in this futuristic concept. Thereby, the same factory, using the same manufacturing production line, can be used to produce multiple types of products.

The flexibility of these flexible work cells depends largely upon the sensing and control functions available to future robots. Vision systems with pattern recognition features and programs that include artificial intelligence capability will be required for maximum flexibility.

Finally, the basic glue that ties this entire manufacturing system together and makes it work as a coordinated entity is the factory communication network. This network includes all of the background functions from raw-material ordering, product scheduling, production scheduling, production-line configuration, robot and work cell programming, to shipping. The system envisioned in Fig. 18-7 even predicts that many more functions will be included in that communications system—in fact, many of those features are already implemented in some industrial systems.

DELIVERY, DISPOSAL, AND TRANSPORT SYSTEMS

As robots improve in their capability and throughput, delivery, disposal, and transport systems must proportionately improve, or we will have underused work cells simply because we cannot move product to and from them quickly enough.

Figure 18-8 illustrates a rather simple delivery, removal, and transport system integrated into a robot workstation. This system has a single conveyor for delivery and disposal, thereby not allowing for rejects or rework paths at this station. Figure 18-9 illustrates a conveyor system bringing parts to several assembly stations. Figure 18-10 illustrates a more elaborate delivery system for an arm-type robot workstation. Several other figures in this chapter illustrate various other delivery, disposal, and transport systems.

Figure 18-8 Robot with workstation and 90° power turner. Note that the workstation assembly raises the workholder up from the conveyor and locks into position. (Courtesy of DORNER Manufacturing Corp.)

SENSING ELEMENTS

There is very little difference between the sensing elements used in the automated process control systems previously covered in this text and those used as robot sensing elements. Robots use all of the position sensors, both linear and rotary, discussed in earlier chapters, with special emphasis on digital incremental linear and shaft angle encoders. Many of the force sensors discussed previously are also used

Figure 18-9 Conveyor system bringing component parts to assembly stations and conveying finished assemblies to the next workstation. (Courtesy of DORNER Manufacturing Corp.)

in robotic systems. These sensors, and most of the others used, do not vary significantly in either theory of operation or design from general automated systems usage to robotic applications. Other types of sensors which are used for robots but which have not been covered elsewhere are laser distance measuring equipment, acoustic range measuring equipment, and even more commonly, visual sensing equipment.

Vision systems (Figs. 18-11 and 18-12) are commonly used in both conventional process control systems and robotic systems. These systems are (conceptually at least) local closed-circuit digital television systems. Special cameras view either the workpiece or the position of the robot end effectors and that image is digitized and passed to the operator and/or the control computer.

The computer is programmed to recognize the digitized pattern of the workpiece or end effector, and the appropriate decisions are made as to when and how to proceed. The field of digital pattern recognition is currently in its infancy, and research is progressing toward recognition of more complex parts and shapes, recognition of the three-dimensional position of the end effector in relation to the

Figure 18-10   Robot with elaborate material delivery and disposal system. (Courtesy of Cincinnati Milacron, Industrial Robot Division.)

workpiece, and eventually, three-dimensional recognition of the positions of both the workpiece and the end effector in relation to the environment.

Pattern recognition techniques and digital simulation of the work area environment are two of the hottest areas of robot and flexible manufacturing system research.

END EFFECTORS

End effectors are the devices fixed to the end of a robot's arm. The objective of the robot control system is to maneuver the end effector to precisely the correct position to perform the intended function. The effectors are specifically designed to hold special tools or special parts; therefore, they vary significantly from one application to another. Figure 18-13 illustrates a few of the more general types of

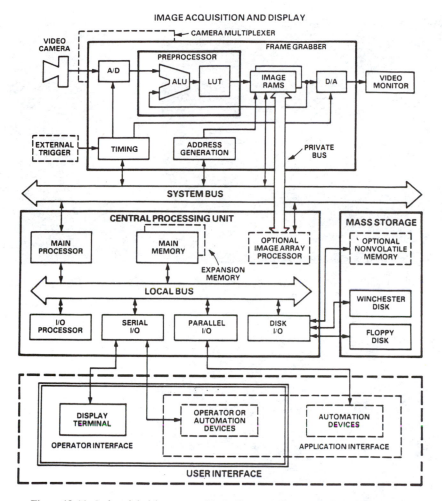

Figure 18-11 Industrial vision system block diagram. (Courtesy of Analog Devices, Inc.)

gripper mechanisms. Typically, these grippers are operated by pneumatic cylinders due to the weight versus holding-power capabilities of these devices. Figure 18-4 also illustrates a selection of several commonly used gripping mechanisms.

Frequently, the end effectors have a mechanical quick-change mechanism where the tools can be changed under automatic control by the robot control system. Thereby the work flexibility of the robot is much increased since it may be able to perform multiple operations without relocating or moving the workpiece.

The end effectors themselves may include actuators which have several degrees of freedom of motion, again increasing the flexibility of the workstation.

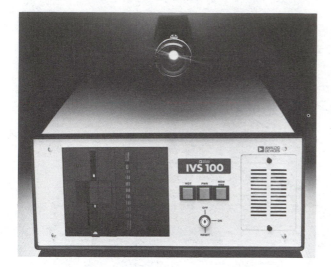

Figure 18-12 Commercial–industrial-grade vision system. (Courtesy of Analog Devices, Inc.)

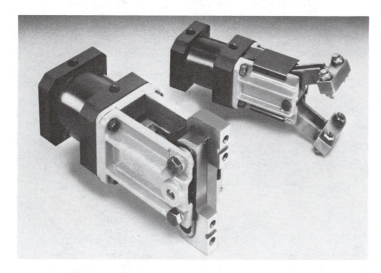

Figure 18-13 Typical robot grippers. (Courtesy of I.S.I. Manufacturing Inc.)

ROBOT APPLICATIONS

Robots, by concept and by design, are the most flexible manufacturing systems ever devised for the factory. Applications are limited only by the imagination and creativity of the control systems designers and the requirements of the workpiece. To compile even a partial listing of the major robot applications would require many pages; therefore, I have opted to list a few and illustrate a few others in Figs. 18-14 and 18-15.

A partial list of robot applications includes the following:

Welding	Painting
Presswork	Processing
Assembly	Die casting
Cutting	Grinding
Forging	Founding
Paletizing	Orienting parts
Transporting	Pick-and-place

They are also used in underwater exploration, as extraterrestrial manipulators, and as human handicapped servomechanisms.

ROBOTS?

The precise distinction of a robot from many other types of industrial automated systems is often confused. There are many devices and systems that perform sequences automatically, like a robot, but for one reason or another do not bear the title *robot*.

Figure 18-16 illustrates one such type of system. This is a dimensional checking system, commonly called a coordinate measuring machine (CMM) throughout the quality control segment of industry. It is basically a robot in that it is computer controlled and responds to a program that causes it to move around and check dimensions of workpieces. It can be programmed to take a complete set of dimensional measurements of a workpiece, compare those measured values to drawing specifications, and pass judgment on the conformance to those specifications, taking into account their associated tolerances.

These machines also typically have a manual mode of operation which is strikingly similar to the teach mode of a robot. The operator moves the dimensional checking probe (on the end of the arm) into contact with the workpiece at preselected locations. At each checking location the computer remembers the coordinates of the probe, calculates the drawing dimensions from those probe positions, and displays them. If those positions were subsequently used to control movement of

Figure 18-14 Industrial robot applications in the automotive industry. [(A) to (C) Courtesy of Cincinnati Milacron, Industrial Robot Division; (D) courtesy of GCA Corporation, Industrial Systems Group.]

the probe to reproduce them, we would have little trouble calling this a robot. In fact, this is the second mode of operation of this type of machine. Basically, the only difference between these machines and the other robots we have been discussing is that they do no work on the workpiece; they simply measure its dimensions.

Figure 18-15 Photo of an automated work cell. The robot in the center services the two production machines on either side supplying workpieces with the proper orientation and disposing of the finished product. Note the elaborate transport system used to service the cell. (Courtesy of Cincinnati Milacron, Industrial Robot Division.)

SUMMARY

A few robot industry statistics may be of interest here. Even though some will be obsolete before this book is printed, they will still serve to give a general concept of the importance of robots in worldwide industry.

The field of industrial robotics can have its beginnings traced back to the early 1930s, but major developments had to wait until the development of low-cost digital computers. Robots remained research and development machines up until the 1970s. The real robot invasion into the factory was not until the later 1970s, when the systems being considered in the early 1970s were actually implemented in the latter half of that decade.

There are about 50 manufacturers in the United States directly involved in robot manufacturing, and over 100 manufacturers in Japan. Depending upon the definition of a robot, there are between 25,000 and 50,000 robots in use worldwide.

Figure 18-16 Coordinate measuring machine. Here the operator is shown manually making the measurements. The machine is also designed to step automatically through complicated measuring sequences in order to check all dimensions of a workpiece. (Courtesy of Federal Products Corporation.)

Japan leads the world in robot usage, employing about half of the industrial robots in use. They use three to four times as many robots as are used in the United States.

It appears that the largest U.S. application for robots, in terms of the number of robots used, is in welding applications. The automotive industry is the leader in robot usage in the United States.

Robot applications require a level of correlation between multiple sensed signals which has not been necessary previously. The robot control system must correlate all of the sensed variable information about the workpiece, the robot's appendages, the work cell environment, and the delivery and disposal systems. This correlation and the software programs necessary to implement it functionally comprise an area of major challenge in robotic research.

Another theoretical element attributed to robots (future ones anyhow) is a level of adaptability which has not been available previously in normal process control systems. This level of adaptive control, or adaptive behavior, will be based upon artificial intelligence programmed into the robot's computers. With this intelligence the robot will be able to react with its environment and have the decision-making ability to adapt its operation to changes in that environment. This area is also the subject of major research effort.

Recent artificial intelligence research by AT&T Bell Labs has produced an integrated circuit chip which implements a type of function called *fuzzy logic*. Fuzzy logic is logic of approximate reasoning, dealing with degrees of certainty rather than being restricted to only two possible decisions (yes or no), as is the case with binary logic. This logic will be used in robots to develop an ability to make intelligent guesses, thereby making the robot much more adaptable to changing conditions.

Robotics is a relatively new field, ripe for research and development efforts, and the necessary element in future industrial progress.

QUESTIONS

1. Referring to the material in this chapter, derive your own definition of an industrial robot.
2. Explain how robotic devices similar to the types offered for sale to the general public through various retail electronics stores fit into the context of robots as used in this chapter.
3. Expand on the list of concepts of robotic devices on page 446.
4. What additional (general) types of functions can you add to the list on page 446?
5. Referring to the list of functions in Question 4, list examples of each of those functions.
6. On page 446 is a list of qualifying attributes which help to justify industrial robots economically.
   A. Add to that list.
   B. Can you give any practical situations or examples where those criteria have in fact been realized?
7. In your own words, explain what a teach mode is when using the term applied in a robotic sense.
8. What criteria would be used to decide how a robot is to be programmed in a particular application (i.e., program versus teach mode)?
9. How do robots differ (in application) from the process control systems discussed in earlier chapters?
10. What inherent properties do robots have that makes them so flexible as to be able to cross applications and industry boundaries with basically the same hardware system?
11. Referring to Fig. 18-5, how many degrees of freedom would a workpiece fixed to the end effector have? If that arm were mounted on an X-Y-Z movable platform, how many degrees of freedom would the workpiece have? How many degrees of freedom would the robot have?
12. Define the term *work envelope* as applied to robots.

GLOSSARY

Degrees of Freedom   Basically, the number of articulated members and the number of distinct motions each is capable of.

End Effector   Mechanical device attached to the end of a robot's arm; its purpose is to hold the tool or workpiece, orient it properly in space, and perform the work task as it is manipulated in space by the robot.

Robot   In an industrial automation context this term refers to a programmable automated mechanical system designed to supplement or replace a human work force and to perform manual work in a seemingly human way. They are typically capable of performing multiple functions or tasks with minor (or no) hardware modifications and with economic and/or environmental benefits.

Teach Mode   A technique of providing the specific applications programming sequence to a robot by either manually forcing the robot through the sequence or by controlling the robot's motion through a teach pendant. In either case the sequence of operations is remembered by the control computer and can be mimicked at a later time.

Teach Pendant   A keyboard device attached to a robot's control computer via an umbilical and through which the robot can be manually moved in response to operator keyboard input. The sequence of operator input motion commands can be remembered and the robot forced to mimic them at a later time.

APPENDIX A

PRACTICAL EQUIVALENCY OF PHYSICAL SYSTEMS

INTRODUCTION

By their very nature, automatic control systems are put together using many different types of components, components that are, based on physical constraints of the process itself, economics and other factors, frequently designed to operate based upon other than exclusively electrical parameters. Therefore, it is not possible to understand how a system operates unless one can understand not only how the electrical portions of the system operate but also basically how the hydraulic, pneumatic, thermal, and mechanical parts of the system operate; and, even more important, how they act together to perform a useful function.

From a realistic point of view, this information is aimed at readers who, first, may not really feel inclined to delve deeply into the fields of pneumatics, hydraulics, thermodynamics, and mechanics, and second, may not have the time to devote, even if they were so inclined. This Appendix, therefore, has the dual purpose of satisfying the requirement for basic understanding of these other systems in order to be better prepared to understand the subject material of this text, and at the same time to accomplish it in such a manner as to make it as brief and as painless as possible.

To digress for a moment, I have heard foreign language professors refer to the fact that most of their students can learn the basics of reading, speaking, and writing a foreign language without ever thinking in that language. The thought process goes on in the student's "native" language. He spontaneously translates what he hears and sees into his native tongue, thinks about it, and then translates his thoughts to the foreign language to express them orally or in writing.

I propose to attempt something of a similar nature here. The idea will be to discuss "foreign" subjects, such as thermodynamics, and purposely help the reader to translate the various parameters in these systems to a "language" with which they may be more familiar, the language of electronics.

PHYSICAL SYSTEM PARAMETERS

Probably the best place to start is with a quick review of the "native tongue," electronics, but (possibly) from a different point of view.

Certainly no electrical circuit is going to be useful without a source of electrons (either a voltage source or a current source). Therefore, in an electrical circuit, the power supply, battery, or source of electrons (whichever you choose to call it) is the prime mover, forcing function, or power source that starts the circuit operating. Without this source of energy, the rest of the components and wiring in an electronic circuit serve no useful purpose, and there is no useful output, regardless of the input.

Given that there is a power supply for any circuit, just what does this power supply actually do? How does it cause the electronic components to "do their trick"? It presents each component with a condition in which there is an imbalance in the number of electrons between each of the component's terminals. This is precisely the condition necessary for each component to exert its own special effect on those electrons as they attempt to establish a condition in which there is no imbalance (or difference in concentration) of electrons between the terminals of the component. In order for the electrons to go from a terminal where there is an excess of them to a terminal where there is a relative scarcity of electrons, they must flow through the particular component, regardless of the type of component.

Therefore, the net effect of this source of power (electrical power in this case) was to create a flow (of electrons in this case). That is its sole function in the circuit. Once energized, all other components in the circuit are left with an imbalance in the relative numbers of electrons present at each of their leads (or terminals) as long as they are operating normally in the circuit (and, of course, the power supply is still connected).

The term used to describe the tendency of components to maintain this difference in concentration of electrons at their terminals is *resistance* to flow. The particular reaction of each component when in this condition (or the particular relationship between relative concentrations of electrons on each of their terminals) is commonly presented as a graph illustrating the relationship between concentration of electrons (voltage) and flow (of current) through the component. This graph is called the *characteristic curve* for the device and is really a curve showing the amount of resistance (or voltage drop) offered by the device to the flow of current under any set of conditions that it has been designed to operate under. These facts about resistance can be applied to any electronic (or electrical for that matter) component.

There are two additional effects (besides resistance) that certain electronic components exert on the flow of current. They are the effects of inductance and capacitance. *Inductance* is a phenomenon exhibited by most components (including some resistors) which has the effect of either opposing or of sustaining the flow of current, or of opposing the *change* in flow of current. This definition will suffice for our purposes. The reason for this brief treatment is that the inductive properties of other physical energy systems are difficult to comprehend and not critical enough to warrant the necessary time and space to explain for the purposes of this book. Therefore, inductive analogies, although they may exist, will not be covered.

The final effect is that of *capacitance,* which is simply the ability of a component to *temporarily* "store" current (flow substance) for later use. Most electronic components exhibit this capability to some degree (again including resistors).

We have now covered the basic effects in electronics from a very qualitative and general point view. It is necessary that you be able to get out of the rut of looking for very exact definitions, equations, and descriptions for the rest of this book. Try to get the general idea rather than to memorize and understand each word as you may have heretofore been taught to do. The difference is between a specific technical text (exact definitions) and a survey text (more general definitions).

To continue to reach the objective of a basic understanding of other than electrical systems, we must first establish the basic concepts that will apply to all of them. In this text they will all be discussed from the point of view of flow: the methods of creating flow (or causing flow), the flow substance itself, and the reactions of the various components which affect that flow (specifically resistance and capacitance).

Probably the best plan of attack would be to take some examples and attempt this translation into electronics of the various foreign system parameters.

Hydraulic/Electronic Analogy

Initially, let us consider an elevated water storage tank, similar to the types used to supply suburban home developments and industrial complexes (refer to Fig. A-1A). Essentially, the water supply system consists of a large empty tank built some distance above the ground. Water is pumped (by a large water pump) from a reservoir up into the top of this tank, and when the tank is full, the pump is turned off. Notice that the pump had to move water against the effects of gravity (among other obstacles) in order to fill the tank. In order to later use this water, the valve on the outlet pipe (V_2) is opened enough to permit the flow of water that is required. When this valve is closed tight, no water will flow from the tank.

Now let us consider a resistor–capacitor circuit as shown in Fig. A-1B. The battery forces current to flow through resistor R_1 in order to charge the capacitor (which temporarily stores current), and when fully charged the battery is disconnected from the circuit by the switch. At a later time resistance R_3 can be reduced in order to permit current to flow to the circuit off the right side of the page. Before proceeding, see if *you* can find analogies between the hydraulic and electrical circuits.

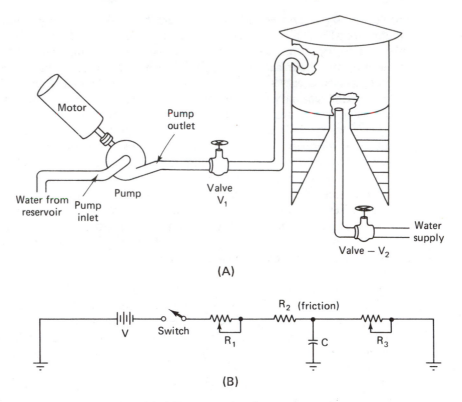

Figure A-1 Town water supply system example.

Surely the pump is the initial "thing" that started the water (the hydraulic flow substance) to flow; so it is analogous to the battery causing electron flow. The flow had to overcome the effect of valve 1 and the effect of friction in order to reach the water tank. This is analogous to the electron flow through resistances R_1 and R_2 in order to charge the capacitor.

The tank stores water in a manner similar to the way in which the capacitor stores electrons. Both have the ability to store the flow substance for further use; therefore, they exhibit the phenomenon of capacitance. The resistance to water flow offered by valve 2 is analogous to the resistance offered to the flow of electrons by R_3. Note that the resistors are shown as variable resistors. This is due to the fact that the hydraulic valves can exert a variable effect on the flow through them in a manner similar to the effect variable resistors have on electron current flowing through them.

As water flows into the tank, the head pressure which supplies water to valve 2 increases in a manner analogous to the way that the capacitor charges up to the value of the battery voltage during its charge cycle (while electron current is flowing from the battery into the capacitor). Therefore, there is a very close analogy between hydraulic pressure and electrical voltage.

A further analogy will be noted in the effect that the valves have on the hydraulic flow through them. As the valves are closed somewhat, the quantity of flow substance (water) is reduced, because they exert a greater resistive effect on the flow through them. Coupled with the reduced flow and increased resistance is a corresponding increase in the pressure drop across the valve. This sequence of events indicates that there must be a sort of Ohm's law for hydraulic circuits—the analogy is that close.

Pneumatic/Hydraulic/Electronic Analogy

Pneumatic systems are very similar to hydraulic systems, with two major exceptions. The most obvious difference is, of course, the flow substance. Rather than hydraulic liquid (such as oil, glycerine, or water) the flow substance is a gas (such as air). This gives rise to the second major difference; hydraulic fluids are not compressible, whereas pneumatic fluids are. By compressibility I mean that a 1-gallon bucket full of water cannot be squeezed into any smaller volume, no matter how hard you squeeze it. A 1-gallon bucket full of air (a pneumatic fluid) can, however, be squeezed into a smaller volume, say a ½-gallon volume. Therefore, in order to store pneumatic fluids, a smaller tank can be used, and the same initial quantity (volume) can be squeezed into the smaller tank.

In the process of squeezing the bucket full of air into a smaller volume, energy was expended. This energy was necessary to overcome the resistance of the air to being compressed. All of this expended energy is actually stored in the compressed air itself in the form of a potential energy, or energy that can later be used.

Remember that in the hydraulic case, it was necessary to store the water high up in the air so that it would flow out by itself at a later time (it has stored in it the same kind of potential energy as a book on a table—potential energy based upon its physical position only). In the pneumatic case it is not necessary to elevate the stored air; the potential energy is stored in the compressed gas itself, which can cause the air to flow "uphill."

Furthermore, as the gas moved through the valve, it had to overcome some resistance (friction), and so a pressure drop was created. Sound familiar? Another physical system with its own equivalent to Ohm's law.

Referring to Fig. A-2A, initially valve V_1 is opened (to offer minimum resistance to the flow of air through it) and valve V_2 is closed. The air is pumped into the tank (by the compressor) and stored there for future use. The analogies between valves and variable resistors, tank and capacitor, and pump and battery are identical to the hydraulic example. The valve V_1 can then be closed and the energy stored in the tank will be available at a later time (by opening valve V_2).

Mechanical/Electronic Analogy

Now let us consider a mechanical system (refer to Fig. A-3). An automobile engine causes a car to move against the effects of friction in bearings and between wheels and road, wind resistance, and against the effects of obstructions (bumps).

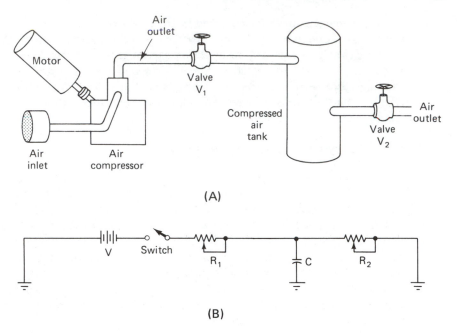

(A)

(B)

Figure A-2 Compressed air system example.

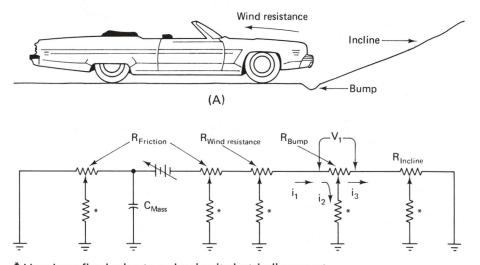

(A)

* Very large fixed value to make circuit electrically correct

(B)

Figure A-3 Moving automobile example.

At this point, can you figure out what the mechanical flow substance is? It is energy. The engine converts gasoline to mechanical energy, which is used up accelerating the car and overcoming the forces listed above which tend to slow it down, all of which tend to use up that energy or create *energy drops* which are analogous to voltage drops across resistors. An energy drop ordinarily would be measured as a *force,* since the energy is converted to force as it is lost from the system.

One effect not specifically mentioned was the effect of mechanical capacitance. Remember that capacitance (as defined previously) is a phenomenon in which the flow substance is temporarily stored in the system. In the mechanical system, energy is the flow substance; it is stored by the mass of the automobile and is called potential energy. Its effect can be visualized by considering an automobile to be traveling up a hill and at the crest of the hill it suddenly loses power. It does not come to a screeching halt at the instant power is lost. Rather, the energy is slowly released through the frictional losses in the system (as described previously) and the vehicle gradually loses its stored energy as it coasts down the other side of the hill in the same way that a capacitor discharges through a resistor.

Before concluding the mechanical system analogy, let us consider an automobile colliding with two objects. The first is a wall consisting of a propped-up sheet of plywood and the second is a stone wall. As the auto collides with the plywood wall, some energy is transferred to the wall, enough energy is transferred to move it, and the auto continues on. It then collides with the stone wall and transfers all of its remaining energy to the wall, attempting to move that wall also. In both cases the energy flowed from the auto to the wall, overcoming its resistance to being physically moved, and only enough energy was transferred to exert the proper force necessary to accomplish this.

In the second case the auto did not have enough energy to move the stone wall, so it transferred all its remaining energy to the wall, in the form of a force, attempting to move it. Certainly this latter amount of force was greater than the force required to move the plywood wall.

Notice that where *energy* flowed from the auto into one of the walls, the amount of *force* developed was directly proportional to the *resistance* offered by the wall. This is exactly analogous to current flowing through resistors and the resulting voltage drops; or there is a mechanical analogy to Ohm's law.

Thermodynamic/Electronic Analogy

Finally, let us consider a thermal system. Let us investigate the thermal situation as applied to a tea pot. Initially, the tea kettle is filled with water and then set in contact with a heat source. Heat (the flow substance) flows from the heat source through the bottom of the tea kettle and then into the water.

The water does not instantaneously change temperature up to the boiling point; rather, there is a relatively long interval during which heat continues to flow into the water, slowly raising its temperature. Conversely, once the heat source is removed, the water does not instantaneously cool down to ambient temperature. Now the water continues to slowly give off the heat it stored while being heated. Obviously, then, the water is exhibiting the phenomenon of thermal (heat) capacitance. It accepted the flow substance and stored it for later use. In fact, it is the mass of the water that is the thermal capacitor.

What about thermal resistance? When you grab the handle of the hot tea kettle, if it has a metal handle, you must use an *insulator* to keep from being burned. Obviously, this thermal insulator material is offering a large resistance to the flow of heat through it, and that is why you will not be burned. Without the insulator, the resistance to heat flow between the metal tea pot handle and your hand is less, and larger quantities of heat flow.

Let us look at this thermal resistor a little closer for a minute. Heat, the flow substance, flows through the insulating material. As it flows, there is a difference in temperature built up on the opposite sides. The higher the value of thermal resistance, the less heat that flows and concurrently the greater the temperature difference between the two sides. This defines the thermal equivalent to Ohm's law, where obviously temperature is analogous to voltage.

At this point I do not think that an electrical circuit is necessary to draw the proper analogies.

SUMMARY OF ANALOGIES

Having studied this very brief review of the analogies among hydraulic, pneumatic, mechanical, and thermodynamic systems, as compared to electronic systems, in your mind you should have noted the important analogies summarized in Table A-1.

At this point it should be obvious that the five energy systems discussed are not five completely different systems. They are, in fact, *physically* different, but they all follow a *common* set of nature's laws. They can all be thought of along common lines, and most certainly each system can be "translated" into any other system for purposes of analysis and/or understanding.

These analogies are so close that design engineers constantly "simulate" the mechanical, pneumatic, hydraulic, and thermodynamic portions of everything from automobile suspension systems to spaceships on electronic analog computers. These "simulations" are used to vary parameters electrically, which would be very difficult or expensive to actually construct and test. For example, can you guess how many spaceships would have to have been made if a trial-and-error process were used to land men on the moon? Simulation and electronic analog computers are discussed in Chapter 9.

There is nothing wrong with thinking of a valve as a pneumatic or hydraulic resistor, mass as a mechanical capacitor, or temperature as a thermal voltage. In fact, these analogies are used every day by scientists and engineers, and I encourage readers to use them to help understand unfamiliar systems.

TABLE A-1

Energy system	Flow substance	Necessary to start flow	Necessary to direct or conduct or confine flow	Opposition to flow	Temporary storage element (capacitance)
Electronic	Electrons	Source of emf (voltage)	Copper wires	Electrical resistance	Electrical capacitor
Hydraulic	Fluid (liquid)	Pump (pressure)	Pipes and tubing	Valve or restriction	Tank, gravity
Pneumatic	Gas	Compressor (pressure)	Pipes or tubing	Valve or restriction	Tank
Mechanical	Energy	Source of energy (force)	Mechanical linkage or contact	Friction or obstructions	Mass-potential energy (anti-gravity) or spring
Thermal	Heat	Heat source (temperature)	Heat transmission medium	Thermal insulation	Mass

MATHEMATICAL RELATIONSHIPS AMONG SYSTEM PARAMETERS

With this background information under your belt, the next logical step in getting a feel for the operation of other-than-electronic systems would be an understanding of the system relationships between specific parameters.

For example, when components are connected together in an electronic circuit (including power supplies), the interrelationships among the individual system components are expressed in terms of voltage and current values at specific points in the circuit (and at specific points in time, if applicable). For the purposes of this appendix, these voltage-current values are determined using very basic mathematical relationships: Ohm's law, and Kirchhoff's voltage and current laws, and other mathematically simple relationships.

Would you be shocked if I told you that with very minor variations, these same laws hold in hydraulic, pneumatic, mechanical, and thermodynamic systems? This appendix started by pointing out the similarities between components of these systems and electronics systems by using common examples. This should have been nothing more than a very brief review of a basic applied physics course. The fact that each system has its own version of Ohm's law may have come as a surprise, but the fact that each of these systems also obeys a law similar to Kirchhoff's laws may not have been previously presented. There are other analogies, power losses, for example; however, the only laws I propose to discuss here are Kirchhoff's voltage and current laws, and their analogies in hydraulic, pneumatic, mechanical, and thermodynamic systems.

Kirchhoff's voltage law states, in effect, that in any closed electronic circuit, the sum of the *voltage drops* around any loop must equal the sum of the voltage rises; or the sum of the *voltages* around any closed circuit is zero. Remember to assign an algebraic sign to each voltage so that they will add properly.

Kirchhoff's current law states, in effect, that the sum of electron currents *flowing into* any point in a circuit must equal the sum of the currents *flowing away* from that point.

In a hydraulic system the analogy is nearly exact. Kirchhoff's voltage law becomes *Kirchhoff's pressure law,* since voltage in an electronic circuit is analogous to pressure in a hydraulic circuit (and for that matter is also analogous to pressure in a pneumatic system).

Therefore, Kirchhoff's voltage law, rewritten for hydraulic and pneumatic systems (circuits), would read: The sum of the *pressure drops* around any closed loop must be equal to the sum of the *pressure rises* (or increases); or the sum of the *pressures* around any closed circuit is zero.

Since there is a close analogy among electron flow, hydraulic fluid flow, and pneumatic (weight) flow, Kirchhoff's current law can be rewritten to cover hydraulic and pneumatic flow. The sum of the hydraulic (or pneumatic) *flows into* any point is equal to the sum of the *flows away* from that point.

At this point refer to Fig. A-4 and see if, using the hydraulic and pneumatic versions of Kirchhoff's voltage and current laws, you can figure out the unknown quantities. Note how pressures are indicated in the same manner that voltages would be, and flows are indicated in the same manner that currents would be.

The mechanical voltage and current (force and energy) versions of Kirchhoff's laws are a bit more difficult to visualize; nevertheless, they do exist. Since the flow substance in a mechanical system is energy, then the mechanical version of Kirchhoff's current law must read: The sum of the energy being delivered to any point must equal the sum of the energy leaving that point. If you think about it awhile, it will become obvious that this is merely a rearrangement of the basic law of physics which states that energy can be neither created nor destroyed—and must always be accounted for.

As an example, refer back to Fig. A-3. Specifically, let us investigate the system as the car hits the bump with the front wheels. Before impact the car has some amount of energy (analogous to i_1 in the Fig. A-3B). As the tire hits the bump, some of the energy that the mass of the auto has possessed is lost or dissipated. This is analogous to the current i_2, which is shunted to ground. From the bump's point of view, this energy was received in the form of a sudden blow or a force. Therefore, energy was lost from the system in the form of a force or voltage drop, analogous to V_1; and the auto continues on with less energy, which is analogous to i_3.

Kirchhoff's mechanical voltage law would read: The summation of *forces* acting at a point must equal zero. This is merely a restatement of the fact that for every action there is an equal but opposite reaction. Refer to Fig. A-5A, which illustrates a tug-of-war between four people, two on each end of the rope. If the flag in the center of the rope is not moving in either direction, what must F_1 be?

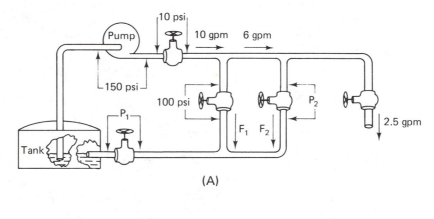

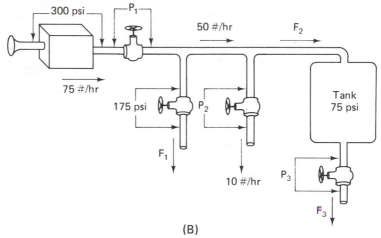

Figure A-4 *Note*: Ambient air pressure is pneumatic "ground"; answers to (A) and (B) appear at the end of this section.

Thermodynamic systems have exact analogies to both of Kirchhoff's laws. Remember that heat is the thermodynamic flow substance, but what was the analogy to voltage? Temperature, of course. Therefore, Kirchhoff's current law analogy would read: The sum of the *heat flows* into any point must equal the sum of the *heat flows* leaving that point. His voltage-law analogy would read: The sum of the *temperature drops* around any closed path must equal the sum of the *temperature rises,* or the sum of the *temperatures* around any closed loop is zero.

Refer to Fig. A-5B and see if you can apply these heat and temperature versions of Kirchhoff's laws to this thermodynamic problem. Note the way that radiated heat is illustrated (heat leaving the three radiators, for example), which is transferred to room air. This air is then recycled to the burner to be reheated.

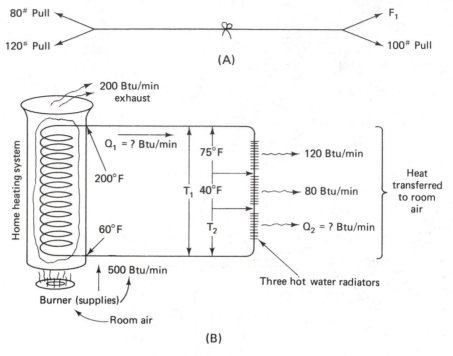

Figure A-5 Hot water heat transfer example.

Figure A-4A Answers	Figure A-4B Answers	Figure A-5 Answers
$P_1 = 150 - 10 - 100 = 40$ psi	$F_1 = 75 - 50 = 25$ lb/hr	$F_1 = 100$-lb pull
$P_2 = P_1 = 100$ psi	$P_1 = 300 - 175 = 125$ psi	$Q_1 = 500 - 200 = 300$ Btu/min
$F_1 = 10 - 6 = 4$ gpm	$P_2 = P_1 = 175$ psi	$Q_2 = 300 - 120 - 80 = 100$ Btu/min
$F_2 = 6 - 2.5 = 3.5$ gpm	$P_3 = 175 - 75 = 100$ psi	$T_1 = 200 - 60 = 140°F$
	$F_2 = 50 - 10 = 40$ lb/hr	$T_2 = 140 - 75 - 40 = 25°F$
	$F_3 = F_2 = 40$ lb/hr	

SUMMARY

Hopefully, this brief review of applied physics will help you to better understand the material that is to follow. I will reiterate that it is impossible to design an industrial control system using electrical components exclusively. Someplace in the system there is a knob or valve that must be twisted, a temperature that must be monitored, or a pressure that must be controlled. Therefore, there is a requirement for devices that perform the function of conversion from thermodynamic, hydraulic, pneumatic, or mechanical parameters into electronic parameters, because the common control element is normally electronic. These devices are called *sensors*. They provide the control element (possibly a computer) with electrical signals that

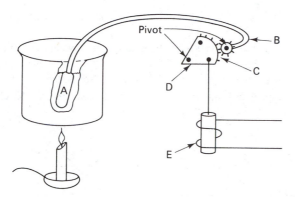

Figure A-6 Complete temperature measurement system.

are proportional to the value of the parameter they represent. In a similar fashion, there are devices which convert from electrical parameters into mechanical, thermodynamic, pneumatic, and hydraulic parameters; these devices are called *transducers*. If the resultant energy is mechanical, the transducer is called an *actuator*.

Both sensors and actuators fall into the category of transducers. *Transducers* are simply devices that convert information (pressure, temperature, force, flow, etc.) from one energy system to another. Therefore, it is an absolute necessity to have an understanding of these other systems in order to understand their conversion and use in control systems.

A brief example of a temperature measurement may help to point out the importance of understanding how systems other than electrical work (Refer to Fig. A-6). Briefly, here are the energy-system conversions necessary to obtain an electrical representation of the temperature in that pot of water.

At point A: thermal to pneumatic
At point B: pneumatic to mechanical (linear motion)
At point C: mechanical (linear) to mechanical (rotational)
At point D: mechanical (rotational) to mechanical (linear)
At point E: mechanical to electrical

This particular system is commonly used, and unless you have some sort of feeling (or understanding) of mechanical, pneumatic, and thermal systems, it would be impossible to analyze the operation of this device to determine if it is properly performing its function in a system.

QUESTIONS

The following questions are designed to impress upon the reader the basic similarities between the parameters in the various energy systems.

1. Reproduce the specific analogies as per Table A-1, without referring to the entries in the table.

2. List several of the basic laws of electricity that apply equally to all the other energy systems discussed herein.

3. Develop a hydraulic example problem illustrating the hydraulic equivalent of Kirchhoff's voltage law.

4. Develop a hydraulic example problem illustrating the hydraulic equivalent of Ohm's law.

5. Develop a thermal example problem illustrating the thermal equivalent of Kirchhoff's current law.

APPENDIX B

INTRODUCTION TO THE BINARY NUMBER SYSTEM

INTRODUCTION

Prior to becoming involved with the technical details of the theory and operation of digital computers (controllers) and analog-to-digital and digital-to-analog conversion techniques, the terms "analog" and "digital" themselves must be understood.

Analog is a term used to imply a *continuous*, uninterrupted, or unbroken sequence of events. Time is an analog quality, in that it always exists; there never is a circumstance under which time stops or is interrupted.

Digital, on the other hand, refers to a sequence of *discrete* events. Each event is completely separated from the previous events and next events. Usually this separation is a period of time. As a comparison between analog and digital, consider the previous statement. Reworded slightly, it says that the digital (discrete) events are normally separated by periods of time; this means that time is continuous (analog), even between the digital events.

As far as the real world is concerned, most naturally occurring phenomena are analog in nature. Our lives are continuous from the instant of birth until death; there is never a period of time in the interim during which we do not exist. Therefore, all life is analog in nature.

Our calendar, on the other hand, is discrete. March 6 occurs only once a year, each one is (normally) separated from any other March 6 by 364 days. It exists only for a period of 24 hours each year. During the interim between these 24-hour periods, it does not exist or has no effect.

The most common application of digital information in the modern world is in the use of binary digital computational equipment. So common is binary information that the term "digital" has acquired the connotation of "binary digital" whenever it appears, even in technical literature.

Since binary digital equipment is in such widespread use, it is absolutely necessary that the modern technician and engineer become familiar with principles upon which the design of this equipment is based. The most basic principle that is common to all this equipment is that it operates using the binary number system.

THE BINARY NUMBER SYSTEM

The number system used in most phases of schoolwork has, for centuries, been the *decimal number system*. It is based upon the existence of 10 different numerical digits (or symbols), 0–9. Certain rules have been developed to facilitate such operations as addition, subtraction, multiplication, and division using this particular number system and these symbols.

Simply because this number system has been taught to us since childhood, many people never realize that any other number system could be used, each system having its own set of rules for addition, subtraction, multiplication, and division. Other number systems in common use include the *octal* (8), *hexadecimal* (16), and *binary* (2) *number systems*. Each of these number systems has its own set of numerical digits; for example, the octal system uses the digits 0–7, the binary system the numbers 0 and 1 (only), and the hexadecimal system the numbers 0–9 plus the letters A–F. Each also has its own set of rules for mathematical operations using those symbols (still called *digits*).

Of primary interest here is the binary number system, since it is the basic numbering system used in all digital computers, even if they are programmed in octal or hexadecimal. The binary number system recognizes only the numbers 0 and 1 (notice that all number systems start with the numeral zero rather than 1). In the decimal number system, using only one digit up to 10 possible events can be counted or represented; however, in the binary number system, only two events can be counted or represented by a single "binary digit."

In the decimal number system, in order to count a quantity greater than 10, two decimal digits are required. Using *all* the possible combinations of the two digits (each ranging from 0 to 9), up to 100 possible events can be counted or represented (00 to 99). Notice that 100 is equal to the base of the decimal number system (10) raised to a power that is the sum of the available number of digits (2 in this case).

In the binary number system, to count more than two events, two binary digits must be used. Using all the possible combinations of these two digits (each ranging from 0 to 1), up to four possible events can be counted or represented (00 to 11). Figure B-1A illustrates all the possible combinations of two binary digits. Notice that four (the total number of possible combinations) is equal to the base of the binary number system (2) raised to a power that is equal to the available number of digits (2 in this case).

(A)	Two	(B)	Three	(C)	Four	
	00		000		0000	1000
	01		001		0001	1001
	10		010		0010	1010
	11		011		0011	1011
			100		0100	1100
			101		0101	1101
			110		0110	1110
			111		0111	1111

Figure B-1 All possible combinations of (up to four) binary digits.

Three decimal numbers will result in 1000 possible unique combinations (000 through 999) of these three digits. Three binary digits will result in two-to-the-third-power possible combinations; or eight. To carry this one step further, whereas four decimal digits can represent 10,000 possible events, four binary digits can represent only 16 possible events (see Fig. B-1 for all possible combinations).

It should be obvious that the binary number system is not as powerful as the decimal number system; so why is it so popular and useful? The answer to this question lies in the ease and speed with which electronic active elements can be made to assume either of two states "cutoff" and "saturation." A computer using a series of transistors which are permitted to operate in only two possible states can be made small, economical, reliable, and extremely fast.

Even though it takes more digits to work with larger numbers, the transistors are economical enough to offset this disadvantage. Furthermore, there just does not happen to be available a 10-state electronic device which could be easily designed into a *decimal* digital computer.

The best place to start learning the binary number system is the same place you started to learn the decimal number system, by counting. In the decimal system the first number is zero; zero is also the first number in the binary number system. The second decimal number is 1; likewise, with the binary number system. Now we run into a problem. The third decimal numeral is 2; however, there is no third binary numeral; the quantity 3 is more than the binary number system can handle using one digit. Therefore, to represent the third state will require two binary digits. To understand how the third event (number) is represented in binary, let us look at how the eleventh number is represented in decimal, for they are handled in an identical manner.

In decimal when the number nine is reached (the tenth count), to add one more to nine, a one is carried to the next most significant place and the nine is changed to a zero. This is the eleventh decimal numeral, since we started with zero, not one. The third binary numeral is therefore identical to the eleventh decimal number, "one-zero," by similar reasoning.

This process continues in a similar manner, the fourth binary numeral being "one-one" and the fifth "one-zero-zero" (three binary digits). Figure B-1 not only illustrates all possible combinations of two, three, and four binary digits but also illustrates them in numerical order from top (the number zero) to bottom (the binary numbers three, seven, and fifteen, respectively).

BINARY ARITHMETIC

As with the decimal number system, the binary number system has rules for the arithmetic processes of addition, subtraction, multiplication, and division. For the purposes of this presentation, only the rules for addition and subtraction will be discussed.

The rules for addition are very simple for the most part: zero plus zero equals zero and one plus zero (or zero plus one) equals one. When it comes to one plus one, the answer is zero, but you must carry a one to the next most significant position in a manner similar to a carry in decimal addition.

Figure B-2 illustrates the procedure for addition of two binary numbers. The top number is the binary equivalent of the decimal number 42, and the bottom number is 12 (decimal) in binary. The answer is the binary equivalent of 54 (decimal). The process of addition requires working from right (least significant digit) to left, adding corresponding digits in each binary number (word) in a manner similar to decimal addition.

The process of binary subtraction is normally accomplished by taking the number to be subtracted and forming the *two's complement* (binary negation) of this number then *adding* it to the other number. The net effect is subtraction of the complemented number from the uncomplemented number.

Refer to Fig. B-3A and follow the process of subtraction of 12 from 42 (decimal), the same numbers used for the addition example. The rule says to take the number to be subtracted and form the two's complement of that number. This is accomplished by first taking the number itself and forming a new number having ones in each position in which the original number had zeros, and zeros in each

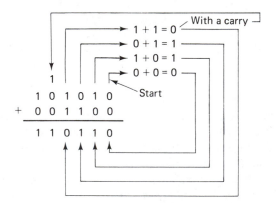

Figure B-2 Binary addition example.

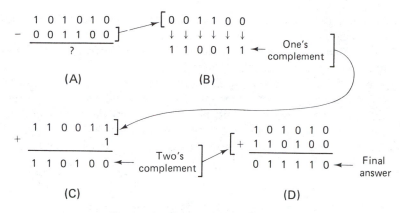

Figure B-3 Binary subtraction example.

position in which the original number had ones. This process is called *complementing* (or *one's complementing*) the original number; and the new number is called the one's complement (or simply the complement) of the original number. This process is illustrated in Fig. B-3B.

The two's complement is now formed by simply adding a binary one to the one's complement, following the rules of addition previously explained. Figure B-3C illustrates this procedure. This number, the two's complement, is the negative equivalent of the original number. Furthermore, if the two's complement of this number is taken, the process will result in the original number again.

The final step in the binary subtraction procedure is to add this negative value to the number from which the positive equivalent was supposed to have been subtracted. This is illustrated in Fig. B-3D. The only point left to note is that there will generally be a carry from addition of the most significant (binary) digits when adding negative numbers; this carry is generally discarded. The result in Fig. B-3D is the binary equivalent of the decimal number 30.

Finally, the difference between positive and negative numbers must be more thoroughly discussed. Refer to Fig. B-4A, where the positive binary equivalent of the number 12 is shown along with several leading (most significant) zeros. As with decimal numbers, the leading zeros have no significance. Note that the one's complement of this number has all leading ones (Fig. B-4B), and that the two's com-

Figure B-4 Positive and negative binary numbers.

plement (which is the negative value of the original number) also has all (most significant) leading ones (Figure B-4C).

Therefore, the most obvious difference between positive and negative binary numbers lies in whether the leading (insignificant) digits are ones or zeros. This fact makes the correct algebraic sign automatically appear when adding binary numbers of opposite sign. Remember, positive numbers have a zero in the most significant position, and negative numbers have a one in the most significant position. The only exception to this rule is when the algebraic sign is understood to be positive; then the leading (most significant digit) zero is simply omitted. This is frequently the situation.

BINARY-TO-DECIMAL NUMBER SYSTEM CONVERSION

Obviously, if you are going to do much working with the binary number system, a means of conversion from binary to decimal and back will be a must. Prior to actually developing the rules for these conversions, the significance of each binary "word" (numbers) must be understood.

Again, this may be easier to understand if it is compared to the decimal system. In the decimal number system the first decimal number to the left of the decimal point carries a multiplier equal to the number itself times 10 to the zero power. In a similar manner, the second number to the left of the decimal point is equal to (or carries the weight of) that specific number multiplied by 10 to the first power. This sequence continues, each succeeding (more significant) number carrying 10 times the weight of the number next to it but closer to the decimal point.

An example might be appropriate at this point. Take the number 0693. (decimal). This number really means to take the least significant digit (3) and multiply it times 10 (the base of the decimal number system) to the zero power (remember *any* number raised to the zero power is equal to 1), and then *add* it to the next most significant digit (9) multiplied by 10 to the first power. Then add this result to the next most significant digit (6) multiplied by 10 to the second power. This sum is then to be added to the most significant digit (0) multiplied by 10 to the third power. This result is your final answer and is illustrated in Fig. B-5A. Notice the effect of the most significant digit (a zero); it has no effect at all on the final result.

Now let us refer to a binary number, the number 01010110101., which has its equivalent of the decimal point, called a *binary point*. The binary point has the same significance as the decimal point, in that it separates the integer and fractional portions of the binary number. The first binary digit to the left of the binary point is the least significant (integer) digit. The binary digit farthest away from the binary point (to the left) is the most significant binary digit.

In a manner analogous to the decimal number system, the least significant binary digit, a 1 in this case, carries the weight of the number itself (1) multiplied by 2 (the base of the binary number system) raised to the zero power. This result (1) is then added to the next most significant binary digit multiplied by 2 raised to

(A) $0693 = 3 \times 10^0 + 9 \times 10^1 + 6 \times 10^2 + 0 \times 10^3$
$= 3 + 90 + 600 + 0000 = 693$

(B) $01010110101._2$

$= 1 \times 2^0 + 0 \times 2^1 + 1 \times 2^2 + 0 \times 2^3 + 1 \times 2^4 + 1 \times 2^5 + 0 \times 2^6 + 1 \times 2^7 + 0 \times 2^8 + 1 \times 2^9 + 0 \times 2^{10}$

(C) $= 1(1) + 0(2) + 1(4) + 0(8) + 1(16) + 1(32) + 0(64) + 1(128) + 0(256) + 1(512) + 0\ (1024)$

$= 1 + 0 + 4 + 0 + 16 + 32 + 0 + 128 + 0 + 512 + 0$

$= 693._{10}$

(D)

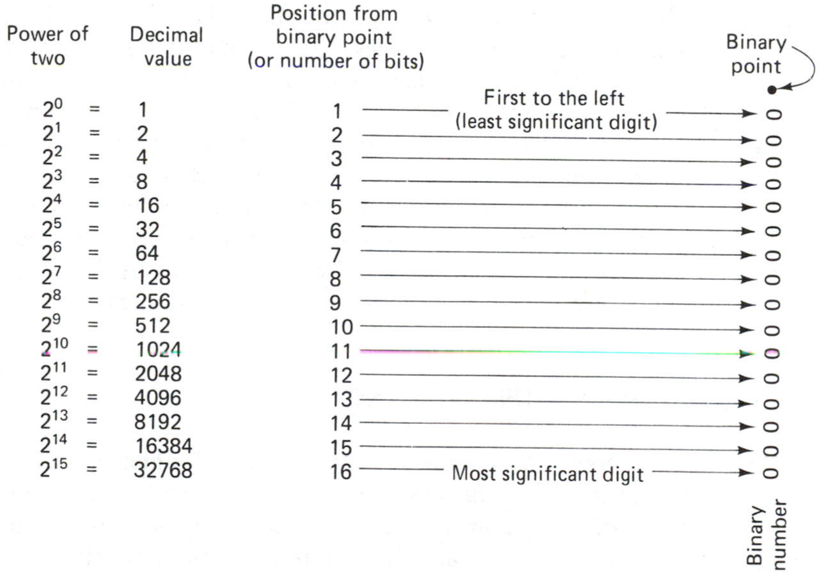

Power of two	Decimal value	Position from binary point (or number of bits)		Binary point
			First to the left (least significant digit)	
$2^0 =$	1	1		0
$2^1 =$	2	2		0
$2^2 =$	4	3		0
$2^3 =$	8	4		0
$2^4 =$	16	5		0
$2^5 =$	32	6		0
$2^6 =$	64	7		0
$2^7 =$	128	8		0
$2^8 =$	256	9		0
$2^9 =$	512	10		0
$2^{10} =$	1024	11		0
$2^{11} =$	2048	12		0
$2^{12} =$	4096	13		0
$2^{13} =$	8192	14		0
$2^{14} =$	16384	15		0
$2^{15} =$	32768	16	Most significant digit	0

Binary number

Figure B-5 Binary-to-decimal conversion.

the first power. Similarly, this result is added to the next most significant binary digit multiplied by 2 to the second power; and so on until the most significant digit has been multiplied by its appropriate power of 2 and added to the rest of the numbers, each multiplied by their appropriate power of two.

Figure B-5B illustrates the entire process for this particular binary number. This discussion illustrates the significance of each binary digit; however, it also illustrates the initial step in the process of binary-to-digital number conversion. Figure B-5C illustrates the rest of the process of binary-to-decimal conversion. Figure

B-5D is a convenient listing of the decimal values of the powers of 2 in order to have this reference for future binary-to-decimal (and decimal-to-binary) number system conversion.

The process of decimal-to-binary conversion is naturally the inverse of the binary-to-decimal conversion process. Again use the decimal number 693 as our example. First refer to Fig. B-5D and pick out the largest power of 2 that is *less than or equal to* our example number. That number is 512, which means that a binary 1 must be present in the tenth binary position to the left of the binary point. A binary number 1 in the tenth position carries a weight of 512 (decimal). Next, subtract 512 from 693, which leaves a remainder of 181 (decimal); then look for the power of 2 that is equal to or less than 181. This number is 128, which is represented by a binary 1 in the eighth position to the left of the binary point. Subtracting 128 from 181, 53 remains to be converted to binary.

At this point the binary number is 01010000000., which is the binary equivalent of the decimal number 640. (512 + 128). The next power of 2 equal to or less than 53 is 32, which is represented by a 1 in the sixth place from the binary point and leaves a remainder of 21. Sixteen is less than 21 and is represented by a 1 in the fifth place from the binary point, leaving a remainder of 5 to be counted. Four is less than 5 and is represented by a 1 in the third digit from the binary point, leaving a remainder of 1. Decimal 1 is represented by a 1 in the least significant position in the binary number. This completes the conversion process.

The binary equivalent of 693 (decimal) therefore has 1s in the tenth, eighth, sixth, fifth, third, and first binary digit positions (to the left of the binary point). The binary equivalent of the decimal number 0693. is therefore 01010110101., which is the binary number used in the binary-to-decimal conversion example.

BINARY-TO-OCTAL NUMBER SYSTEM CONVERSION

Many digital computers are programmed in octal. What this means is that, for simplicity in writing and speaking, the binary codes are *spoken* of as though they were base eight numbers rather than binary codes. In fact, the computers use only binary for any purposes whatsoever insofar as their internal logic, arithmetic operations, and data-handling hardware is concerned. The octal number system is used *only* externally to the computer for the convenience of the program writer. It is merely a very convenient shorthand means of "talking" about long strings of binary numbers.

The large majority of process control computers have 16-bit word sizes, that is, all of their internal binary processing, data transfers and storage are performed on 16 binary bits simultaneously (or in parallel). Therefore, every time the computer programmer refers to a binary computer word, he must speak of a string of 16 1s and 0s. This is inconvenient, confusing, and downright inefficient. Therefore, several techniques have been developed simply to make it easier to communicate binary codes between human beings or between human being and machine (computer); the

Binary number	000_2	010_2	011_2	101_2	110_2	111_2	' denotes
Coding (4-2-1)	4 2 1	4 2 1	4 2 1	4 2 1	4 2 1	4 2 1	an octal
Octal Number	'0	'2	'3	'5	'6	'7	number

Figure B-6 Conversion of three binary digits into octal numbers.

adaptation of the octal number system is one of these techniques. (The use of the hexadecimal number system is another commonly used technique, as will be explained in the next section.)

To convert from binary to octal, simply group the binary digits into groups of three, starting at the binary point and working toward the most significant bit (MSB). Then, apply a "4, 2, 1" coding to each group of three and, using this coding, convert each group of three to decimal form. This is illustrated for several examples in Fig. B-6. Since the only decimal digits possible are the ones from zero through seven, this conversion is not the decimal number system but a system based upon the existence of only eight different digits (0–7), called the *octal number system*.

What has been gained by this conversion? Instead of having to speak about three binary digits when we refer to a 3-bit binary number, we can now, by inspection, convert it to octal form and speak of only one octal digit, and communicate that number more easily to another person (who is also knowledgeable about octal-to-binary conversions). This really is not too much of an advantage for a single group of 3 binary bits, but consider what happens when we talk of strings of 16 binary bits. Figure B-7 illustrates the conversion into octal form of the same number that was used in the binary-to-decimal conversion example. The octal equivalent is much more convenient to communicate with than the binary form is.

Actually, the ideal would frequently be to convert to decimal form and communicate directly in that form. However, that conversion process is much more complicated and time-consuming and is therefore done only when there is a more pressing need for the decimal value. Octal is more convenient to use than binary and easier to convert to than decimal; that is why it is used so frequently.

What about octal-to-decimal conversion? Certainly, a set of rules can be developed to convert between octal and decimal—but that would be another set of rules to remember. Since the conversion between octal and binary is so simple, the

	MSB									LSB	
16-bit binary number to be converted	0		000		001		010		110		101	Binary point
Grouped into 3's			'		'		'		'		'	
Conversion code	1		4 2 1		4 2 1		4 2 1		4 2 1		4 2 1	Octal point
Converted to octal	0		0		1		2		6		5	. point

Figure B-7 Conversion of 16-bit binary number into octal.

easiest procedure is to convert either your octal number, or decimal number, first into binary and then into the other number system.

BINARY-TO-HEXADECIMAL NUMBER SYSTEM CONVERSION

As mentioned previously, many computers are programmed in octal. The computers that are not programmed in octal are programmed in hexadecimal. As we have mentioned before, the computer itself does nothing in hexadecimal; it is used only externally from the computer for the convenience of the programmer. It is also used as a convenient means of referring to long strings of binary numbers. It, therefore, competes in every respect with the use of the octal number system in these applications.

Since they are both used for the same purpose, there must be a reason for the preference of one or the other in specific applications. In fact, it is easier to convert from binary to the octal number system than to the hexadecimal number system. However, it requires six octal digits to represent a 16-bit binary word in octal form, whereas it only requires four hexadecimal digits to represent the same 16-bit binary word. Therein lies the basic criterion for the choice between the two. Originally, octal was used almost exclusively for programming small computers, but recently hexadecimal has come into very popular use, especially with microcomputers.

Therefore, we also have the need to convert between binary and hexadecimal; conversion between octal and hexadecimal or decimal and hexadecimal would be accomplished via an intermediate conversion to binary. To convert a binary number to hexadecimal form, start at the binary point and group the binary digits into groups of four (remember that octal grouped them into threes). Then convert each of these groups of four binary digits into hexadecimal form according to the conversion chart in Fig. B-8. Note that the conversion is basically similar to octal conversion, except that for the four binary digits, an "8, 4, 2, 1" coding is used for all the digits from 0 to 9. Then the beginning letters of the alphabet (A–F) are used as symbols for the six binary-code combinations left over after the number nine. We are, therefore, using a number system based upon the existence of a total of 16

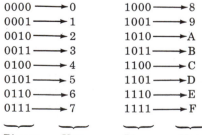

Figure B-8   Binary-to-hexadecimal codes.

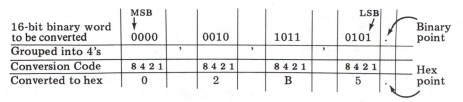

	MSB							LSB		Binary
16-bit binary word to be converted	0000		0010		1011		0101	.		Binary point
Grouped into 4's		,		,		,				
Conversion Code	8 4 2 1		8 4 2 1		8 4 2 1		8 4 2 1			Hex
Converted to hex	0		2		B		5	.		point

Figure B-9 Conversion of 16-bit binary number into hexadecimal.

arithmetic "digits." Figure B-9 illustrates the same binary number used in the binary-to-decimal conversion example as it would be converted into hexadecimal form.

BINARY-TO-BINARY CODED DECIMAL CONVERSION

One final very common convenience code will be presented; not that there aren't others, but the ones presented here are the most commonly encountered ones in practice. The *binary coded decimal* (BCD) *conversion* is very similar to the hexadecimal. The single difference is that only the characters 0–9 are allowed. The binary codes corresponding to the hexadecimal characters A–F are not permitted in BCD conversions; the existence of these binary codes causes an error condition and an invalid conversion.

BCD coding is commonly used for binary codes either being sent to a computer from a numeric (only) peripheral device (input device) or from the computer to a numeric (output) peripheral. It is commonly used for displays associated with analog-to-digital converters and is extensively used in the electronic (hand-held) calculators that are in such popular use. Figure B-10 compares the four coding schemes discussed here: binary, octal hexadecimal, and BCD.

Character   Binary codes

```
0        0   000
1        0   001
2        0   010
3        0   011          Octal    BCD
4        0   100
5        0   101
6        0   110
7        0   111                        Hex
8        1   000
9        1   001
(A)      1   010
(B)      1   011
(C)      1   100
(D)      1   101
(E)      1   110
(F)      1   111
```

Figure B-10 Comparison of several popular coding schemes.

Numerous other coding schemes have been developed since the advent of the binary digital computer, each of which has it applications. The *gray codes* are used for reliable parallel data communications channels, where errors are probable. The ASCII code is the standard used with many devices, including the common teletype units associated with the majority of small computers, although the *Baudot code* is also used with teletypes. Certain companies that manufacture input/output devices have developed their own codes, which suit their particular requirements. There are so many codes that it would be impossible to cover them here; however, many of the more (commercially) popular ones have at least been mentioned.

QUESTIONS

Convert the following binary numbers to decimal, then from binary to octal, hexadecimal, and BCD (if possible).

1. 0001 0110 1110 0111.

2. 1100 1100 0011 0011.

3. 0101 0101 0101 0101.

4. 0000 1111 0000 1111.

5. 0101 1011 1010 1101.

Convert the following numbers to decimal form.

6. 011011 (binary)

7. 7B9A (hexadecimal)

8. 7468 (BCD)

9. 0714 (octal)

The next four questions refer to the following two binary numbers. Both are unsigned binary integers.

    (A) 10011011.
    (B) 01010001.

10. Add them using binary arithmetic.

11. Subtract (B) from (A) using binary arithmetic.

12. If (A) had been a signed binary number, would it have been a positive or a negative number?

13. Repeat Question 12 for (B).

14. What is the arithmetic significance of shifing a number right or left one place?

15. How many pieces of discrete information can be uniquely represented by a binary word consisting of 8 bits?

16. Repeat Question 15 for a 16-bit binary word.

INDEX